High Temperature Properties and Thermal Decomposition of Inorganic Salts with Oxyanions

High Temperature Properties and Thermal Decomposition of Inorganic Salts with Oxyanions

Kurt H. Stern

CRC Press
Boca Raton London New York Washington, D.C.

Library of Congress Cataloging-in-Publication Data

Stern, K. H. (Kurt H.)
High temperature properties and thermal decomposition of inorganic salts with oxyanions / by Kurt H. Stern.
p. cm.
Includes bibliographical references.
ISBN 0-8493-0256-0 (alk. paper)
1. Salts—Effect of temperature on. 2. Salts—Deterioration. I. Title.

QD189 .S75 2000
546′.34—dc21 00-041427
CIP

Direct all inquiries to CRC Press LLC, 2000 N.W. Corporate Blvd., Boca Raton, Florida 33431.

International Standard Book Number 0-8493-0256-0
Library of Congress Card Number 00-041427
Printed in the United States of America 1 2 3 4 5 6 7 8 9 0
Printed on acid-free paper

Preface

The origin of this book is the four monographs on the thermal decomposition of inorganic salts with oxyanions which the author wrote under the auspices of the Standard Reference Data Program between 1968 and 1972. The aim of that work was to provide a reference for scientists and engineers who need to use these compounds at high temperatures, but who do not know whether a particular compound even exists at a particular temperature. Since the original work is now out of print and more information has become available in the thirty years since the monographs were published, it seemed useful to prepare an updated version.

The aim is still to provide useful and critically evaluated information which the potential user may not have the time or inclination to find for himself. This information is useful in a variety of contexts, from questions of basic chemistry to technologically related questions involving the corrosion of metals in molten salts or the preparation of molten salt baths as heat-transfer media.

On the assumption that some information is better than none, both qualitative and quantitative material have been included. The latter includes thermodynamics of the decomposition reactions and their rates, as well as phase transitions and densities of the solid and liquid salts. Although some of the qualitative material may be quite old, it has been included if it is the only information available.

A discussion of the thermodynamics used in this book, which is essential for understanding the meaning of the tables, as well as a discussion of the kinetics of solid-state decomposition, is included in Chapter 1, which also includes a discussion of the sources of the data used.

Kurt H. Stern

The Author

Dr. Kurt H. Stern obtained his B.A. degree in chemistry from Drew University in Madison, NJ in 1948, his M.S. degree in physical chemistry from the University of Michigan in 1950, and a Ph.D. from Clark University under Prof. Arthur Martell in 1953. He then joined the faculty of the University of Arkansas, where he and his students carried out research in non-aqueous and molten salt electrochemistry. In 1959, while spending a sabbatical year at the National Bureau of Standards in Washington, he was asked to join the staff and to start a program in molten salt electrochemistry. While at NBS he became involved in the formation of the National Standards Reference Data Program under whose auspices he wrote the four monographs on which the present book is based. In 1969 he moved to the Naval Research Laboratory where he continued his research, most recently on the electrodeposition of refractory compounds, such as carbides and silicides, from molten salts. He has published approximately 100 research articles and monographs on his work. Concurrently with his research, he has served on the faculty of the National Institutes of Health Graduate School.

Since his retirement from active research in 1991, he has completed a book on protective metallurgical and ceramic coatings (Chapman and Hall, 1996).

Acknowledgments

I would like to thank my wife Faith for her great patience and perseverance in dealing with the problems of a semi-literate computer-competent husband and author, for rescuing me several times from semi-fatal computer errors, for developing the format for the tables, and for the artwork involved in the preparation of many of the figures.

Thanks also to my son Alan who several times managed to retrieve the manuscript from the electronic domain to which I had unintentionally consigned it by inadvertent key strokes.

Thanks to Dr. Irwin Singer of the Naval Research Laboratory for assistance with thermochemical software which greatly increased the speed of preparation of the many tables in the book.

Thanks to Dr. Elizabeth Mayfield of the Mathematics Department of Hood College, Frederick, MD, who kindly calculated the partial pressures for nitrate decomposition in Chapter 5.

Finally, thanks to several librarians at both the Library of Congress and the Naval Research Laboratory, particularly Harriet Oxley at NRL who helped me with literature searches and copyright problems.

Any errors can be blamed on the author, who would be glad to hear of them from readers, so corrections can be made in possible future printings.

Contents

Chapter 8

Oxysalts of Bromine

Chapter 9

Oxysalts of Iodine

1 Introduction

1.1 GENERAL

In many areas of chemistry and chemical technology there exists a need to know the properties of salts at high temperatures. These areas include chemical synthesis, metallurgy, corrosion by salts, and catalysis.

Questions often relate to the thermal stability of the salt, usually phrased in terms of the following questions: Up to what temperature is the salt stable? What happens to it above that temperature — does it melt or decompose in either the solid or liquid state? What are its decomposition products? If it is placed in a closed container, how much will the pressure build up?

This is true whether one needs to know the properties of a particular salt, or whether one wishes to select salts fitting particular criteria, such as melting points or decomposition pressures.

The kind of information desired proceeds from the qualitative — congruent melting point or decomposition products to the quantitative — melting points, decomposition pressures, and rates of decomposition, i.e., thermodynamics and kinetics.

The topics covered in this book include the above kinds of information for oxyanions of the elements which form discrete anions: carbon, sulfur, nitrogen, and the halogens including:

Carbon: carbonates
Sulfur: sulfites, sulfates
Nitrogen: nitrites, nitrates
Chlorine: chlorites, chlorates, perchlorates
Bromine: bromites, bromates, perbromates
Iodine: iodates, periodates

In order to avoid complicating factors (the subject is already complicated enough) coverage has been restricted to anhydrous salts and monatomic cations, e.g., no ammonium and acid salts are included. Even with these restrictions, nearly 300 salts fit these criteria.

There is also a large body of literature dealing with additives, such as catalysts, the effect of radiation and mechanical treatment, which affects both the surface and internal structure of crystals, which are not treated in detail in this book.

Between 1968 and 1972 the author published four monographs covering this subject under the auspices of the National Standard Reference Data Program (NSRDS) (see below). This material is no longer readily available and needs revision and updating.

This book generally follows the format of the monographs. Each class of salts, e.g., sulfates, is treated in a separate chapter. Since a major question is always "Does this salt exist?" each section includes a periodic table which provides this information. A blank space does not mean the salt does not (or cannot) exist, but only that nothing is known of its existence, perhaps because no one has tried to synthesize it.

The discussion of each salt includes both qualitative and quantitative information, what is known of its behavior above ambient temperature, and quantitative data: phase transitions, densities, and thermodynamic properties. When possible, high temperature properties, enthalpies and entropies at 298.15 K and above, are listed. When these are available, they are used to calculate equilibrium constants and partial pressures of the component gases. (For a discussion of this topic see Chapter 1.2). Kinetic data for decomposition reactions, such as rates and activation energies, are also included, but this information is inherently less reliable since these parameters depend not only on the nature of the material, but also on its physical state, e.g., particle size and defect structure. This topic is also discussed in a separate section.

1.2 EQUILIBRIUM CALCULATIONS

The main method for deciding what happens to substances when they are heated is thermodynamic calculations. Before they can be carried out, however, the relevant decomposition reaction(s) must be known. For some substances this is straightforward; for example, the decomposition of a carbonate always results in the formation of a metal oxide, nearly always in the solid state, and CO_2. For other classes of salts the problem is more difficult. For sulfates the gaseous decomposition products may be SO_3, or mixtures of this gas, SO_2, and O_2, depending on the equilibrium $SO_3 = SO_2 + 1/2O_2$ at the temperature of interest. Some oxyhalogens decompose to halides, others to oxides, and some to a mixture of both. Correspondingly, the gaseous products will also be different in the two cases. Thus it is essential to decide first what the reactions are for which calculations are to be performed. In most of these cases, where simultaneous equilibria obtain, calculations of partial pressures usually require the solution of simultaneous equations. Examples will be given in the appropriate sections.

In order to calculate the equilibrium constant for a particular reaction, thermodynamic functions of reactants and products as a function of temperature must be known. The calculation of these functions is discussed in detail in the work of Wagman et al. cited below.

The ΔG of the reaction is then calculated in one of two ways: (1) the appropriate addition and substraction of ΔGf for reactants and products, or (2) the calculation of Gibbs energy functions for reactants and products from enthalpy and entropy increments

$$gef = \frac{G_T^0 - H_{298}^0}{T} = \frac{H_T^0 - H_{298}^0}{T} - (S_T^0 - S_{298}^0) - S_{298}^0 \quad (1.1)$$

Finally, the equilibrium constant is calculated from

$$G_T = T\Delta gef + \Delta H_{298}^0 \text{ and} \quad (1.2)$$

$$\Delta G = RT \ln K \quad (1.3)$$

These methods were used in this work when necessary. However, in many cases the equilibrium constants could be calculated by a computer program developed by Outukompu Research, which uses data stored in their program. These data are usually the most recent available. References used in this program are listed below. When data above 298 K were not available, data at 298 K are given. These are taken from Wagman, D. D.; Evans, W. H.; Parker, V. B.; Schumm, R. H.; Halow, I.; Bailey, S. M.; Churney, K. L.; and Nuttall, R. L., The NBS tables of chemical thermodynamic properties, *J. Phys. Chem. Reference Data* 11, Suppl. 2, 1982

HSC Chemistry for Windows

This program stores data for approximately 7800 compounds. Data come from a variety of sources. Those chiefly used for calculations in this work include

Barin, I., *Thermochemical Data of Pure Substances*, VCH Velagsgesellschaft, Weinheim, 1989 and 1993.

Barin, I., Knacke, O., and Kubaschewski, O., *Thermodynamic Properties of Inorganic Substances*, Springer-Verlag, Berlin, 1973, Supplement 1977.

Chase, M. W., Davis, C. A., Downey, J. R., Frurip, D. J., McDonald, R. A., and Syverud, A. N., JANAF thermochemical tables, *J. Phys. Chem. Reference Data,* 14, Supplement 1, 1985.

Knacke, O., Kubaschewski, O., and Hesselman, K., Thermochemical Properties of Inorganic Substances, Springer Verlag, Berlin, 1991.

In addition, some tables come from a 1997 edition of HSC. References for these tables appear at the end of each chapter.

Previous work by the present author is given in the following monographs:

High Temperature Properties and Decomposition of Inorganic Salts

Part 1. Sulfates, Stern, K. H. and Weise, E. L., NSRDS – NBS, 1966.

Part 2. Carbonates, Stern, K. H. and Weise, E. L., NSRDS – NBS, 1969.

Part 3. Nitrates and Nitrites, Stern, K. H., *J. Phys. Chem. Ref. Data*, 1, 3, 1972.

Part 4. Oxysalts of the Halogens, Stern, K. H., *J. Phys. Chem. Ref. Data*, 3, 2, 1974.

When data from the HSC program were used, ΔHf°, ΔGf°, and S° have been listed for reactants and products as a function of temperature. Gaseous equilibria are listed in separate tables in each section, e.g., $SO_2 + 1/2\ O_2 = SO_3$ for sulfates. In many cases the equilibrium constants for the decomposition reactions can be calculated directly from data in the program. When this was not possible, other methods, using the equations given above, were used.

When the equilibrium constants are known, the partial pressures of the individual gases can be calculated under certain restrictive conditions, e.g., closed system, pressures of some gases are fixed, etc. For example, if one starts with a closed system that is initially pumped out and the salt allowed to decompose so that pressure builds until equilibrium is reached, it is possible to calculate the partial pressures of the gases in the system. Thus, for carbonates, the pressure of CO_2 equals the equilibrium constant. For other classes of salts the calculations can be much more complicated. Appropriate methods are discussed in the individual chapters.

1.3 PHASE TRANSITIONS

Solid–solid phase transitions have been much studied and are generally reliable if they were carried out appropriately, e.g., if equilibrium was approached from above and below.

Melting points are usually obtained by rapid heating of the salt to minimize decomposition, but this cannot always be avoided. The lower values should therefore be regarded as lower limits. Above the melting point most salts decompose so rapidly that boiling points cannot be measured. When more recent or reliable data were not available, phase transitions were taken from the following NBS publication: Rossini, F. D., Wagman, D. D., Evans, W. H., Levine, S., and Jaffe, I. Selected Values of Chemical Thermodynamic Properties, NBS Circular 500, U.S. Govt. Printing Office, Washington, D.C. 1952.

1.4 DENSITY

The densities of solid salts are generally determined directly by pycnometry, or are calculated from crystallographic (X-ray diffraction) data. In the latter method atomic packing must be assumed. Since no allowance is made for crystal imperfections, calculated densities are generally higher than experimental values by as much as 10%. Since most crystallographers do not indicate the temperature for which the density is calculated, in this work the experimental values have been given when they were available. However, in some cases the experimental values seem to depend on the method of preparation, perhaps because of variations in the defect density. Thus, although the density of a given sample can certainly be determined to four or five significant figures, not too much significance should be attached to figures beyond the third. This is particularly true because the units of the reported density values are frequently in doubt. In some cases they are not stated at all; in others it is not clear whether the units are g/cm^3 or g/ml; in still others they are specific gravity values and may be referred to water at 4°C or at some other temperature. The difference between cubic centimeters and milliliters is so small (the ratio is 1.000028) as to be negligible. However, the difference in the density of water between 4 and 20° is in the third significant figure and is no longer negligible for better measurements.

In this work the units reported by the original authors are given. When no units are given it can generally be assumed that g/ml is meant.

There are some density data for solid salts above ambient temperature which are given in the form of thermal expansion coefficients. These have been listed when they seemed reliable. Above the melting point, density data are scarce. Most are available for alkali halides but those available for salts are taken from the critically evaluated compilation: Janz, G.J., Thermodynamics and transport properties for molten salts. correlation equations for critically evaluated density, surface tension, electrical conductance, and viscosity data, *J. Phys. Chem. Reference Data*, 17, Suppl. 2, 1988.

1.5 KINETICS OF THERMAL DECOMPOSITION

1.5.1 General

In contrast to the thermodynamics of decomposition, where a few parameters permit the calculation of the equilibrium properties of the system, the determination of decomposition rates is largely an experimental problem, i.e., there are no standard kinetic data from which these rates can be calculated. This is particularly true for the decomposition reactions of solids which are "topochemical," i.e., where the rate depends on structural factors. One reason for this situation is that it does not yet seem possible to prepare duplicate samples of any solid inorganic salt that are identical in all the properties that may determine the rate of decomposition, e.g., the dislocation density of the crystals.

Nevertheless, the kinetics of decomposition are of sufficient interest, for both theoretical and practical reasons, that it seems worthwhile to include the results of at least the more carefully done studies. In this section we summarize features common to all decompositions and the factors affecting decomposition rates. The discussion is essentially limited to solids, since most experimental and theoretical work has been done in this area. It should be pointed out that many of the theoretical foundations were laid several decades ago, and that newer developments deal mostly with refinements. Kinetics of the individual compounds are treated in their respective sections.

1.5.2 Mechanism of Endothermic Solid-State Decomposition

Many authors have derived equations for the rate of decomposition from theoretical models. The aim of these studies is to find functional relations between some measurable quantities of the system, such as weight and time, which can be tested against experimental results. Although the usual caveat against taking agreement between the predictions of the model and experimental results as evidence for the correctness of the model applies, nevertheless such agreement, together with other experimental evidence, e.g., microscopic examination, is frequently very helpful in at least narrowing the possibilities.

In this section some of the models which have been proposed are briefly summarized. Greater detail can be found in books by Garner,[1] Young,[2] and Tompkins[3] and in the original references.

Most models try to account for the shape of the experimental α(= fraction decomposed) vs. time(t) curves. These are generally S-shaped, with a relatively slow induction period, a steep portion during which the bulk of the material decomposes, and a slow final section. Experimentally determined curves differ primarily in the relative length of the induction period. This induction period is commonly identified with the formation of nuclei of the new phase B imbedded in A; these nuclei grow, by spreading both over the surface and into the bulk of the decomposing particle. After the surface is covered by B, this phase grows into the bulk of the decomposing particle. Decomposition thus occurs only at the reactant/product phase boundary, as has been argued by Langmuir from phase rule considerations.[4] There seems to be general agreement that the evolution of gas C during endothermic solid decomposition is not a rate-limiting step in the process because the product B is sufficiently porous to permit C to escape readily. However, if the gas is generated in the bulk of the crystal, e.g., by ionizing radiation, and must diffuse along grain boundaries, the rate of decomposition may depend markedly on the diffusion constant of the gas. This case has been discussed by Gafner.[5]

Although the measured decomposition rates are not rate-limited by diffusion of gas out of the crystal, it is clear that in the case of readily reversible decompositions, such as those of the carbonates, the measured rate will depend on the difference between the equilibrium pressure p_e and the actual pressure p at the reaction interface. Thus, although the functional dependence of α on t may be independent of $p_e - p$, the actual rates are not. It is largely for this reason that it is not possible to tabulate "standard reference data" rate constants, since the pressure p is not usually known. Instead, one finds decompositions carried out in streams of various "sweep gases," which may or may not remove product gases from the reaction interface, or in a vacuum. In the latter case, the question as to whether the manometrically measured pressure equals the pressure at the interface needs to be answered. Small sample size and a porous oxide favor this condition.

1.5.2.1 Kinetic Equations

The most complete classification of solid decompositions has been worked out by Jacobs and Tompkins.[6] When the decomposition begins at some points on the A lattice where the energy is most favorable, small fragments of B are imbedded in A. These may at first retain the lattice parameters of A and will therefore possess some strain energy as a result of this deformation. Jacobs and Tompkins show that the Gibbs energy accompanying the formation of a fragment of B containing m molecules is

$$\Delta G = am^{2/3} - bm$$

where a is proportional to the strain energy and b is the negative of the bulk Gibbs energy per molecule. ΔG passes through a maximum at $m = m^*$ when a fragment has the critical size to be in equilibrium with its surroundings. Smaller fragments are unstable and revert to A; larger fragments are stable nuclei and grow. Whether decomposition proceeds from the growth of a few nuclei or whether many small nuclei are formed depends on the relative magnitudes of the activation energies for nucleus formation ($\Delta G_1^{\ddagger}$) and the growth stage $\Delta G_2^{\ddagger}$. If $\Delta G_2^{\ddagger} < \Delta G_1^{\ddagger}$, growth of existing nuclei predominates over formation of new ones. If $\Delta G_2^{\ddagger} = \Delta G_1^{\ddagger}$, many small nuclei are formed, none of which grow to visible size. In that case the induction period is much reduced. The whole surface then nucleates rapidly and the kinetics of the remainder of the reaction depends on the rate at which the interface progresses into the crystal.

We next consider various mechanisms for each of the stages in the decomposition.

1.5.2.1.1 Nucleus Formation

Jacobs and Tompkins[6] consider only structure-sensitive nucleation, which occurs at definite sites in the lattice where the activation energy is least, such as lattice defects and dislocations. The rate of nucleus formation thus depends on both the defect density and on the activation energy.

Jacobs and Tompkins distinguish two cases:

a. The decomposition of a single molecule leads to the formation of a nucleus. The probability of this unimolecular decomposition is

$$k_1 = \Gamma \exp(\Delta G_1^{\ddagger} / RT) \tag{1.4}$$

where Γ is the frequency of lattice vibrations and $\Delta G_1^{\ddagger}$ is the activational Gibbs energy for nucleus formation. The rate of nucleus formation at time t then is

$$dN/dt = k_1 N_0 \exp(-k_1 t) \tag{1.5}$$

or

$$N = N_0 [1 - \exp(-k_1 t)]$$

where N_0 is the total number of potential nucleus-forming sites. The possible loss of nucleus- forming sites through their ingestion by growing nuclei is neglected. In the early stages of the reaction and especially for large $\Delta G_1^{\ddagger}$ and consequently small k_1

$$N \cong k_1 N_0 t \tag{1.6}$$

so that the number of nuclei increases linearly with time.

b. A power law results from two possible mechanisms: (1) a stable nucleus is formed in a bimolecular combination of two active intermediates, and (2) several decompositions are required to form a stable nucleus. The first of these possibilities may require the surface migration of several B molecules until they coalesce into a stable nucleus. Some aspects of this phenomenon have been discussed by MacDonald.[7]

1.5.2.2 Growth of Nuclei

As each nucleus is formed, it grows, generally hemispherically, i.e., it spreads across the surface of the crystal and into its interior. Since the laws for the formation of nuclei are different from those for their growth, the overall decomposition rate is likely to be complicated if the two rates are of comparable magnitude. This is particularly true if the complications resulting from overlapping nuclei are taken into account. As nuclei grow, they ingest or overlap sites which would otherwise yield nuclei. Also, as nuclei grow, they impinge on one another and cease to grow along the line of contact. Both of these effects result in rates different from those predicted from models without these refinements.

The simplest decomposition kinetics, commonly observed for carbonates, result from a model in which nucleation and subsequent surface growth are extremely rapid. The rate of decomposition is then determined by the progression of the reaction interface into the crystal. Rate laws have been derived for various geometries, but the ones for spherical particles are most useful and have been given in several equivalent forms by various authors. In this derivation it is assumed that the reaction interface proceeds at a constant velocity into the crystal; i.e., for a constant interfacial area, the rate of decomposition is a constant. For a spherical particle, the reaction proceeds along a constantly diminishing interface. For such particles of initial radius R_0, the fraction decomposed at time t is

$$\frac{4/3\pi R_0^3 - 4/3\pi(R_0 - k_2 t)^3}{4/3\pi R_0^3} = 1 - \left(1 - \frac{k_2 t}{R_0}\right)^3 = \frac{3k_2 t}{R_0} - \frac{3k_2^2 t^2}{R_0^2} + \frac{k_2^3 t}{R_0^3} \tag{1.7}$$

and

$$\frac{d\alpha}{dt} = \frac{3k_2}{R_0} - \frac{6k_2^3 t}{R_1^2} + \frac{3k_2^3 t^3}{R_0^3} \tag{1.8}$$

These equations were first derived by Hume and Colvin.[8]

The same, but superficially different, equation was derived by Fischbeck and Schnaidt[9] in a form now frequently used:

$$d\alpha/dt = k_1 A \tag{1.9}$$

where A is the total area of the reacting interface. For a spherical particle $A = k'_2\ (1 - \alpha)^{2/3}$, where k'_2 depends on R_0, and hence

$$\frac{d\alpha}{dt} = k'_2(1 - \alpha)^{2/3} \tag{1.10}$$

In this form the equation is frequently referred to as the 2/3 power law. In terms of mass change it can be written as[19]

$$-\frac{dm}{dt} = k''_2 m^{2/3} \tag{1.11}$$

where k''_2 again depends on R_0. Equation 1.10 is usually given in integrated form

$$k'_2 t = 1-(1 - \alpha)^{1/3} \tag{1.12}$$

The dependence of rate on particle size can be included by substitution $k_2 = k'_2/R_0$.

The second limiting case arises when the rate of linear propagation of the reacting interface is so great that each particle decomposes as soon as it has a nucleus. For this case, Hume and Colvin give

$$\text{Rate} = k_0 N_t$$

i.e.,

$$dN/dt = k_0\ N_t \tag{1.13}$$

where N_t is the number of particles at time t. Integration from N_0 to N_t gives

$$\ln(N_t/N_0) = k_0 t \tag{1.14}$$

a first-order reaction. This behavior is most likely to be approached by very small particles.

In general, the kinetics of decomposition is more complex than in the two limiting cases just outlined. Thus, if nucleation is not extremely rapid, the number of nuclei will increase in time. Each of these nuclei will then begin to grow at a different time. Some of them will spread over sites at which nucleation would have occurred at some future time. These sites are referred to as "phantom nuclei" by Jacobs and Tompkins.[6] This problem has been studied in great detail by

Mampel.[10] Both Jacobs and Tompkins and Mampel assume the rate of nuclei formation to be first order,

$$dN/dt = k_1 N_0 \exp(-k_1 t) \tag{1.15}$$

where N_0 is the total number of possible sites. For the induction period, Mampel finds a t^4 law for α, a relation derived by Jacobs and Tompkins for random nucleation and short times, and by Fischbeck and Spingler[11] for the case of no overlap, a constant rate of nuclei formation, and a large total number of nuclei.

For large radii ($R > k_2 t$) Mampel finds

$$\frac{k_2 t}{R} \approx 1-(1-\alpha)^{1/3} \tag{1.16}$$

an equation (1.12) previously given for the contracting sphere case. For small radii the fraction decomposed is

$$X = 1 - ce^{-kt} \tag{1.17}$$

where c is independent of t, but does depend on R in a rather complicated way. The rates are dependent on particle size and both small and large radii give lower rates than intermediate radii.

A generalized equation has been derived by Erofeev[12] in terms of probability theory. In his treatment,

$$\alpha = 1 - \left(-\int_0^t p \, dt \right) \tag{1.18}$$

where p is the probability that a molecule will react in the time interval dt.

The general solution of (1.18) for thermal decomposition is

$$\alpha = 1 - \exp(-kt^n) \tag{1.19}$$

where the constant n depends on the shape of the nucleus and the number of electrons necessary for the formation of a stable nucleus. For example, cylindrical nuclei (the centers of formation are edges or surface cracks) give $n = 3$, flat nuclei give $n = 4$. Thus the evaluation of n, most easily carried out by transforming (1.19) into

$$\ln[-\ln(1-\alpha)] = \ln k + n \ln t \tag{1.20}$$

and plotting the left side of (1.20) vs. $\ln t$, gives some information about the nucleation mechanism.

1.5.2.3 Surface Area Changes Decomposition

Since the molar volume of the product oxides is less than that of the corresponding salts, decomposition results in a porous structure through which the product gases readily escape. Although not directly related to decomposition kinetics, the resulting increase in surface area during decomposition is of some interest in the production of "active solids" and has been treated by several authors. Accordingly, we briefly review it here.

The oxide is first formed at sites on the salt lattice and is therefore in a strained state. Gregg[13] has described these as small crystallites or micelles. The strained pseudo-lattice will tend to recrystallize into a stable oxide lattice with a resulting decrease in surface area. Three mechanisms for this recrystallization are generally recognized. At temperatures $T/T_m < 0.2$, where T_m is the melting point of the oxide, adhesion predominates. This mechanism results from contact between parts of the micelles which are atomically contiguous. For $0.2 < T/T_m < 0.35$, surface diffusion predominates. Here mobile atoms or ions in the surface layer leave their position on the lattice to migrate to sites of lower energies, i.e., to stable oxide lattice positions. Above $T/T_m = 0.5$, sintering, i.e., lattice or bulk diffusion, is the most significant process leading to decreased surface area. Thus, as decomposition proceeds, there are basically two processes leading to changes in surface area: formation of the metastable oxide produces an increase in surface, and formation of stable oxide decreases the area. Consequently, the "surface area" vs. "fraction decomposed" plot exhibits a maximum before decomposition is complete. Nicholson[14] has developed a mathematical model which agrees well with experimental observation. If the purpose of the decomposition is the preparation of an "active solid," the reaction should not be carried to completion in order to achieve the desired maximum surface area.

1.5.3 The Effect of Pressure on Rate

Most of the work on this topic has been carried out on carbonates because only one gas is produced and the analysis is thus inherently simpler. For other salts, several gases are present. Although the principles are basically the same, the discussion has been given here in terms of carbonates.

When the pressure of CO_2 in a carbonate–oxide system is equal to the equilibrium pressure p_e, no net reaction occurs. When $p < p_e$, the thermodynamic driving force favors oxide formation; conversely, when $p > p_e$, carbonate formation is favored. In the actual system the favored reaction may not occur, however, because kinetic factors prevent it. Particularly when p is not far from p_e, the reaction may not proceed because some rate-limiting process, such as nucleus formation, is proceeding too slowly. The resulting spurious equilibria[15] give rise to hysteresis effects, i.e., decomposition stops for some $p < p_e$, recombination stops for some $p > p_e$. It is for this reason that this work relies largely on thermodynamic methods for the calculation of equilibrium pressures.

When the carbonate and oxide are present as pure phases, the value of p_e is completely unambiguous, i.e., at a given temperature $p_e = K_p$, the equilibrium constant for the reaction $MCO_3 = MO + CO_2$. However, when the condensed phase consists of a solid or liquid solution, p_e no longer has a unique value since

$$K = (a_{MO} / a_{MCO3}) p_e \tag{1.21}$$

and p_e now depends on the solution composition. Since the decomposition rate is generally some function of $p_e - p$, we consider only systems without solution formation. We have also omitted discussion of the effect of "foreign" gases, i.e., gases other than CO_2, on the rate. Such gases, e.g., O_2, H_2O, frequently function as catalysts or inhibitors, but these effects are difficult to relate to well-defined properties of the system.[16–18] In order to study the dependence of the reaction rate on CO_2 pressure, it is necessary that the pressure remains fixed throughout the course of the decomposition while more CO_2 is evolving. Since the significant pressure is that at the reaction interface, it is possible that in a closed system where one depends on diffusion to remove the excess CO_2, the pressure at the interface will exceed the nominal value, particularly if the sample size is large and the reaction is rapid. The effect of a pressure gradient has been investigated[19] by the application of Fick's law. For both slow and fast reactions the rate is proportional to $p_e - p_0$ where p_0 is the pressure at the reaction site but the proportionality constants differ in the two cases. One way to avoid the problem of pressure gradients is to use sweep gases made from a known mixture of CO_2

and some inert gas. Independence of decomposition rate from flow rate is generally taken as evidence that the reaction is not diffusion controlled, e.g.,[20] if the surface area is constant, the decomposition rate is proportional to $p_e - p_0$, i.e.,

$$v = k(p_0 - p) \tag{1.22}$$

Such a relation was also found to hold for both cylindrical pellets of calcite and powdered $CaCO_3$.[21] Cremer and Nitsch,[22] in studying the decomposition of $CaCO_3$, found that for samples which followed a 2/3 power rate law the pressure dependence of the rate (in $m^{1/3}t^{-1}$) was given by

$$v = k(1/p - 1/p_0) \tag{1.23}$$

A similar equation has been derived[20] from a two-step model in which the CaO is assumed to be formed first in some "active" state CaO* which occupies a constant fraction of the surface. The reactions then are

$$\underset{1-\Theta}{CaCO_3} \underset{k_2}{\overset{k_1}{\rightleftharpoons}} \underset{\Theta}{CaO^*} + \underset{P}{CO_2}$$

$$CaO^* \underset{k_4}{\overset{k_3}{\rightleftharpoons}} CaO$$

both of which are reversible. The rate

$$R = -dw/dt = \frac{k_0k_1k_3 - Pk_0k_2k_4}{k_1 + k_3 + k_4 + k_2P} \tag{1.24}$$

where the constant k_0 is introduced to convert the surface area to surface concentration in molecules per square centimeters, to account for the roughness factor, and the fact that only a portion of the total surface area may be potentially reactive. Equation 1.24 can be written as

$$R = \frac{1 - CP}{VP + D} \tag{1.25}$$

where $P = 0$, $R = R_0 = 1/D$, $P = P_e$, $R = R_0$, and $C = 1/P_e$
Hence

$$R = \frac{1 - P/P_e}{BP + 1/R_0} \tag{1.26}$$

When $1/R_0 >> BP$,

$$R = 1/BP - 1/P_e$$

which is essentially the equation given by Cremer and Nitsch.[22]

1.5.4 Activation Energy

The activation energy E^{+}, as defined by the usual Arrhenius equation

$$k = Ae^{-E+/RT}$$

is usually at least as great as the thermodynamic enthalpy for the reaction, ΔH, though it is frequently greater. When $E^{+} = \Delta H$, the reaction is often described as occurring "without activation." The question as to whether $E^{+} = \Delta H$ for the endothermic carbonate decomposition was first considered by Fischbeck and Schnaidt.[23] Shortly thereafter, Zawadzki and Bretsnajder[15] showed that for a reversible reaction of the type A(s) = B(s) + C(g), the experimental value of E^{+} for the forward reaction, as determined from the temperature dependence of k, must necessarily increase with rising pressure and will have its minimum value in the absence of the backward reaction, i.e., in a vacuum. This dependence was verified for the decomposition of $CaCO_3$, $CdCO_3$, and Ag_2CO_3. In a vacuum the temperature dependence of k was the same as that of the equilibrium pressure p_e. i.e., $E = \Delta H$. The dependence of E^{+} on the pressure was reexamined by Pavlyuchenko and Prodan,[24] who distinguished two cases: (1) if E^{+} is determined from the temperature dependence of the rate, with the pressure p held constant at the various temperatures, then E^{+} increases as $p \rightarrow p_e$. (2) If, however, the ratio p/p_e is held constant as the temperature is changed, then E^{+} will be independent of the actual numerical value of the ratio and will have the same value as in a vacuum. This conclusion was reached by a theoretical argument and verified by data on the decomposition of $CdCO_3$. Since condition (1) is the more usual experimentally, particularly in the older literature, it is not surprising that widely varying values of E^{+} for the same reaction have been reported[25] by authors who failed to recognize the dependence of E^{+} on pressure.

The atomic mechanism responsible for monomolecular reactions, including thermal decompositions, was first discussed by Polanyi and Wigner.[26] Their model assumes that decomposition occurs when, due to energy fluctuations in the bonds of the molecule, the bond strength is exceeded; or more precisely, the bond energy "resides in harmonic vibrations and that decomposition occurs when their amplitude is exceeded." The resulting expression for the first-order Polanyi–Wigner rate constant is

$$k = \nu \exp(-E^{+}/RT) \tag{1.27}$$

where ν is the atomic frequency of vibration and E^{+} is the activation energy. Thus, a molecule having energy E^{+} would leave the surface.

Shannon[27] analyzed in some detail the theory as it applies to the thermal decomposition of solids. He found that of the 31 reactions for which he compared experimental rate constants with those calculated from the Polanyi–Wigner equation, only a third showed order-of-magnitude agreement. In Shannon's view, the lack of agreement stems from neglecting rotational and other vibrational degrees of freedom.

Shannon calculated the rate constant for thermal decomposition of a solid from absolute reaction rate theory. The resulting equation is of the same form as Equation 1.27, but ν is replaced by a partition function ratio:

$$k = (kT/h)(Q^{*}/Q) \exp(-E/kT) \tag{1.28}$$

where Q^{*} is the complete partition function of the activated complex, excluding that for the reaction coordinate, and Q is the complete partition complex for the reactant. Detailed calculations of the partition functions were carried out for the decomposition of $CaCO_3$ and $MgCO_3$, using known values of the rotational and vibrational partition functions and several models for the activated complex. The best agreement with experiment was obtained for a model in which the carbonate ion in the lattice is free to rotate and the CO_2 molecule leaves directly from the surface of the solid, rather than from a mobile surface layer.

It should be noted that over the past 50 years studies of solid decomposition kinetics have progressed from the application of equations which were originally derived for gases to more and more detailed studies on molecular events occurring at the reactant-product interface. This point has been emphasized by Boldyrev[28] who, with his co-workers, has amassed a large body of evidence that shows the importance of defect structure on kinetics. This structure may be affected by radiation or by mechanical forces, such as grinding, which affect both the surface and the internal structure of crystals.

1.5.5 The Effect of Preparation Method on Rate

Various authors have recognized that the rate of thermal decomposition may depend on the method of preparation and the thermal "history" of the sample, but systematic studies in this area are inherently difficult. However, there has been progress in this field, called "tribochemical decomposition." Boldyrev[28] has distinguished between physical and chemical effects. The main process in the former is the introduction of defects by diminution, e.g., grinding. The latter arise from chemical processes at the contact points of solid particles. Boldyrev[28] has provided a framework for the analysis of these effects and tested his hypotheses with an experimental study of some nitrate and bromate decompositions. Additional problems arise if impurities present in the samples at very low concentrations, such as H_2O, catalyze the decomposition.[29] In that case, the resulting change in mechanism will affect not only the rate but also the activation energy.

If the carbonate is a single crystal, the reaction will propagate from the surface to the interior. For polycrystalline materials the reaction may additionally propagate along grain boundaries.[30]

A systematic study of the effect of sample constitution on the decomposition rate was carried out by Cremer and Nitsch.[22] They decomposed four calcite samples at 850°C and CO_2 pressures ranging from 120 to 220 torr. The CaO product exhibited increasing grain size with increasing pressure. The four CaO samples were then reacted with CO_2 at 850°C and 429 torr to produce four $CaCO_3$ samples with similarly increasing grain size. When these four samples were then decomposed under identical conditions (850°, 200 torr CO_2) the rate of decomposition increased in inverse order of crystal size. The disappearance of the induction period with decreasing crystal size was particularly noticeable. These results are explained in terms of more rapid nucleus formation at lower pressures, leading to smaller CaO crystals and consequently a larger number of active sites in the $CaCO_3$ formed subsequently.

It is evident from these results that the course of the decomposition is affected by the microstructure of the sample and that reproducible preparation and characterization of nominally identical materials are important prerequisites for understanding and systematizing decomposition kinetics.

1.5.6 The Effect of Heat Transfer on Rate

An endothermic solid-state decomposition proceeds as the product–reactant interface advances into the interior of the sample. The interface can, however, advance only if the necessary heat of reaction is applied to it. Since the source of heat is outside the sample, the rate of heat transfer to the interface may become rate-determining if the inherent rate of reaction exceeds the rate of heat transport. The rate of heat transport depends not only on the properties of the product through which heat must be transported, but also on the general experimental arrangement. It is for this reason that one expects and finds the literature to be conflicting.

For example, Narsimhan[30] derived an equation for the thermal decomposition rate based on the idea that the entire particle rapidly reaches the decomposition temperature, so there are no internal temperature gradients. All the heat reaching the reaction interface is used up in the decomposition, so the reaction rate depends on the rate of heat transport. For both spherical and cylindrical particles, good agreement with some literature values is found. Under these conditions the chemical steps are no longer rate-determining. It can thus be seen that in any solid-state reaction whose enthalpy is appreciable, the establishment, maintenance, and measurement of the temperature at the reaction interface is a complicated problem. Thus, careful studies of the decomposition of $PbCO_3$[31] and several other inorganic salts[32] have shown that the induction period may disappear if the entire sample is rapidly brought to the reaction temperature. There is, therefore, some question now as to whether the commonly reported induction periods in the decomposition of so many substances are artifacts of the experimental arrangements. Considerable care in the experiment's design is certainly required to insure a constant and known temperature at the reaction interface throughout the course of the decomposition.

1.5.7 The Influence of Crystal Defect Structure on Decomposition Rate

The idea that the defect structure of a solid reactant affects the rate of decomposition seems to be generally accepted; but of all the factors influencing the kinetics, this is the one most difficult to characterize quantitatively. Boldyrev and his many co-workers[28] have made considerable progress in elucidating the factors which affect the decomposition rate of solids; however, at the level of detail required for the understanding of a particular reaction, it seems difficult to make connections to the standard kinetic equations derived by earlier workers.

Searcy and co-workers[32–34] have addressed the problems that arise from the use of large and/or powdered samples by measuring decomposition from specific faces of small single crystals, using the Langmuir method in which weight loss in a vacuum is measured. The authors cite several advantages of this approach: (1) the possibility that intergranular diffusion may be slow enough to influence the measured rate is eliminated; (2) the area and morphology of the interface at which the reaction proceeds can be precisely measured as a function of time; (3) the ratio of the measured flux of gaseous reaction product which leaves the reactant surface to the maximum (equilibrium) flux which can be predicted from equilibrium data and the kinetic theory of gases, provides a potentially valuable parameter for describing and understanding reaction kinetics, just as it does for congruent vaporization.

The maximum flux is given by the Hertz—Knudsen–Langmuir (HKL) equation

$$J_{max} = \frac{P_{eq}}{(2\pi MRT)^{1/2}} = \frac{\exp(-G_d^0/RT)}{(2\pi MRT)^{1/2}} \tag{1.29}$$

The ratio of the experimental flow J to J_{max}, α_v, the decomposition coefficient, may be useful in assessing the decomposition mechanism. However, account must be taken of formation of the solid product in a metastable state. The formation of this material may be the rate-limiting step. Beruto and Searcy[33,34] have carried out a study of $CaCO_3$ decomposition using these principles. They have also studied the nature of the CaO formed. Thus, the field of solid decomposition kinetics seems to be moving toward a detailed examination of the products formed, with less emphasis on the overall kinetic equations.

REFERENCES

1. Garner, W.E., *Chemistry of the Solid State*, Academic Press, New York, 1955.
2. Young, D. A., *Decomposition of Solids*, Pergamon Press, Oxford, 1966.
3. Tompkins, F. C., Decomposition Reactions, in Reactivity of Solids, *Treatise on Solid State Chemistry*, Vol 4, Hannay, N. B., Ed., Plenum Press, New York, 1976.
4. Langmuir, I., *J. Am. Chem. Soc.*, 38, 2263, 1916.
5. Gafner, G., *Trans. Faraday Soc.*, 55, 981, 1959.

6. Jacobs, P. W. M. and Tompkins, I., Ch. 7 in reference 1.
7. MacDonald, J. Y., *Trans. Faraday Soc.*, 47, 860, 1951.
8. Hume, J. and Colvin, J., *Phil. Mag.*, 8, 589, 1929.
9. Fischbeck, K. and Schnaidt, K., *Z. Elektrochem.*, 38, 649, 1932.
10. Mampel, K., *Z. Phys. Chem.*, A187, 43, 235, 1940.
11. Fischbeck, K. and Spingler, H., *Z. Anorg. Chem.*, 241, 209, 1939.
12. Erofeev, B. V., *Dokl. Akad. Nauk SSSR*, 52, 511, 1946.
13. Gregg, S. J., *J. Chem. Soc.*, 3940, 1953.
14. Nicholson, D., *Trans. Faraday Soc.*, 61, 990, 1965.
15. Zawadzki, J. and Bretsnajder, S., *Z. Elektrochem.*, 41, 721, 1938.
16. Bischoff, F., *Z. Anorg. Chem.*, 252, 10, 1942; ibid. 262, 18, 1950.
17. Huttig, G. F. and Hermann, E., *Kolloid Z.*, 92, 9, 1940.
18. Huttig, G. F. and Strial, K., *Z. Anorg. Chem.*, 289, 249, 1932.
19. Barret, P. and Bonnefair, L., *Bull. Soc. Chim. France,* 576, 1961.
20. Hyatt, E. P., Cutler, I. B., and Wadsworth, M. E., *J. Am. Ceram. Soc.,* 41, 70, 1958.
21. Ingraham, T. R. and Marier, P., *Can. J. Chem. Eng.*, 41, 170, 1963.
22. Cremer, E. and Nitsch, E., *Z. Elektrochem.*, 66, 697, 1962.
23. Fischbeck, K. and Schnaidt, K., *Z. Elektrochem.*, 38, 649, 1932.
24. Pavlyuchenko, M. M. and Prodan, E. A., *5th International Symp. Reactivity of Solids*, Schwab, G. M., Ed., Elsevier, Amsterdam, 407, 1965.
25. Stern, K.H. and Weise, E.L., *High Temperature Properties of Inorganic Salts, Part 1, Sulfates*, NSRDS-NBS - 7, 1966.
26. Polanyi, M. and Wigner, E., *Z. Phys. Chem.*, A 139, 439, 1928.
27. Shannon, R. D., *Trans. Faraday Soc.,* 60, 1902, 1964.
28. Boldyrev, V. V., *Thermochim. Acta,* 100, 315, 1986; *Z. Phys. Chem. (Leipzig),* 256, 342, 1975.
29. Prodan, E. A. and Pavlyuchenko, M. M., *Geterogennye Khim. Rektsii Inst. Obshch. i Neorg. Khim. Akad. Nauk Belorusski SSSR,* 64, 1965.
30. Narsimhan, G. V., *Chem. Eng. Sci.*, 16, 7, 1961.
31. Zhabrova, G. M. and Gordeeva, V. A., *Izvest. Akad. Nauk SSSR Otdel Khim. Nauk,* 569, 1959.
32. Basu, T. K. and Searcy, A.W., *J. Chem. Soc. Faraday Trans.,* I 72, 1889, 1976.
33. Beruto, D. and Searcy, A.W., *J. Chem. Soc. Faraday,* 70, 2145, 1974.
34. Searcy, A.W. and Beruto, D., *J. Phys. Chem.*, 80, 425, 1976.

2 Carbonates

2.1 INTRODUCTION

2.1.1 STRUCTURE OF THE CARBONATE ION

The carbonate ion is a well-defined entity in which the three oxygens are arranged in the same plane as the carbon atom at a distance of 1.30 + 0.01 A with bond angles of 120°. This value of the C–O distance, calculated by Pauling[1] on the basis of the resonance structures shown in Figure 2.1, is in good agreement with experimental measurements on calcite,[2,3] ranging from 1.29 to 1.31 A. The resonance energy is 176 kJ per mole.

2.1.2 GENERAL FEATURES OF CARBONATE DECOMPOSITION

All carbonates decompose with the evolution of CO_2, the decomposition product being either the corresponding oxide or a basic carbonate, the latter decomposing with further evolution of CO_2 as the temperature is raised. The oxides of some metals are unstable in the same temperature range as the corresponding carbonates, so the decomposition occurs as simultaneous reactions

$$MCO_3 = MO + CO_2(g)$$

$$MO = M(g) + 1/2\ O_2(g)$$

Consequently, the gas phase contains CO_2, M, and O_2. The oxide equilibria have not been calculated, but the text for particular elements indicates when they are important, such as for the higher atomic alkali weight metals.

The melting points of pure carbonates are generally not measurable directly, though they may be estimated by extrapolation from phase diagrams or by rapid heating, particularly under high CO_2 pressures which minimizes decomposition.

The dissociation $CO_2 = CO + 1/2\ O_2$ is too slight at temperatures relevant for carbonate decomposition, but a separate table for this reaction has been provided for up to 2400 K.

The thermal stability of a polyatomic anion decreases with increasing polarizing power of the associated cation, since polarization of the anion structure leads to the distortion and consequent weakening of the intra-ionic bonds. This topic has been discussed in detail elsewhere.[4] Here we provide Figure 2.2 which plots ΔHf_{298}, a measure of thermal stability, against $r^{1/2}/Z^*$, a measure of cation polarizing power, for the carbonates. It can be seen that the plot is approximately linear and thus can be used for predicting one parameter when the other is known. Similar plots can be constructed for the other anions in this book,[4] but will not be reproduced.

The discussion of the high-temperature properties and thermal decomposition of carbonates in this chapter, a plan which will be followed for other classes of compounds in subsequent chapters, is given for each element. A brief discussion is followed by quantitative data on phase transitions, densities, and thermodynamic parameters for the carbonate, the corresponding oxide, and the decomposition reaction $MCO_3 = MO + CO_2$. If the carbonate and oxide are solids and form no solid solution, the equilibrium constant is the pressure of CO_2 which would be obtained if one begins with an evacuated container and lets the system come to equilibrium. In this case, the pressure is a unique function of the temperature. As pointed out in Chapter 1, this is no longer true

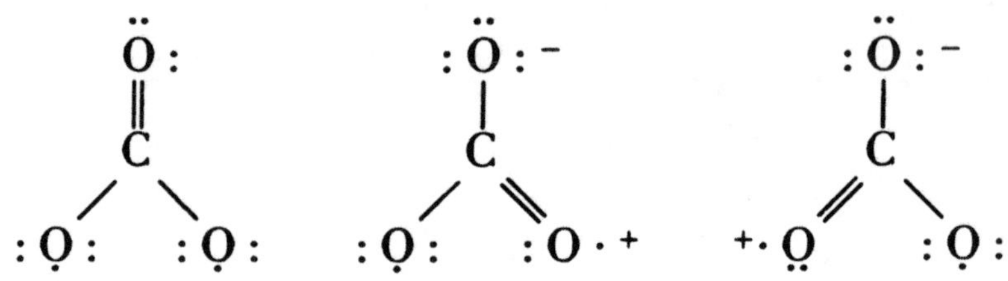

FIGURE 2.1 Structure of the carbonate (CO_3) ion. (From Pauling, L., *The Nature of the Chemical Bond*, 3rd ed., Cornell University Press, Ithaca, NY, 1960. With permission.)

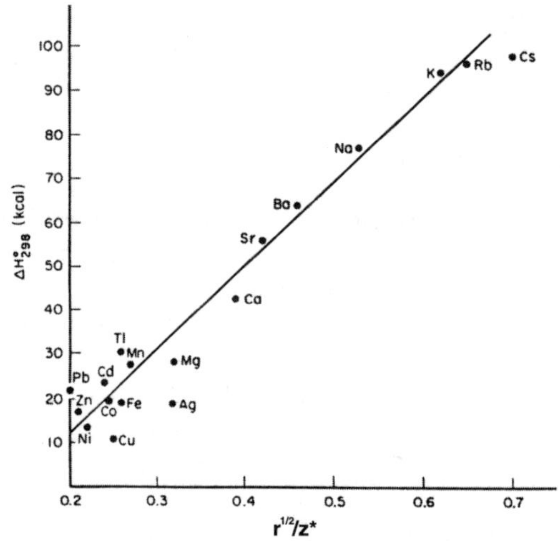

FIGURE 2.2 ΔH_{298} vs. $r^{1/2}/Z^*$ for the carbonates.

if the oxide and carbonate form a solid or liquid solution. In that case, K = (a_{MO} / a_{MO_3}). P, and the pressure of CO_2 depends on the condensed phase composition.

One aspect of carbonate decomposition that should be kept in mind is the quality of the "anhydrous" starting materials. In many cases they were prepared by heating hydrates that were not completely dehydrated, and the residual water may affect the decomposition, particularly the kinetics.

2.1.3 Phase Transitions

This topic was discussed in Chapter 1.3. Transitions of carbonates are listed under the individual salts. When they are taken from the NBS compilation given in Chapter 1, the reference is 1.3.

2.1.4 Density

This topic was treated in detail in 1.4. Densities for individual carbonates are given for each salt.

2.2 CO, CO_2, AND O_2

The CO_2 produced by the decomposition of a carbonate dissociates partly to CO and O_2. However, at temperatures for which most of the carbonate equilibria have been calculated, the decomposition of CO_2 is very slight. Therefore, the decomposition equilibria for carbonates have been given in terms of CO_2 only. However, for those wishing to refine the calculations, thermodynamic functions for the three gases and the equilibrium constant for the reaction $CO_2 = CO + O_2$ have been listed for up to 2400 K in Tables 2.1 to 2.4. It should also be noted that if a carbonate is decomposed in air (0.21 atm O_2) the decomposition will be partially repressed.

2.3 THE INDIVIDUAL CARBONATES

The known carbonates are shown in Figure 2.3.

2.3.1 Barium

The effect of the $BaCO_3$ transitions at 1079 and 1241 K on the calculated equilibrium pressure is virtually nil. The solid transitions (Table 2.5) of $BaCO_3$ exhibit pronounced hysteresis, i.e., different temperatures are obtained, depending on whether the transition is approached from above or from below.[6] The melting point can only be measured under high CO_2 pressure. ΔH and ΔS for the transitions are taken from Chapter 1.3. Densities (Table 2.6) are only known for the solid. Thermodynamic functions for $BaCO_3$ and BaO and the equilibrium $BaCO_3 = BaO + CO_2$ are given in Tables 2.7 to 2.9. The salt is evidently quite stable to 1200 K.

2.3.2 Beryllium

Normal anhydrous $BeCO_3$ may not exist at ambient temperatures. Dehydration of the hydrate at elevated temperatures probably forms basic carbonates.[10] The values of ΔHf° for $BeCO_3$ are calculated from a measured value of ΔH for the reaction $BeO + CO_2 = BeCO_3$[10] and the current value for BeO. The known thermodynamic values are summarized in Table 2.10.

2.3.3 Cadmium

The density (Table 2.11) must be regarded as very uncertain since there are no 20th century measurements, and De Schulten[12] reports a value of 4.960.

Thermodynamic data (Tables 2.12 to 2.14) indicate that the salt is unstable above 500 K.

The rate of decomposition has been measured as a function of CO_2 pressure by Centnerszwer and Brusz[14] and Prodan and Pavlyuchenko.[15] The decomposition yields CdO without the formation of intermediates. In vacuum the activation energy is 151 kJ. The rate of decomposition depends on the method of preparation. The kinetics follows the contracting sphere model.

2.3.4 Calcium

The aragonite–calcite transition has been studied as a function of temperature and pressure.[16,17] At ambient pressure the transition occurs near 730 K.[18,19] In addition, a transition from vaterite to calcite occurs in the range 620 to 670 K, with small impurity concentrations producing large changes in the transition temperature.[19] This information is summarized in Table 2.17. Thermodynamic data (Tables 2.18 to 2.20) indicate that $CaCO_3$ is stable to 1100–1200 K. The density is given in Table 2.15, and the expansion coefficient of $CaCO_3$ in Table 2.16.

2.3.5 Cesium

The melting point of Cs_2CO_3 is 1065 K.[23] Transitions are listed in Table 2.21. Very little is known of its decomposition, but the thermodynamic data (Tables 2.22 to 2.24) indicate considerable thermal stability.

2.3.6 Cobalt

Experimental information on the thermal decomposition of $CoCO_3$ is scarce. A number of basic oxides have been reported,[24] but their existence cannot be regarded as proved. In a more recent

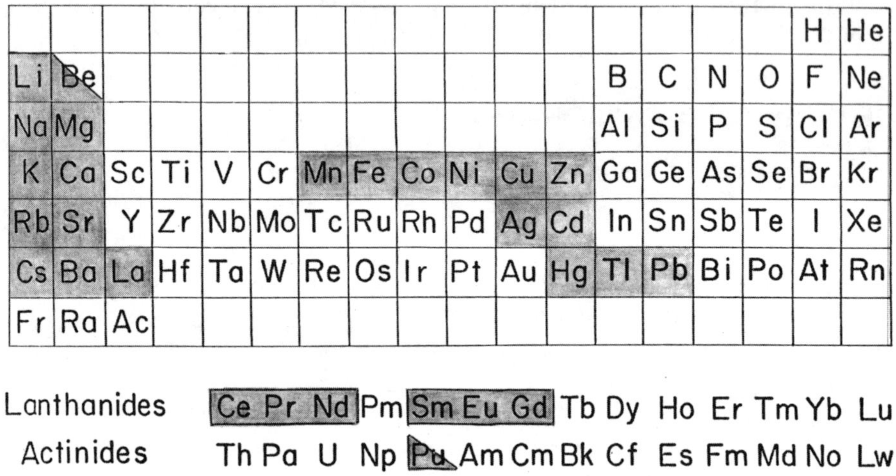

FIGURE 2.3 The known carbonates: full square, existence known; half square, existence uncertain.

study[25] no intermediates were reported, but the decomposition pressure exhibited an unexplained break in the log p vs. 1/T plot.

In view of several density values in the 4.1 to 4.2 range (Table 2.25), the value of 2.818 for spherocobaltite given in [13] is probably in error.

Thermodynamic data (Tables 2.26 to 2.28) indicate that $CoCO_3$ is unstable toward decomposition above 400 K.

2.3.7 Copper

Little is known of $CuCO_3$, but thermodynamic data have been published (Tables 2.29 to 2.31); however, they are based on extrapolations. The decomposition reaction (Table 2.31) is spontaneous from ambient temperature upward, but the salt may be stable for kinetic reasons.

2.3.8 Iron

Anhydrous $FeCO_3$ can be prepared by heating the 1/2 hydrate in CO_2 at high pressure (40 to 50 atm) to near 300°C. The thermal decomposition always leads to a higher-valent oxide, since the CO_2 liberated immediately oxidizes the FeO to Fe_2O_3.[28] The reaction may also be written

$$3\ FeCO_3 = Fe_2O_3 + 2\ CO_2 + CO$$

The oxidation occurs so fast that even in vacuum no FeO is formed. Thermodynamic data that are available at 298 K are given in Table 2.34; density in Table 2.32. Data on the thermal expansion of $FeCO_3$ are given in Table 2.33.

2.3.9 Lead

Several studies[32–35] indicate that decomposition of $PbCO_3$ results in the formation of several oxycarbonates, but which ones are formed depends on the atmosphere — CO_2 and its pressure, air, or nitrogen — and temperature. The following intermediates have been reported: $PbO.PbCO_3$, $3\ PbO.5PbCO_3$, $2\ PbO.PbCO_3$, and $3\ PbO.PbCO_3$.

Because of the variety of conditions employed, the reported equilibrium pressures and temperatures between the various oxycarbonates probably do not have much thermodynamic significance. Therefore, only thermodynamic parameters for decomposition to PbO have been listed (Tables 2.35

to 2.37). These thermodynamic results are consistent with experimental studies which indicate that decomposition occurs between 500 and 700 K.

Since there is so much disagreement about the intermediates formed, the kinetics measured by various workers can also not be considered reliable. Not only are the decomposition products not known reliably, but it has been reported that $PbCO_3$ specimens of different particle size decompose in different ways under identical conditions.

The density at ambient temperature has been reported as 6.533[7] and 6.582.[8a]

2.3.10 Lithium

The melting point of $LiCO_3$ is still uncertain. The measurements[36,37] are considerably lower than the value of 1008 K given in Chapter 1.3, but differ from each other by 6 degrees. In both these studies the salt was protected from decomposition by being kept under a pressure of CO_2 high enough to prevent weight changes. ΔH_m given by Janz et al.[38] and Rolin and Recapet[39] differ by 7 kJ. These data are collected in Table 2.39. Density data are given in Table 2.38.

Equilibrium pressures of CO_2 above liquid Li_2CO_3 were measured by Janz and Lorenz,[40] using a dynamic method in which the temperature was varied until weight changes (losses on heating, gains on cooling) were noted. The resulting pressures were larger by several orders of magnitude than K values determined from thermodynamic data. This discrepancy can be accounted for by remembering that in a liquid decomposition P depends on liquid composition (see Chapter 1.2), and Li_2O certainly dissolves in the liquid salt. This also means that above the melting point, K values in Table 2.42 no longer equal partial pressures. Thermodynamic data are given in Tables 2.40 to 2.42.

2.3.11 Magnesium

In vacuum, $MgCO_3$ decomposes to MgO at temperatures as low as 600 K, the solids exhibiting X-ray lines corresponding to $MgCO_3$ and MgO only.[41] In an earlier work Centnerszwer and Bruzs[42] reported the following equilibria:

$$MgCO_3 \leftrightarrow MgO.MgCO_3 \leftrightarrow 3\ MgO.MgCO_3 \leftrightarrow 4\ MgO + CO_2$$

These equilibria are based entirely on pressure measurements and weight loss data. Until structural evidence for the existence of these compounds is produced, their existence must be regarded as doubtful.

The decomposition follows "rapid nucleation and contracting sphere" kinetics with an activation energy of 150 kJ, somewhat higher than the enthalpy of decomposition.[43–45] Densities are given in Table 2.43, thermal expansion coefficient in Table 2.44, and thermodynamic data in Tables 2.45 to 2.47. From data in Table 2.47, it is apparent that $MgCO_3$ is unstable above 700 K, in agreement with quoted experimental studies.

2.3.12 Manganese

The thermal decomposition of $MnCO_3$ is complicated by the easy interconvertability of the various manganese oxides. When $MnCO_3$ is decomposed in O_2-containing atmospheres, the oxide formed depends on the temperature. In addition, a number of non-stoichiometric oxides are also formed,[46] the particular oxide formed depending primarily on the temperature and partial pressure of oxygen. In the absence of oxygen, the decomposition product is MnO, at least up to 600–700 K.[47] Above this temperature range MnO may be oxidized by the CO_2 formed according to Mahanty and Ray[48]

$$MnO + 1/3CO_2 = 1/3Mn_3O_4 + 2\ CO_2$$

The decomposition can thus also be thought of as

$$3\ MnCO_3 = Mn_3O_4 + 2\ CO_2 + CO$$

The actual temperature at which such transformations occur depends, therefore, on the efficiency of CO_2 removal and on the partial pressure of CO_2. The thermodynamics of many reactions involving manganese compounds have been given by Mah.[49] The predominant mode of decomposition of $MnCO_3$ is $MnCO_3 = MnO + CO_2$. We have, therefore, given equilibrium constants for this reaction and for the reaction $3\ MnCO_3 = Mn_3O_4 + 2\ CO_2 + CO$. Thermodynamic functions for other oxides are listed in Mah and Coughlin.[49,50] There are many studies of $MnCO_3$ decomposition, e.g., Lamure and Iltis and Mu and Perlmutter,[52,53] but the results seem to depend largely on experimental conditions: atmosphere, temperature, and the condition of the carbonate. We have, therefore, included only the calculated thermodynamic results Tables 2.49 to 2.51). For both reactions the calculated results show that decomposition becomes significant in the 700 to 800 K range, in agreement with experimental results. The kinetics follows the contracting sphere model.

The density at 294 K has been given as 3.125[13] and 3.633[29] (Table 2.48).

2.3.13 Mercury

2.3.13.1 $HgCO_3$

Most authors indicate that the compound does not exist, but contrary evidence has been presented by Erdoes[87] who gives dissociation pressures for the reaction

$$HgCO_3 = HgO + CO_2(g)$$

Thermodynamic data for the salt are given in Table 2.53, and for decomposition in Table 2.54. These indicate decomposition above 500 K.

2.3.13.2 Hg_2CO_3

This compound has been reported to be very sensitive to light and to decompose near 100°C.[88] Thermodynamic data for the salt are given in Table 2.55. Data for the decomposition (Table 2.56) are in approximate agreement with this.

2.3.14 Nickel

Yellow and green forms of $NiCO_3$ have been reported, with densities of 4.36 and 4.395, but they may be identical. The salt decomposes directly to NiO, decomposition being complete at 300 K (Tables 2.57 to 2.59).[53]

2.3.15 Potassium

Transitions are listed in Table 2.61, densities in Table 2.60, and thermodynamic functions in Tables 2.63 to 2.65. The equilibrium calculations have been carried out for $K_2CO_3 = K_2O(s) + CO_2(g)$, although both $K_2O(g)$ and KO(g) have been reported to be minor constituents above decomposing K_2CO_3.[54]

2.3.16 Rare Earths

Although hydrates of rare earth carbonates at ambient temperature are well-known, their dehydration most likely leads to a series of basic carbonates. However, anhydrous carbonates of La, Ce, Pr, Nd, Sm, Eu, and Gd have been reported by Head and Holley,[89] who carefully decomposed the hydrates by TGA and removed the theoretical amounts of water.

2.3.17 Rubidium

Very little information, aside from thermodynamic data, is available. These (Tables 2.65 to 2.67) indicate that the salt is stable even to 1500 K. The melting point is 1143 K.[56]

2.3.18 Silver

The thermal decomposition of Ag_2CO_3 is complicated by the fact that the decomposition product, Ag_2O, decomposes further ($Ag_2O = 2\ Ag + 1/2\ O_2$; for thermodynamic data see Otto[57]). Therefore, weight loss experiments can only be carried out in a rather narrow temperature range, < 520 K, where Ag_2CO_3 decomposes but Ag_2O does not. Alternatively, at higher temperatures the reaction can be followed by absorbing CO_2 and O_2 in separate traps and measuring their respective weight changes.

The kinetics are most consistent with a model of rapid nucleation followed by contracting sphere kinetics.[58] However, the actual rates are extremely dependent on the method of preparation. Kadlets and Dubinin[35] studied this effect by preparing Ag_2CO_3 using two different methods: one by precipitation from concentrated solutions which precipitated rapidly, the other by precipitation from very dilute solution so the salt first formed as a supersaturated solution. Although both preparations exhibited the same X-ray diffraction pattern, the second preparation decomposes much more rapidly than the first, an effect which the authors attribute to a higher defect density. Direct evidence is lacking, however. We have given data both for the decomposition of Ag_2CO_3 and of Ag_2O (Tables 2.68 to 2.71).

The density of the salt at ambient temperature has been reported as 6.077[13] and 6.131[59] by X-ray diffraction.

2.3.19 Sodium

The literature on the phase transitions of Na_2CO_3 is somewhat contradictory. The transitions at 629 and 759 K, but not the one at 891 K, were observed by Reisman,[37] as was the melting point. Two other transitions, at 593 and 723 K, were observed by Ginzburg[60] and Popov,[61] and very similar values also by Jaffray and Martin.[62] ΔH_m values of Janz et al.,[55] and Rolin and Recarpet[39] differ by 5 kJ; the average value has been listed in Table 2.73, as has all other information on transitions. According to Khlapova[63] who studied the polymorphism of Na_2CO_3 by DTA, the actual transition temperature will vary somewhat, depending on sample preparation and treatment, in some cases by 10 or 15 degrees. If this is so, solid–solid transitions would require careful reinvestigation.

Because the salt is stable well into the liquid range, reliable density values could be measured. These are listed in Table 2.72.

Decomposition pressures were measured by Janz and Lorenz,[36] but the same comments apply to them as were made for Li_2CO_3. The difficulties encountered in measuring equilibrium pressures above the liquid salt by the effusion method have been described in great detail by Motzfeld.[64] He showed that above the melting point, CO_2, Na(g), and O_2 vaporize from the melt, but the activation energy for evaporation of CO_2 is quite high. Consequently, both the observed pressure and the steady-state composition of the melt depend on the size of the effusion orifice. In this work the pressures of these three gases have not been given, but they can be calculated from the data for the dissociation of the oxide given in Chase et al.[65] and the tables in this work. All of the thermodynamic data pertaining to the dissociation of Na_2CO_3 are given in Tables 2.74 to 2.76.

2.3.20 Strontium

The decomposition of $SrCO_3$ resembles that of $BaCO_3$. Wanmaker and Radielovic[66] studied the decomposition in air gravimetrically and reported that it commenced at 1120 K and ended at 1450 K. Up to 1220 K the decomposition rate was constant; at higher temperatures a change from zero-order to first-order kinetics occurred as the reaction proceeded beyond $\alpha = 0.5$. A kinetic study of the effect of processing parameters between 1170 and 1270 K shows that the slow step is the diffusion of the reaction components to the product layer.[69]

The density of the salt as strontianite at 298 K was reported as 3.736,[51] and for the precipitated salt as 3.585.[67]

Phase changes are given in Table 2.77, and thermodynamic data related to decomposition in Tables 2.78 to 2.80.

2.3.21 Thallium

There seem to be no high-temperature thermodynamic data for Tl_2CO_3. Data for 298.15 K are given in Table 2.82

Transitions are listed in Table 2.81. The density at 298 K has been given as 7.11[13] and as 7.23 from X-ray diffraction.[73]

2.3.22 Zinc

The decomposition of $ZnCO_3$ has been studied by several authors. Huttig, Meller, and Lehman[75] used natural $ZnCO_3$, smithsonite. Bretsnajder and Cibor[76] used synthetic material. In neither case was the material pure $ZnCO_3$. There is also some indication that the decomposition is not entirely reversible.[77,78] It is generally agreed that for most of the concentration range contracting sphere kinetics are followed, but the details of the induction period are not certain. Gedalla[79] analyzed the kinetics of heating curves obtained in air in terms of mechanism, but could not reach a definite conclusion. Various values have been reported for the activation energy, but it is not certain whether the value at zero pressure equals the reaction enthalpy. Thermodynamic data related to the decomposition are given in Tables 2.83 to 2.85.

Ambient temperature densities have been reported as 4.4,[13] 4.348,[7] and 4.333[8a] for X-ray diffraction.

TABLE 2.1
Thermodynamic Data for CO_2(g)

T K	ΔHf° kJ mol⁻¹	S° J mol⁻¹ deg⁻¹	ΔGf° kJ mol⁻¹	Reference
298.15	–393.51	213.77	–457.24	Barin 1993
400.00	–389.45	225.44	–479.62	Barin 1993
500.00	–385.03	235.28	–502.67	Barin 1993
600.00	–380.35	243.81	–526.63	Barin 1993
700.00	–375.46	251.34	–551.40	Barin 1993
800.00	–370.41	258.08	–576.88	Barin 1993
900.00	–365.23	264.19	–602.99	Barin 1993
1000.00	–359.91	269.78	–629.70	Barin 1993
1100.00	–354.49	274.95	–656.94	Barin 1993
1200.00	–348.97	279.76	–684.67	Barin 1993
1300.00	–343.35	284.25	–712.88	Barin 1993
1400.00	–337.65	288.47	–741.51	Barin 1993
1500.00	–331.88	292.46	–770.56	Barin 1993
1600.00	–326.04	296.23	–800.00	Barin 1993
1700.00	–320.13	299.81	–829.80	Barin 1993
1800.00	–314.17	303.21	–859.95	Barin 1993
1900.00	–308.16	306.46	–890.44	Barin 1993
2000.00	–302.10	309.57	–921.24	Barin 1993
2100.00	–296.01	312.54	–952.35	Barin 1993
2200.00	–289.89	315.39	–983.74	Barin 1993
2300.00	–283.75	318.12	–1015.42	Barin 1993
2400.00	–277.58	320.74	–1047.37	Barin 1993

TABLE 2.2
Thermodynamic Data of CO(g)

T K	ΔHf° kJ mol⁻¹	S° J mol⁻¹ deg⁻¹	ΔGf° kJ mol⁻¹	Reference
298.15	–110.54	197.66	–169.47	Barin 1989
400.00	–107.55	206.28	–190.07	Barin 1989
500.00	–104.59	212.90	–211.04	Barin 1989
600.00	–101.57	218.40	–232.61	Barin 1989
700.00	–98.50	223.14	–254.69	Barin 1989
800.00	–95.37	227.31	–277.22	Barin 1989
900.00	–92.19	231.06	–300.14	Barin 1989
1000.00	–88.95	234.47	–323.42	Barin 1989
1100.00	–85.66	237.60	–347.02	Barin 1989
1200.00	–82.33	240.51	–370.91	Barin 1989
1300.00	–78.94	243.21	–395.12	Barin 1989
1400.00	–75.52	245.75	–419.57	Barin 1989
1500.00	–72.05	248.14	–444.26	Barin 1989
1600.00	–68.55	250.40	–469.19	Barin 1989
1700.00	–65.01	252.55	–494.34	Barin 1989
1800.00	–61.44	254.59	–519.70	Barin 1989
1900.00	–57.83	256.54	–545.26	Barin 1989
2000.00	–54.21	258.40	–571.00	Barin 1989
2100.00	–50.56	260.18	–596.93	Barin 1989
2200.00	–46.88	261.89	–623.04	Barin 1989
2300.00	–43.19	263.53	–649.31	Barin 1989
2400.00	–39.49	265.11	–675.74	Barin 1989

TABLE 2.3
Thermodynamic Data of O_2(g)

T K	ΔHf° kJ mol⁻¹	S° J mol⁻¹ deg⁻¹	ΔGf° kJ mol⁻¹	Reference
298.15	0.00	205.15	–61.17	Barin 1993
400.00	3.06	213.96	–82.53	Barin 1993
500.00	6.21	220.96	–104.28	Barin 1993
600.00	9.44	226.87	–126.68	Barin 1993
700.00	12.74	231.95	–149.63	Barin 1993
800.00	16.09	236.42	–173.05	Barin 1993
900.00	19.45	240.42	–196.90	Barin 1993
1000.00	22.92	244.04	–221.12	Barin 1993
1100.00	26.39	247.35	–245.70	Barin 1993
1200.00	29.91	250.41	–270.59	Barin 1993
1300.00	33.45	253.25	–295.77	Barin 1993
1400.00	37.03	255.90	–321.23	Barin 1993
1500.00	40.65	258.39	–346.95	Barin 1993
1600.00	44.29	260.74	–372.90	Barin 1993
1700.00	47.96	262.97	–399.09	Barin 1993
1800.00	51.66	265.09	–425.49	Barin 1993
1900.00	55.39	267.10	–452.10	Barin 1993
2000.00	59.15	269.03	–478.91	Barin 1993
2100.00	62.93	270.87	–505.91	Barin 1993
2200.00	66.74	272.65	–533.08	Barin 1993
2300.00	70.56	274.35	–560.43	Barin 1993
2400.00	74.42	275.99	–587.95	Barin 1993

TABLE 2.4
CO_2 (g) = CO(g) + _ O_2 (g)

T K	ΔH° kJ	ΔS° J deg^{-1}	ΔG° kJ	K
298.15	282.96	86.47	257.18	8.682E – 046
400.00	283.42	87.82	248.29	3.745E – 033
500.00	283.55	88.11	239.49	9.513E – 026
600.00	283.50	88.02	230.68	8.233E – 021
700.00	283.33	87.77	221.89	2.759E – 017
800.00	283.09	87.45	213.13	1.210E – 014
900.00	282.78	87.08	204.41	1.367E – 012
1000.00	282.42	86.71	195.72	5.971E – 011
1100.00	282.03	86.33	187.06	1.307E – 009
1200.00	281.60	85.96	178.45	1.705E – 008
1300.00	281.14	85.59	169.87	1.492E – 007
1400.00	280.65	85.23	161.33	9.554E – 007
1500.00	280.15	84.88	152.83	4.761E – 006
1600.00	279.63	84.55	144.35	1.936E – 005
1700.00	279.10	84.23	135.92	6.660E – 005
1800.00	278.56	83.92	127.51	1.993E – 004
1900.00	278.02	83.63	119.13	5.304E – 004
2000.00	277.47	83.34	110.78	1.278E – 003
2100.00	276.92	83.08	102.46	2.826E – 003
2200.00	276.38	82.82	94.17	5.808E – 003
2300.00	275.84	82.58	85.90	1.120E – 002
2400.00	275.30	82.36	77.65	2.041E – 002

TABLE 2.5
Transitions of $BaCO_3$

Phase Change	T(K)	ΔH kJ/mol	ΔS J/(mol*K)	Reference
c, orthorhombic(γ) →	1076	14.9	13.7	1.3
c, hexagonal(β) ← c, ortho-rhombic	1023			6
c, hexagonal → c, cubic(α)	1249	3.0		6
c, hexagonal(β) ← c, cubic	1229			
c(α) → l	1653			1.3

TABLE 2.6
Density of $BaCO_3$

Phase	T(K)	d	Reference
c	293	4.287	7
c (witherite)	299	4.308 (X-ray)	8a
c (cubic)	1348	3.889 (X-ray)	8c

TABLE 2.7
Thermodynamic Data of $BaCO_3$

T K	ΔHf° kJ mol⁻¹	S° J mol⁻¹ deg⁻¹	ΔGf° kJ mol⁻¹	Reference
298.15	–1216.29	112.13	–1249.72	Barin 1993
400.00	–1206.89	139.15	–1262.55	Barin 1989
500.00	–1196.66	161.96	–1277.64	Barin 1989
600.00	–1185.65	182.00	–1294.85	Barin 1989
700.00	–1173.99	199.97	–1313.97	Barin 1989
800.00	–1161.71	216.36	–1334.80	Barin 1989
900.00	–1148.85	231.50	–1357.20	Barin 1989
1000.00	–1135.42	245.65	–1381.06	Barin 1989
1100.00	–1103.60	275.49	–1406.64	Barin 1989
1200.00	–1088.12	288.96	–1434.88	Barin 1989
1300.00	–1069.02	304.26	–1464.56	Barin 1989
1400.00	–1052.70	316.36	–1495.60	Barin 1989
1500.00	–1036.38	327.61	–1527.80	Barin 1989

TABLE 2.8
Thermodynamic Data of BaO

T K	ΔHf° kJ mol⁻¹	S° J mol⁻¹ deg⁻¹	ΔGf° kJ mol⁻¹	Reference
298.15	–553.54	70.29	–574.50	Barin 1993
400.00	–548.57	84.62	–582.42	Barin 1993
500.00	–543.47	96.00	–591.47	Barin 1993
600.00	–538.21	105.58	–601.56	Barin 1993
700.00	–532.83	113.87	–612.54	Barin 1993
800.00	–527.35	121.19	–624.30	Barin 1993
900.00	–521.77	127.76	–636.75	Barin 1993
1000.00	–516.11	133.72	–649.83	Barin 1993
1100.00	–510.36	139.20	–663.48	Barin 1993
1200.00	–504.53	144.28	–677.66	Barin 1993
1300.00	–498.62	149.01	–692.32	Barin 1993
1400.00	–492.63	153.44	–707.45	Barin 1993
1500.00	–486.56	157.62	–723.00	Barin 1993

TABLE 2.9
$BaCO_3$ = BaO + CO_2(g)

T K	ΔH° kJ	ΔS° J deg^{-1}	ΔG° kJ	K
298.15	269.24	171.93	217.98	6.421E – 039
400.00	268.88	170.91	200.51	6.506E – 027
500.00	268.16	169.32	183.50	6.734E – 020
600.00	267.10	167.39	166.66	3.087E – 015
700.00	265.69	165.23	150.03	6.364E – 012
800.00	263.95	162.90	133.62	1.882E – 009
900.00	261.85	160.44	117.45	1.523E – 007
1000.00	259.40	157.86	101.54	4.963E – 006
1100.00	238.76	138.66	86.23	8.035E – 005
1200.00	234.63	135.07	72.55	6.949E – 004
1300.00	227.05	128.99	59.36	4.118E – 003
1400.00	222.42	125.56	46.64	1.819E – 002
1500.00	217.93	122.46	34.24	6.422E – 002

TABLE 2.10
Thermodynamic Data of $BeCO_3$

	ΔH (kJ mol^{-1})	Reference	ΔS (J mol^{-1} deg^{-1})	Reference
$BeCO_3$	–102.5	1.2	—	
BeO	–606.09	1.2	14.15	1.2

TABLE 2.11
Density of $CdCO_3$

Phase	T(K)	d	Reference
c	293	4.259 g ml^{-1}	13

TABLE 2.12
Thermodynamic Data of $CdCO_3$

T K	ΔHf° kJ mol^{-1}	S° J mol^{-1} deg^{-1}	ΔGf° kJ mol^{-1}	Reference
298.15	–751.87	92.47	–779.43	Barin et al. 1977
400.00	–742.79	118.55	–790.21	Barin et al. 1977
500.00	–732.55	141.35	–803.23	Barin et al. 1977
600.00	–720.99	162.39	–818.42	Barin et al. 1977
700.00	–708.11	182.21	–835.66	Barin et al. 1977
800.00	–693.92	201.14	–854.84	Barin et al. 1977
900.00	–678.41	219.40	–875.87	Barin et al. 1977
1000.00	–661.58	237.12	–898.70	Barin et al. 1977

TABLE 2.13
Thermodynamic Data of CdO

T K	ΔHf° kJ mol^{-1}	S° J mol^{-1} deg^{-1}	ΔGf° kJ mol^{-1}	Reference
298.15	–258.99	54.81	–275.33	Barin et al. 1977
400.00	–254.37	68.12	–281.62	Barin et al. 1977
500.00	–249.60	78.75	–288.98	Barin et al. 1977
600.00	–244.69	87.70	–297.31	Barin et al. 1977
700.00	–239.67	95.46	–306.48	Barin et al. 1977
800.00	–234.55	102.28	–316.37	Barin et al. 1977
900.00	–229.35	108.40	–326.91	Barin et al. 1977
1000.00	–224.08	113.96	–338.03	Barin et al. 1977

TABLE 2.14
$CdCO_3$ = CdO + CO_2(g)

T K	ΔH° kJ	ΔS° J deg^{-1}	ΔG° kJ	K
298.15	99.37	176.11	46.86	6.156E – 009
400.00	98.97	175.00	28.97	1.645E – 004
500.00	97.92	176.68	11.58	6.172E – 002
600.00	95.96	169.13	–5.52	3.025E + 002
700.00	92.99	164.57	–22.21	4.548E + 001
800.00	88.96	159.21	–38.41	3.222E + 002
900.00	83.83	153.19	–54.03	1.369E + 003
1000.00	77.59	146.62	–69.03	4.036E + 003

TABLE 2.15
Density of $CaCO_3$

Phase	T(K)	d	Reference
c, calcite	273	2.771 g ml^{-1}	13
c, calcite	298	2.713	7
c, calcite	299	2.711 (X-ray)	8
c, aragonite	297	2.930	7

TABLE 2.16
Thermal Expansion Coefficient of $CaCO_3$

Calcite	α(2° to 81°) = 25.1353 + 0.0118c opt. axis and –5.5782t + 0.00138t ⊥ opt. axis α^{40} = 26.21 + 0.160(t – 40) opt. axis; and = –5.40 0.0087(t – 40), ⊥ opt. axis α(50° to 60°) = 1.447
Aragonite	α^1 = 34.60 + 0.0337t 20° – 70° α_2 = 17.19 + 0.0368t α_3 = 10.16 + 0.0064t

TABLE 2.17
Transitions of $CaCO_3$

Phase Change	T(K)	ΔH kJ/mol	ΔS J/(mol*K)	Reference
c (aragonite) → c (calcite)	728			18
c (vaterite) → c (calcite)	623–673			19

TABLE 2.18
Thermodynamic Data of $CaCO_3$ (calcite)

T K	ΔHf° kJ mol^{-1}	S° J mol^{-1} deg^{-1}	ΔGf° kJ mol^{-1}	Reference
298.15	–1206.92	92.90	–1234.62	Barin 1993
400.00	–1197.65	119.55	–1245.47	Barin 1993
500.00	–104.55	142.07	–1258.58	Barin 1993
600.00	–1176.82	161.62	–1273.79	Barin 1993
700.00	–1165.61	178.88	–1290.83	Barin 1993
800.00	–1154.01	194.37	–1309.50	Barin 1993
900.00	–1142.05	208.46	–1329.66	Barin 1993
1000.00	–1129.76	221.49	–1351.16	Barin 1993
1100.00	–1117.15	233.41	–1373.91	Barin 1993
1200.00	–1104.25	244.64	–1397.81	Barin 1993
1300.00	–1091.05	255.20	–1422.81	Barin 1993
1400.00	–1077.56	265.20	–1448.83	Barin 1993
1500.00	–1063.79	274.70	–1475.83	Barin 1993

TABLE 2.19
Thermodynamic Data of CaO

T K	ΔHf° kJ mol^{-1}	S° J mol^{-1} deg^{-1}	ΔGf° kJ mol^{-1}	Reference
298.15	–635.09	38.07	–646.44	Barin 1993
400.00	–630.47	51.38	–651.02	Barin 1993
500.00	–625.65	62.12	–656.71	Barin 1993
600.00	–620.67	71.20	–663.39	Barin 1993
700.00	–615.58	79.04	–670.91	Barin 1993
800.00	–610.40	85.96	–679.17	Barin 1993
900.00	–605.15	92.14	–688.08	Barin 1993
1000.00	–599.84	97.74	–697.58	Barin 1993
1100.00	–594.47	102.86	–707.61	Barin 1993
1200.00	–589.04	107.58	–718.13	Barin 1993
1300.00	–583.55	111.97	–729.11	Barin 1993
1400.00	–578.20	116.07	–740.52	Barin 1993
1500.00	–572.43	119.92	–752.32	Barin 1993

TABLE 2.20
$CaCO_3 = CaO + CO_2(g)$

T K	ΔH° kJ	ΔS° J deg^{-1}	ΔG° kJ	K
298.15	178.33	158.94	130.94	1.143E – 023
400.00	177.74	157.27	114.83	1.008E – 015
500.00	176.87	155.33	99.20	4.323E – 011
600.00	175.80	153.39	83.76	5.095E – 008
700.00	174.57	151.50	68.52	7.701E – 006
800.00	173.19	149.66	53.46	3.228E – 004
900.00	171.67	147.87	38.59	5.759E – 003
1000.00	170.01	146.12	23.89	5.651E – 002
1100.00	168.20	144.40	9.36	3.593E – 001
1200.00	166.25	142.70	–4.99	1.65E + 000
1300.00	164.15	141.02	–19.18	5.898E + 000
1400.00	161.89	139.35	–33.20	1.733E + 001
1500.00	159.47	137.68	–47.05	4.350E + 000

TABLE 2.21
Transitions of Cs_2CO_3

Phase Change	T(K)	ΔH kJ/mol	ΔS J/(mol*K)	Reference
c → l	1065			23

TABLE 2.22
Thermodynamic Data of Cs_2CO_3

T K	ΔHf° kJ mol^{-1}	S° J mol^{-1} deg^{-1}	ΔGf° kJ mol^{-1}	Reference
298.15	–1147.25	204.47	–1208.22	Barin et al. 1977
400.00	–1133.97	242.72	–1231.05	Barin et al. 1977
500.00	–1119.85	274.18	–1256.94	Barin et al. 1977
600.00	–1104.86	301.48	–1285.75	Barin et al. 1977
700.00	–1089.07	325.81	–1317.14	Barin et al. 1977
800.00	–1072.52	347.89	–1350.84	Barin et al. 1977
900.00	–1055.24	368.24	–1386.66	Barin et al. 1977
1000.00	–1037.23	387.21	–1424.44	Barin et al. 1977
1100.00	–1018.51	405.05	–1464.06	Barin et al. 1977
1200.00	–999.07	421.95	–1505.41	Barin et al. 1977
1300.00	–978.94	438.07	–1548.43	Barin et al. 1977
1400.00	–958.09	453.51	–1593.01	Barin et al. 1977
1500.00	–936.55	468.38	–1639.11	Barin et al. 1977

TABLE 2.23
Thermodynamic Data of Cs_2O

T K	ΔHf° kJ mol⁻¹	S° J mol⁻¹ deg⁻¹	ΔGf° kJ mol⁻¹	Reference
298.15	–345.98	146.86	–389.76	Knacke et al. 1991
400.00	–338.06	169.67	–405.93	Knacke et al. 1991
500.00	–329.95	187.75	–423.83	Knacke et al. 1991
600.00	–321.51	203.14	–443.39	Knacke et al. 1991
700.00	–312.73	216.66	–464.39	Knacke et al. 1991
800.00	–303.63	228.81	–486.68	Knacke et al. 1991
900.00	–294.19	239.92	–510.12	Knacke et al. 1991
1000.00	–284.42	250.21	–534.63	Knacke et al. 1991
1100.00	–274.33	259.83	–560.14	Knacke et al. 1991
1200.00	–263.90	268.90	–586.58	Knacke et al. 1991
1300.00	–253.15	277.51	–613.90	Knacke et al. 1991
1400.00	–242.06	285.72	–642.07	Knacke et al. 1991
1500.00	–230.64	285.72	–671.04	Knacke et al. 1991

TABLE 2.24
$Cs_2CO_3 = Cs_2O + CO_2(g)$

T K	ΔH° kJ	ΔS° J deg⁻¹	ΔG° kJ	K
298.15	–407.77	156.16	–361.22	5.144E – 064
400.00	–406.46	152.39	–345.50	7.553E – 046
500.00	–404.87	148.85	–330.44	2.992E – 035
600.00	–403.01	145.46	–315.73	3.245E – 028
700.00	–400.88	142.19	–301.35	3.246E – 023
800.00	–398.48	139.00	–287.29	1.740E – 019
900.00	–395.82	135.87	–273.55	1.326E – 016
1000.00	–392.89	132.78	–260.11	2.582E – 014
1100.00	–389.69	129.73	–246.99	1.865E – 012
1200.00	–386.21	126.70	–234.17	6.400E – 011
1300.00	–382.44	123.69	–221.65	1.240E – 009
1400.00	–378.38	120.68	–209.43	1.533E – 008
1500.00	–374.03	117.68	–197.51	1.323E – 007

TABLE 2.25
Density of $CoCO_3$

Phase	T(K)	d	Reference
c		4.07	26
c		4.24 (X-ray)	27
c (spherocobaltite)	299	4.214 (X-ray)	8

TABLE 2.26
Thermodynamic Data of $CoCO_3$

T K	ΔHf° kJ mol⁻¹	S° J mol⁻¹ deg⁻¹	ΔGf° kJ mol⁻¹	Reference
298.15	–713.00	87.86	–739.29	Barin et al. 1977
400.00	–704.12	113.40	–749.48	Barin et al. 1977
500.00	–694.40	135.06	–761.93	Barin et al. 1977
600.00	–683.99	154.02	–775.40	Barin et al. 1977
700.00	–673.02	170.92	–792.66	Barin et al. 1977
800.00	–661.55	186.23	–810.53	Barin et al. 1977
900.00	–649.62	200.27	–829.86	Barin et al. 1977
1000.00	–637.26	213.30	–850.55	Barin et al. 1977
1100.00	–624.46	225.48	–872.50	Barin et al. 1977
1200.00	–611.25	236.97	–895.62	Barin et al. 1977

TABLE 2.27
Thermodynamic Data of CoO

T K	ΔHf° kJ mol⁻¹	S° J mol⁻¹ deg⁻¹	ΔGf° kJ mol⁻¹	Reference
298.15	–237.94	52.97	–253.74	Barin 1989
400.00	–232.42	68.93	–259.99	Barin 1989
500.00	–227.04	80.92	–267.50	Barin 1989
600.00	–221.67	90.71	–276.10	Barin 1989
700.00	–216.27	99.04	–285.60	Barin 1989
800.00	–210.82	106.31	–295.87	Barin 1989
900.00	–205.31	112.80	–306.83	Barin 1989
1000.00	–199.73	116.68	–318.41	Barin 1989
1100.00	–194.07	124.08	–330.55	Barin 1989
1200.00	–188.31	129.09	–343.21	Barin 1989

TABLE 2.28
$CoCO_3 = CoO + CO_2(g)$

T K	ΔH° kJ	ΔS° J deg⁻¹	ΔG° kJ	K
298.15	81.55	178.87	28.22	1.137E – 005
400.00	82.26	180.97	9.87	5.143E – 002
500.00	82.323	181.14	–8.25	7.271E + 000
600.00	81.97	180.51	–26.33	1.962E + 002
700.00	81.28	179.46	–44.33	2.035E + 003
800.00	80.31	178.16	–62.22	1.155E + 004
900.00	79.08	176.72	–79.96	4.377E + 004
1000.00	77.61	175.17	–97.56	1.248E + 005
1100.00	75.91	173,54	–114.99	2.890E + 005
1200.00	73.98	171.87	–132.26	5.724E + 005

TABLE 2.29
Thermodynamic Data of $CuCO_3$

T K	Hf⁰ kJ mol⁻¹	S° J mol⁻¹ deg⁻¹	ΔGf° kJ mol⁻¹	Reference
298.15	–596.22	87.86	–622.42	Barin et al. 1977
400.00	–587.00	114.38	–632.75	Barin et al. 1977
500.00	–576.94	136.79	–645.34	Barin et al. 1977
600.00	–566.20	156.36	–660.01	Barin et al. 1977
700.00	–554.89	173.78	–676.54	Barin et al. 1977
800.00	–543.09	189.53	–694.71	Barin et al. 1977
900.00	–530.83	203.97	–714.40	Barin et al. 1977
1000.00	–518.13	217.35	–735.47	Barin et al. 1977

TABLE 2.30
Thermodynamic Data of CuO

T K	ΔHf° kJ mol⁻¹	S° J mol⁻¹ deg⁻¹	ΔGf° kJ mol⁻¹	Reference
298.15	–156.06	42.59	–168.76	Knacke et al. 1991
400.00	–151.50	55.73	–173.79	Knacke et al. 1991
500.00	–146.69	66.46	–179.92	Knacke et al. 1991
600.00	–141.67	75.59	–187.03	Knacke et al. 1991
700.00	–136.51	83.55	–195.00	Knacke et al. 1991
800.00	–131.23	90.60	–203.71	Knacke et al. 1991
900.00	–125.85	96.94	–213.09	Knacke et al. 1991
1000.00	–120.37	102.72	–223.08	Knacke et al. 1991

TABLE 2.31
$CuCO_3 = Cu + CO_2(g)$

T K	ΔH° kJ	ΔS° J deg⁻¹	ΔG° kJ	K
298.15	46.65	168.50	–3.59	4.249E + 000
400.00	46.05	166.79	–20.66	4.994E + 002
500.00	45.22	164.95	–37.25	7.794E + 003
600.00	44.18	163.04	–53.65	4.688E + 004
700.00	42.92	161.11	–69.86	1.634E + 005
800.00	41.45	159.14	–85.87	4.048E + 005
900.00	39.76	157.16	–101.69	7.983E + 005
1000.00	37.85	155.15	–117.30	1.342E + 006

TABLE 2.32
Density of $FeCO_3$

Phase	T(K)	d	Reference
c, siderite	293.9	0.851	29

TABLE 2.33
Thermal Expansion Coefficient of $FeCO_3$

		Reference
$FeCO_3$, siderite	α^{40} = 19.2 with $\Delta(20° - 70°)$ = 25.5 and α^{40} = 6.05 with 13 $\Delta(20°- 70°)$ = 917⊥	13

TABLE 2.34
Thermodynamic Data of $FeCO_3$ (298.15K)

	$\Delta Hf°$ kJ mol^{-1}	Reference	S° J mol^{-1} deg^{-1}	Reference
$FeCO_3$	–740.6	1.2	92.9	30
FeO	–266.5	1.2	57.49	31
Fe_3O_4	–1117.	30	146.4	31
Fe_2O_3	–822.1	30	87.4	31

TABLE 2.35
Thermodynamic Data of $PbCO_3$

T K	$\Delta Hf°$ kJ mol^{-1}	S° J mol^{-1} deg^{-1}	$\Delta Gf°$ kJ mol^{-1}	Reference
298.15	–699.10	131.00	–738.16	Barin 1993
400.00	–689.58	158.39	–752.93	Barin 1993
500.00	–679.02	181.90	–769.97	Barin 1993
600.00	–667.27	203.29	–789.24	Barin 1993
700.00	654.32	223.23	–810.58	Barin 1993
800.00	640.17	242.11	–833.85	Barin 1993
900.00	624.83	260.16	–858.97	Barin 1993
1000.00	608.29	277.58	–885.87	Barin 1993

TABLE 2.36
Thermodynamic Data of PbO

T K	$\Delta Hf°$ kJ mol^{-1}	S° J mol^{-1} deg^{-1}	$\Delta Gf°$ kJ mol^{-1}	Reference
298.15	–218.06	68.70	–238.55	Barin 1993
400.00	–213.25	82.57	–246.28	Barin 1993
500.00	–208.30	93.61	–255.10	Barin 1993
600.00	–203.17	102.96	–264.94	Barin 1993
700.00	–197.88	111.10	–275.65	Barin 1993
800.00	–192.45	118.36	–287.13	Barin 1993
900.00	–186.88	124.92	–299.30	Barin 1993
1000.00	–181.17	130.93	–312.10	Barin 1993

TABLE 2.37
$PbCO_3 = PbO + CO_2(g)$

T K	ΔH° kJ	ΔS° J deg⁻¹	ΔG° kJ	K
298.15	87.53	151.47	42.37	3.766E – 008
400.00	86.88	149.62	27.04	2.946E – 004
500.00	85.69	146.99	12.20	5.316E – 002
600.00	83.75	143.48	–2.332	1.596E + 000
700.00	80.98	139.21	–16.47	1.695E + 001
800.00	77.31	134.35	–30.15	9.311E + 001
900.00	72.73	128.94	–43.32	3.270E + 002
1000.00	67.21	123.13	–55.93	8.349E + 002

TABLE 2.38
Density of $LiCO_3$

Phase	T(K)	d	Reference
c	290.7	2.111	13
l	1012.2	1.8246	40
	1027.4	1.8189	
	1029.7	1.8190	
	1052.1	1.8119	
	1069.7	1.8036	
	1082.7	1.7987	
	1105.1	1.9710	
	1120.1	1.7843	

TABLE 2.39
Transitions of $LiCO_3$

Phase Change	T(K)	ΔH kJ/mol	ΔS J/(mol*K)	Reference
c → l	993			13
	999	42		40
	1008			13

TABLE 2.40
Thermodynamic Data of Li_2CO_3

T K	ΔHf° kJ mol⁻¹	S° J mol⁻¹ deg⁻¹	ΔGf° kJ mol⁻¹	Reference
298.15	–1216.04	90.17	–1242.92	Barin 1989
400.00	–1205.41	120.71	–1253.70	Barin 1989
500.00	–1193.38	147.49	–1267.13	Barin 1989
600.00	–1179.50	172.74	–1283.15	Barin 1989
700.00	–1162.89	198.123	–1301.65	Barin 1989
800.00	–1147.90	218.22	–1322.48	Barin 1989
900.00	–1131.10	237.99	–1345.29	Barin 1989
1000.00	–1067.79	302.59	–1370.78	Barin 1989
1100.00	–1049.24	320.27	–1401.54	Barin 1989
1200.00	–1030.70	336.40	–1434.38	Barin 1993
1300.00	–1012.16	351.25	–1468.78	Barin 1989
1400.00	–993.61	364.99	–1504.60	Barin 1989
1500.00	–975.07	377.78	–1541.74	Barin 1989

TABLE 2.41
Thermodynamic Data of Li_2O

T K	ΔHf° kJ mol⁻¹	S° J mol⁻¹ deg⁻¹	ΔGf° kJ mol⁻¹	Reference
298.15	–598.73	37.89	–610.03	Barin 1993
400.00	–592.64	55.40	–614.79	Barin 1993
500.00	–585.83	70.57	–621.11	Barin 1993
600.00	–578.52	83.87	–628.84	Barin 1993
700.00	–570.85	95.68	–637.83	Barin 1993
800.00	–562.90	106.31	–647.94	Barin 1993
900.00	–554.69	115.97	–659.06	Barin 1993
1000.00	–546.24	124.87	–671.11	Barin 1993
1100.00	–537.58	133.12	–684.02	Barin 1993
1200.00	–528.72	140.83	–697.72	Barin 1993
1300.00	–519.65	148.09	–712.17	Barin 1993
1400.00	–510.38	154.96	–727.32	Barin 1993
1500.00	–500.93	161.48	–743.15	Barin 1993

TABLE 2.42
$Li_2CO_3 = Li_2O + CO_2(g)$

T K	ΔH° kJ	ΔS° J deg^{-1}	ΔG° kJ	K
298.15	223.80	161.49	175.66	1.673E – 031
400.00	223.33	160.12	159.28	1.578E – 021
500.00	222.52	158.35	143.35	1.055E – 015
600.00	220.63	154.94	127.67	7.667E – 012
700.00	216.58	148.80	112.42	4.077E – 009
800.00	214.59	146.16	97.66	4.197E – 007
900.00	211.19	142.17	83.23	1.475E – 005
1000.00	161.63	92.06	69.58	2.320E – 004
1100.00	157.17	87.80	60.59	1.326E – 003
1200.00	153. 02	84.19	51.99	5.452E – 003
1300.00	149.16	81.09	43.73	1.748E – 002
1400.00	145.58	78.44	35.76	4.630E – 002
1500.00	142.26	76.15	28.04	1.056E – 001

TABLE 2.43
Density of $MgCO_3$

Phase	T(K)	d	Reference
c	294.6	2.980	29
c	298.0	3.037	13
c	298.0	3.009 (X-ray)	86

TABLE 2.44
Thermal Expansion Coefficient of $MgCO_3$

$MgCO_3$, magnesite	α^{40} = 21.3 with Δ(20° – 70°) = 33.9‖ and α^{40} = 5.99 with Δ(20° – 70°) = 24.3⊥

TABLE 2.45
Thermodynamic Data of $MgCO_3$

T K	ΔHf° kJ mol^{-1}	S° J mol^{-1} deg^{-1}	ΔGf° kJ mol^{-1}	Reference
298.15	1095.80	65.70	–1115.39	Knacke et al. 1991
400.00	–1087.30	90.13	–1123.35	Knacke et al. 1991
500.00	–1077.78	111.33	–1133.44	Knacke et al. 1991
600.00	–1067.39	130.24	–1145.54	Knacke et al. 1991
700.00	–1056.26	147.38	–1159.43	Knacke et al. 1991
800.00	–1044.45	163.14	–1174.97	Knacke et al. 1991
900.00	–1032.00	177.81	–1192.02	Knacke et al. 1991
1000.00	–1018.91	191.58	–1210.50	Knacke et al. 1991

TABLE 2.46
Thermodynamic Data of MgO (periclase)

T K	ΔHf° kJ mol⁻¹	S° J mol⁻¹ deg⁻¹	ΔGf° kJ mol⁻¹	Reference
298.15	–601.24	26.92	–609.27	Knacke et al. 1991
400.00	–597.13	38.73	–612.63	Knacke et al. 1991
500.00	–592.69	48.64	–617.01	Knacke et al. 1991
600.00	–588.02	57.15	–622.31	Knacke et al. 1991
700.00	–583.21	64.57	–628.40	Knacke et al. 1991
800.00	–578.29	71.14	–635.19	Knacke et al. 1991
900.00	–573.28	77.03	–642.61	Knacke et al. 1991
1000.00	–568.20	82.38	–650.58	Knacke et al. 1991

TABLE 2.47
$MgCO_3 = MgO + CO_2(g)$

T K	ΔH° kJ	ΔS° J deg⁻¹	ΔG° kJ	K
298.15	101.05	174.99	48.88	2.729E – 009
400.00	100.72	174.05	31.10	8.682E – 005
500.00	100.06	172.59	13.76	3.648E – 002
600.00	99.02	170.72	–3.41	1.979E + 000
700.00	97.60	168.52	–20.37	3.312E + 001
800.00	95.76	166.07	–37.10	2.646E + 002
900.00	93.49	163.41	–53.58	1.288E + 003
1000.00	90.80	160.58	–69.78	4.417E + 003

TABLE 2.48
Density of $MnCO_3$

Phase	T(K)	d	Reference
c	293	3.125 gml⁻¹	13
c, rhodochrosite	294.3	3.633	51

TABLE 2.49
Thermodynamic Data of $MnCO_3$

T K	ΔHf° kJ mol⁻¹	S° J mol⁻¹ deg⁻¹	ΔGf° kJ mol⁻¹	Reference
298.15	–894.10	85.80	–219.68	Barin 1989
400.00	–885.03	111.89	–929.80	Barin 1989
500.00	–875.05	134.11	–942.11	Barin 1989
600.00	–864.37	153.58	–956.51	Barin 1989
700.00	–853.10	170.93	–972.75	Barin 1989

TABLE 2.50
Thermodynamic Data of MnO (manganosite)

T K	ΔHf° kJ mol⁻¹	S° J mol⁻¹ deg⁻¹	ΔGf° kJ mol⁻¹	Reference
298.15	–385.22	59.71	–403.02	Barin 1993
400.00	–380.51	73.28	–409.82	Barin 1993
500.00	–375.68	84.05	–417.71	Barin 1993
600.00	–370.71	93.11	–426.58	Barin 1993
700.00	–365.62	100.95	–436.29	Barin 1993

TABLE 2.51
$MnCO_3 = MnO + CO_2(g)$

T K	ΔH° kJ	ΔS° J deg⁻¹	ΔG° kJ	K
298.15	115.37	187.68	59.42	3.885E – 011
400.00	115.06	186.82	40.34	5.399E – 006
500.00	114.34	185.21	21.73	5.366E – 003
600.00	113.31	183.34	3.30	5.159E – 001
700.00	112.02	181.36	–14.93	1.302E + 001

TABLE 2.52
$3\ MnCO_3 = Mn_3O_4 + 2CO_2(g) + CO(g)$

T K	ΔH° kJ	ΔS° J deg⁻¹	ΔG° kJ	K
298.15	396.95	523.39	240.90	6.149E – 043
400.00	395.92	520.52	187.71	3.062E – 025
500.00	393.73	515.69	135.89	6.349E – 015
600.00	390.78	510.33	84.59	4.321E – 008
700.00	387.22	504.84	33.83	2.990E – 003
800.00	383.11	499.37	–16.38	1.174E + 001

TABLE 2.53
Thermodynamic Data of $HgCO_3$

T K	ΔHf° kJ mol⁻¹	S° J mol⁻¹ deg⁻¹	ΔGf° kJ mol⁻¹	Reference
298.15	–553.29	184.10	–608.18	Karapet 70
400.00	–553.29	184.10	–626.93	Karapet 70
500.00	–553.29	184.10	–645.34	Karapet 70
600.00	–553.29	184.10	–663.75	Karapet 70

TABLE 2.54
$HgCO_3 = HgO + CO_2(g)$

T K	ΔH° kJ	ΔS° J deg⁻¹	ΔG° kJ	K
298.15	69.00	99.94	39.20	1.354E – 007
400.00	77.77	125.19	27.70	2.414E – 004
500.00	87.19	146.18	14.10	3.363E – 002
600.00	97.17	164.36	–1.45	1.336E + 000

TABLE 2.55
Thermodynamic Data of Hg_2CO_3

T K	ΔHf° kJ mol^{-1}	S° J mol^{-1} deg^{-1}	ΔGf° kJ mol^{-1}	Reference
298.15	–553.50	180.00	–607.17	CRC 94
400.00	–542.98	210.37	–627.12	CRC 94
500.00	–532.64	233.43	–649.36	CRC 94

TABLE 2.56
$Hg_2CO_3 = Hg_2O + CO_2(g)$

T K	ΔH° kJ	ΔS° J deg^{-1}	ΔG° kJ	K
298.15	68.76	165.15	19.55	3.761E – 004
400.00	65.72	156.27	3.22	3.800E – 001
500.00	63.27	150.77	–12.12	1.844E + 001

TABLE 2.57
Thermodynamic Data of $NiCO_3$ (zaratite)

T K	ΔHf° kJ mol^{-1}	S° J mol^{-1} deg^{-1}	ΔGf° kJ mol^{-1}	Reference
298.15	–694.19	86.19	–720.24	Barin 1989
400.00	–685.20	113.07	–730.43	Barin 1989
500.00	–675.22	135.31	–742.88	Barin 1989
600.00	–664.65	154.58	–757.39	Barin 1989
700.00	–653.56	171.66	–773.72	Barin 1989

TABLE 2.58
Thermodynamic Data of NiO

T K	ΔHf° kJ mol^{-1}	S° J mol^{-1} deg^{-1}	ΔGf° kJ mol^{-1}	Reference
298.15	–239.70	37.99	–251.03	Barin 1989
400.00	–234.79	52.11	–255.64	Barin 1989
500.00	–229.08	64.80	–261.48	Barin 1989
600.00	–223.17	75.61	–268.54	Barin 1989
700.00	–217.88	83.78	–276.52	Barin 1989

TABLE 2.59
$NiCO_3 = NiO + CO_2(g)$

T K	ΔH° kJ	ΔS° J deg^{-1}	ΔG° kJ	K
298.15	61.34	165.57	11.97	7.983E – 003
400.00	60.97	164.48	–4.83	4.269E + 000
500.00	61.11	164.76	–21.27	1.668E + 002
600.00	61.12	164.84	–37.78	1.947E + 003
700.00	60.22	163.46	–54.20	1.109E + 004

TABLE 2.60
Density of K_2CO_3

Phase	T(K)	d	Reference
c	293	2.330	13, Vol. III
l	1180.8	1.8922 gcm^{-1}	40
	1185.3	1.8904	
	1193.9	1.8864	
	1195.9	1.8848	
	1206.6	1.8824	
	1213.5	1.8778	
	1220.0	1.8749	
	1223.6	1.8739	
	1234.1	1.8684	
	1245.3	1.8640	
	1257.4	1.8584	
	1270.0	1.8527	
	1283.3	1.8467	

TABLE 2.61
Transitions of K_2CO_3

Phase Change	T(K)	ΔH kJ/mol	ΔS J/(mol˙K)	Reference
c,IV → c,III	523			1.2
c,III → c,II	791			40
c,II → c,I	895			40
c,I → l	1171			23, 39, 55

TABLE 2.62
Thermodynamic Data of K_2CO_3

T K	ΔHf° kJ mol⁻¹	S° J mol⁻¹ deg⁻¹	ΔGf° kJ mol⁻¹	Reference
298.15	–1150.18	155.52	–1196.55	Knacke et al. 1991
400.00	–1137.78	191.20	–1214.26	Knacke et al. 1991
500.00	–1124.34	221.14	–1234.91	Knacke et al. 1991
600.00	–1109.81	247.60	–1258.37	Knacke et al. 1991
700.00	–1094.27	271.54	–1284.35	Knacke et al. 1991
800.00	–1077.04	293.59	–1312.62	Knacke et al. 1991
900.00	–1060.27	314.17	–1343.02	Knacke et al. 1991
1000.00	–1041.84	333.58	–1375.41	Knacke et al. 1991
1100.00	–1022.47	352.03	–1409.70	Knacke et al. 1991
1200.00	–974.48	393.27	–1446.40	Knacke et al. 1991
1300.00	–953.56	419.02	–1486.58	Knacke et al. 1991
1400.00	–932.64	425.52	–1528.37	Knacke et al. 1991
1500.00	–911.72	439.96	–1571.65	Knacke et al. 1991

TABLE 2.63
Thermodynamic Data of K_2O

T K	ΔHf° kJ mol⁻¹	S° J mol⁻¹ deg⁻¹	ΔGf° kJ mol⁻¹	Reference
298.15	–361.50	102.01	–391.91	Barin 1989
400.00	–353.66	124.59	–403.50	Barin 1989
500.00	–345.59	142.59	–416.88	Barin 1989
600.00	–337.25	157.79	–431.92	Barin 1989
700.00	–328.68	170.99	–448.37	Barin 1989
800.00	–319.90	182.71	–466.07	Barin 1989
900.00	–310.93	193.27	–484.88	Barin 1989
1000.00	–301.78	202.92	–504.69	Barin 1989
1100.00	–264.06	239.78	–527.82	Barin 1989
1200.00	–253.36	249.09	–552.27	Barin 1989
1300.00	–242.66	257.65	–577.61	Barin 1989
1400.00	–231.96	265.58	–603.78	Barin 1989
1500.00	–221.26	272.97	–630.71	Barin 1989

TABLE 2.64
$K_2CO_3 = K_2O + CO_2(g)$

T K	ΔH° kJ	ΔS° J deg^{-1}	ΔG° kJ	K
298.15	395.18	160.36	347.40	1.356E – 061
400.00	394.68	158.83	331.14	5.670E – 044
500.00	393.72	156.72	315.36	1.127E – 033
600.00	392.22	154.00	299.82	7.875E – 027
700.00	390.13	150.79	284.58	5.793E – 022
800.00	387.43	147.20	269.67	2.458E – 018
900.00	384.11	143.29	255.15	1.550E – 015
1000.00	380.15	139.12	241.03	2.565E – 013
1100.00	403.91	162.70	224.94	2.077E – 011
1200.00	372.15	135.57	209.46	7.615E – 010
1300.00	367.54	131.88	196.09	1.319E – 008
1400.00	363.02	128.53	183.07	1.475E – 007
1500.00	358.57	125.47	170.38	1.166E – 006

TABLE 2.65
Thermodynamic Data of Rb_2CO_3

T K	ΔHf° kJ mol^{-1}	S° J mol^{-1} deg^{-1}	ΔGf° kJ mol^{-1}	Reference
298.15	–1136.00	181.33	–1190.07	Barin 1993
400.00	–1123.30	217.89	–1210.45	Barin 1993
500.00	–1109.66	248.28	–1233.80	Barin 1993
600.00	–1093.78	277.10	–1260.04	Barin 1993
700.00	–1078.25	301.02	–1288.96	Barin 1993
800.00	–1061.85	322.90	–1320.17	Barin 1993
900.00	–1044.60	343.21	–1353.49	Barin 1993
1000.00	–1026.51	362.26	–1388.77	Barin 1993
1100.00	–1007.60	380.28	–1425.91	Barin 1993
1200.00	–962.90	426.23	–1474.38	Barin 1993
1300.00	–944.07	441.30	–1517.76	Barin 1993
1400.00	–925.25	455.25	–1562.60	Barin 1993
1500.00	–906.42	468.24	–1608.78	Barin 1993

TABLE 2.66
Thermodynamic Data of Rb_2O

T K	ΔHf° kJ mol⁻¹	S° J mol⁻¹ deg⁻¹	ΔGf° kJ mol⁻¹	Reference
298.15	–339.00	125.52	–376.42	Barin 1993
400.00	–331.28	147.76	–390.39	Barin 1993
500.00	–323.37	165.41	–406.07	Barin 1993
600.00	–314.28	181.98	–423.47	Barin 1993
700.00	–301.52	202.03	–442.94	Barin 1993
800.00	–271.64	240.86	–464.33	Barin 1993
900.00	–262.44	251.70	–488.97	Barin 1993
1000.00	–253.23	261.40	–514.64	Barin 1993
1100.00	–244.03	270.18	–541.22	Barin 1993
1200.00	–234.83	278.19	–568.65	Barin 1993
1300.00	–225.62	285.55	–596.84	Barin 1993
1400.00	–216.42	292.37	–625.74	Barin 1993
1500.00	–207.21	298.73	–655.30	Barin 1993

TABLE 2.67
$Rb_2CO_3 = Rb_2O + CO_2(g)$

T K	ΔH° kJ	ΔS° J deg⁻¹	ΔG° kJ	K
298.15	403.50	157.96	356.40	3.587E – 063
400.00	402.57	155.31	340.44	3.458E – 045
500.00	401.26	152.40	325.06	1.093E – 034
600.00	399.15	148.69	309.93	1.037E – 027
700.00	401.27	152.35	294.63	1.030E – 022
800.00	419.79	176.04	278.97	6.081E – 019
900.00	416.94	172.68	261.53	6.610E – 016
1000.00	413.37	168.93	244.44	1.701E – 013
1100.00	409.08	164.85	227.75	1.528E – 011
1200.00	379.11	131.71	221.06	2.381E – 010
1300.00	375.10	128.50	208.05	4.363E – 009
1400.00	371.18	125.59	195.35	5.139E – 008
1500.00	367.33	122.94	182.92	4.261E – 007

TABLE 2.68
Thermodynamic Data of Ag_2CO_3

T K	ΔHf° kJ mol⁻¹	S° J mol⁻¹ deg⁻¹	ΔGf° kJ mol⁻¹	Reference
298.15	–505.80	167.40	–555.71	Barin et al. 1977
400.00	–493.87	201.74	–574.57	Barin et al. 1977
500.00	–481.07	230.27	–596.20	Barin et al. 1977
600.00	–467.18	255.56	–620.51	Barin et al. 1977
700.00	–452.21	278.61	–647.24	Barin et al. 1977
800.00	–436.17	300.02	–676.18	Barin et al. 1977

TABLE 2.69
Thermodynamic Data of Ag_2O

T K	ΔHf° kJ mol^{-1}	S° J mol^{-1} deg^{-1}	ΔGf° kJ mol^{-1}	Reference
298.15	–31.05	121.30	–67.214	Knacke et al. 1991
400.00	–24.01	141.57	–80.64	Knacke et al. 1991
500.00	–16.54	158.21	–95.65	Knacke et al. 1991

TABLE 2.70
$Ag_2CO_3 = Ag_2O + CO_2(g)$

T K	ΔH° kJ	ΔS° J deg^{-1}	ΔG° kJ	K
298.15	81.25	167.67	31.26	3.338E – 006
400.00	80.42	165.27	14.31	1.352E – 002
500.00	79.49	163.22	–2.12	1.663E + 000
600.00	78.24	160.94	–18.33	3.940E + 001
700.00	76.55	158.35	–34.29	3.623E + 002

TABLE 2.71
$Ag_2O = 2Ag + 1/2\ O_2(g)$

T K	ΔH° kJ	ΔS° J deg^{-1}	ΔG° kJ	K
298.15	31.05	66.63	11.19	1.098E – 002
400.00	30.75	65.81	4.43	2.639E – 001
500.00	30.08	64.32	–2.08	1.649E + 000
600.00	29.08	62.50	–8.42	5.411E + 000
700.00	27.78	60.50	–14.57	1.223E + 001
800.00	26.21	58.41	–20.52	2.187E + 001
900.00	24.38	56.25	–26.25	3.341E + 001
1000.00	22.30	54.07	–31.77	4.567E + 001

TABLE 2.72
Density of Na_2CO_3

Phase	T(K)	d	Reference
c	293	2.533 gml^{-1}	13
l	1137.9	1.9685 gcm^{-3}	36
	1146.2	1.9666	
	1165.5	1.9576	
	1178.8	1.9508	
	1184.7	1.9477	
	1223.3	1.9288	
	1233.2	1.9264	
	1244.8	1.9211	
	1260.3	1.9142	
	1277.0	1.9080	

TABLE 2.73
Transitions of Na_2CO_3

Phase Change	T(K)	ΔH kJ/mol	ΔS J/(mol*K)	Reference
c,IV → c,III	629			1.3
c,III → c,II	759			
c,II → c,I	891			
c,I → l	1129	31		36, 38, 39

TABLE 2.74
Thermodynamic Data of Na_2CO_3

T K	ΔHf° kJ mol⁻¹	S° J mol⁻¹ deg⁻¹	ΔGf° kJ mol⁻¹	Reference
298.15	–1130.77	138.78	–1172.15	Knacke et al. 1991
400.00	–1118.78	175.27	–1188.09	Knacke et al. 1991
500.00	–142.35	202.97	–1206.92	Knacke et al. 1991
600.00	–1090.19	230.71	–1238.61	Knacke et al. 1991
700.00	–1072.68	257.66	–1253.04	Knacke et al. 1991
800.00	–1056.18	279.78	–1280.01	Knacke et al. 1991
900.00	–1040.20	298.59	–1308.93	Knacke et al. 1991
1000.00	–1022.93	316.77	–1339.70	Knacke et al. 1991
1100.00	–1004.37	334.75	–1372.27	Knacke et al. 1991
1200.00	–955.66	377.44	–1408.59	Knacke et al. 1991
1300.00	–936.71	392.61	–1447.10	Knacke et al. 1991
1400.00	–917.75	406.66	–1487.07	Knacke et al. 1991
1500.00	–898.80	419.74	–1528.40	Knacke et al. 1991

TABLE 2.75
Thermodynamic Data of Na_2O

T K	ΔHf° kJ mol⁻¹	S° J mol⁻¹ deg⁻¹	ΔGf° kJ mol⁻¹	Reference
298.15	–417.98	75.04	–440.36	Barin 1989
400.00	–410.55	96.43	–449.12	Barin 1989
500.00	–402.65	114.05	–459.67	Barin 1989
600.00	–394.31	129.23	–471.85	Barin 1989
700.00	–385.63	142.61	–485.46	Barin 1989
800.00	–376.66	154.59	–500.33	Barin 1989
900.00	–367.44	165.44	–516.34	Barin 1989
1000.00	–358.01	175.37	–533.38	Barin 1989
1100.00	–346.66	186.24	–551.52	Barin 1989
1200.00	–336.92	194.71	–570.57	Barin 1989
1300.00	–315.14	212.19	–590.98	Barin 1989
1400.00	–305.21	219.54	–612.57	Barin 1989
1500.00	–247.08	260.68	–638.10	Barin 1989

TABLE 2.76
$Na_2CO_3 = Na_2O + CO_2(g)$

T K	ΔH° kJ	ΔS° J deg^{-1}	ΔG° kJ	K
298.15	319.28	150.03	274.55	7.868E – 049
400.00	318.78	148.60	259.35	1.350E – 034
500.00	317.76	146.35	244.59	2.794E – 026
600.00	315.53	142.33	230.13	9.191E – 021
700.00	311.59	136.29	216.19	7.357E – 017
800.00	309.11	132.89	202.80	5.717E – 014
900.00	307.54	131.04	189.60	9.885E – 012
1000.00	305.01	128.39	176.62	5.935E – 012
1100.00	303.22	126.74	163.82	1.661E – 008
1200.00	269.78	97.03	153.35	2.111E – 007
1300.00	278.21	103.82	143.25	1.753E – 006
1400.00	274.89	101.36	132.99	1.091E – 005
1500.00	319.84	133.40	119.74	6.761E – 005

TABLE 2.77
Transitions of $SrCO_3$

Phase Change	T(K)	ΔH kJ/mol	ΔS J/(mol*K)	Reference
c, orthorhombic → c, hexagonal	1198	17		1, 6, 65
c, hexagonal → c, cubic	1770			65
c → l	1770			1, 1.3

TABLE 2.78
Thermodynamic Data of $SrCO_3$

T K	ΔHf° kJ mol^{-1}	S° J mol^{-1} deg^{-1}	ΔGf° kJ mol^{-1}	Reference
298.15	–1219.85	97.07	–1248.79	Barin et al. 1977
400.00	–1210.66	123.50	–1260.06	Barin et al. 1977
500.00	–1200.79	145.48	–1273.54	Barin et al. 1977
600.00	–1190.34	164.54	–1289.06	Barin et al. 1977
700.00	–1179.38	181.41	–1306.37	Barin et al. 1977
800.00	–1167.99	196.62	–1325.28	Barin et al. 1977
900.00	–1156.18	210.52	–1345.65	Barin et al. 1977
1000.00	–1143.97	223.38	–1367.35	Barin et al. 1977
1100.00	–1131.38	235.38	–1390.30	Barin et al. 1977
1200.00	–1098.71	263.12	–1415.46	Knacke et al. 1991
1300.00	–1084.48	274.51	–1441.35	Knacke et al. 1991
1400.00	–1070.26	285.05	–1469.33	Knacke et al. 1991
1500.00	–1056.03	294.87	–1498.33	Knacke et al. 1991

TABLE 2.79
Thermodynamic Data of SrO

T K	ΔHf° kJ mol⁻¹	S° J mol⁻¹ deg⁻¹	ΔGf° kJ mol⁻¹	Reference
298.15	–592.04	55.52	–608.59	Barin 1989
400.00	–587.24	69.34	–614.98	Barin 1989
500.00	–582.28	80.40	–622.48	Barin 1989
600.00	–577.15	89.74	–631.00	Barin 1989
700.00	–571.89	97.86	–640.39	Barin 1989
800.00	–566.51	105.04	–650.54	Barin 1989
900.00	–561.03	111.49	–661.37	Barin 1989
1000.00	–555.46	117.36	–672.82	Barin 1989
1100.00	–549.81	122.74	–684.83	Barin 1989
1200.00	–544.09	127.72	–697.35	Barin 1989
1300.00	–538.29	132.37	–710.36	Barin 1989
1400.00	–532.41	136.72	–723.82	Barin 1989
1500.00	–526.47	140.82	–737.70	Barin 1989

TABLE 2.80
$SrCO_3 = SrO + CO_2(g)$

T K	ΔH° kJ	ΔS° J deg⁻¹	ΔG° kJ	K
298.15	234.30	172.22	182.96	8.795E – 033
400.00	233.97	171.28	165.46	2.462E – 022
500.00	233.48	170.19	148.39	3.140E – 016
600.00	232.84	169.02	131.43	3.609E – 012
700.00	232.04	167.79	114.59	2.811E – 009
800.00	231.07	166.50	97.87	4.066E – 007
900.00	229.93	165.16	81.29	1.914E – 005
1000.00	228.60	163.76	64.84	4.100E – 004
1100.00	227.08	162.31	48.54	4.955E – 003
1200.00	205.66	144.36	32.43	3.875E – 002
1300.00	202.85	142.11	18.11	1.872E – 001
1400.00	200.19	140.14	4.00	7.092E – 001
1500.00	197.69	138.41	–9.926	2.217E + 000

TABLE 2.81
Transitions of Tl_2CO_3

Phase Change	T(K)	ΔH kJ/mol	ΔS J/(mol˙K)	Reference
c,II → c,I	591			74
c,I → l	546	18	36	1
	536			73

TABLE 2.82
Thermodynamic Data (298.15 K) of Tl_2CO_3

	ΔHf° kJ mol^{-1}	S° J mol^{-1} deg^{-1}	Reference
Tl_2CO_3	–700.0	–155.2	1.3
Tl_2O	–178.7	–126.	

TABLE 2.83
Thermodynamic Data of $ZnCO_3$

T K	ΔHf° kJ mol^{-1}	S° J mol^{-1} deg^{-1}	ΔGf° kJ mol^{-1}	Reference
298.15	–812.78	82.40	–837.35	Barin 1993
400.00	–803.91	107.90	–847.35	Barin 1993
500.00	–793.80	130.39	–859.00	Barin 1993
600.00	–782.32	151.29	–873.09	Barin 1993
700.00	–769.45	171.09	–889.22	Barin 1993

TABLE 2.84
Thermodynamic Data of ZnO (zincite)

T K	ΔHf° kJ mol^{-1}	S° J mol^{-1} deg^{-1}	ΔGf° kJ mol^{-1}	Reference
298.15	–350.46	43.64	–363.47	Barin 1993
400.00	–346.07	56.28	–368.58	Barin 1993
500.00	–341.50	66.48	–374.74	Barin 1993
600.00	–336.75	75.13	–381.83	Barin 1993
700.00	–331.88	82.63	–389.72	Barin 1993

TABLE 2.85
$ZnCO_3 = ZnO + CO_2(g)$

T K	ΔH° kJ	ΔS° J deg^{-1}	ΔG° kJ	K
298.15	68.82	175.01	16.64	1.217E – 003
400.00	68.39	173.82	–1.14	1.407E + 000
500.00	67.28	171.37	–18.41	8.378E + 001
600.00	65.22	167.65	–35.37	1.200E + 003
700.00	62.11	162.88	–51.90	7.469E + 003

REFERENCES

1. Pauling, L., *The Nature of the Chemical Bond*, third ed., Cornell University Press, Ithaca, N.Y., 1960, 282–283
2. Elliott, N., *J. Am. Chem. Soc.,* 59, 1380, 1937.
3. Suss, R. L. and Donohue, *J. Acta Cryst.,* 10, 567, 1957.
4. Stern, K. H., *J. Chem. Educ.,* 46, 645, 1969.
5. Mu, J. and Perlmutter, D., *Thermochim. Acta,* 49, 207, 1981.
6. Lander, J. J., *J. Chem. Phys.,* 17, 829, 1949.
7. Anderson, C. T., *J. Am. Chem. Soc.,* 56, 340, 1934.
8. Swanson, H. E., *Standard X-Ray Diffraction Powder Patterns*, NBS Circular 539 (a) with Fuyet, R. K., part 2, 1953; (b) with Gilfrich, N. T. and Cook, M. I., part 7, 1957.
9. Kelley, K. K., *Contributions to the Data on Theoretical Metallurgy IV. Metal Carbonates — Correlations and Applications of Thermodynamic Properties*, U.S. Bur. Mines Bull. 384. 1935; Bull. 601, 1962.
10. Kapustinskii, A. F. and Stakhanova, M. S., *Dokl. Akad.,* 57, 575, 1947.
11. Andrussov, L., *Z. Phys. Chem.,* 115, 273, 1925.
12. De Schulten, A., *Bull. Soc. Chim.,* 19, 34, 1898; *Bull. Soc. Min.* 20, 195, 1897.
13. *International Critical Tables*, McGraw Hill, New York, 1928.
14. Centnerszwer, M. and Brusz, B., *Z. Phys. Chem.,* 119, 405, 1926.
15. Prodan, E. A. and Pavlyuchenko, M., *Get. Khim. Reaktsii,* 79, 1961.
16. Clark, S. P., *J. Am. Mineralogist,* 42, 564, 1957.
17. MacDonald, G. J. F., *Am. Mineralogist,* 42, 564, 1957.
18. Gruver, R. M., *J. Am. Ceram. Soc.,* 33, 171, 1950.
19. Subba Rao, G. V., Natarajan, M., and Rao, C. N. R., *J. Am. Ceram. Soc.,* 51, 171, 1968.
20. Ingraham, T. R. and Marier, P., *Canad. J. Chem. Eng.,* 41, 170, 1963.
21. Kappel, H. and Huttig, G. F., *Kolloid Z.*, 91, 117, 1940.
22. Auffredic, J. P. and Vallet, C. P. R., *Acad. Sci. Paris,* Ser. 263, 652, 1966.
23. Reisman, A., *J. Am. Chem. Soc.,* 80, 3558, 1958.
24. Krustinsons, J., *Z. Elektrochem.*, 39, 193, 1933.
25. Buzdov, K. A., *Zhur. Neorg. Khim.*, 11, 1251, 1966.
26. Saslawski, I. I., *Z. Anorg. Chem.*, 146, 319, 1925.
27. Ferrari, A., and Colla, C., *Atti Acad. Lincei*, 10, 594, 1929.
28. Berg, L. G. and Buzdov, K. A., *Zhur. Neorg. Khim.*, 6, 2003, 1961.
29. Anderson, C. T., *J. Am. Chem. Soc.,* 56, 894, 1934.
30. Kubaschewski, O. and Evans, E., *Metallurgical Thermochemistry,* 3rd ed., Plenum Press, New York, 1958.
31. Kelley, K. K. and King, E. G., *Contributions to Data on Theoretical Metallurgy XIV. Entropies of the Elements and Inorganic Compounds*, U.S. Bur. Mines Bull. 592, 1961.
32. Centnerszwer, M., Falk, G., and Awerbuch, A., *Z. Phys. Chem.,* 115, 29, 1925.
33. Glasner, A. and Hodara, I., *Bull. Res. Council (Israel)*, 7A, 66, 1958.
34. Centnerszwer, M. and Awerbuch, A., *Z. Phys. Chem.,* 123, 127, 1926.
35. Kadlets, O. and Dubinin, M. M., *Izvest. Akad. Nauk.,* SSSR, Otd. Khim. Nauk., 390, 1961.
36. Janz, G. J. and Lorenz, M. R., *J. Electrochem. Soc.,* 108, 1052, 1961.
37. Reisman, A., *J. Am. Chem. Soc.,* 80, 3558, 1958.
38. Janz, G. J., Neuenschwander, E., and Kelly, F. J., *Trans. Faraday Soc.,* 59, 841, 1963.
39. Rolin, M. and Recapet, J. M., *Bull. Soc. Chim.,* 2504, 1964.
40. Janz, G. J. and Lorenz, M. R., *J. Electrochem. Soc.,* 108, 1052, 1961.
41. Gordeeva, V. A., Zhabrova, G. M., and Kushnerer, N. Y., *Dokl. Akad. Nauk. SSSR,* 129, 835, 1959.
42. Centnerszwer, M. and Bruzs, B., *Z. Phys. Chem.,* 114, 237, 1924; ibid. 115, 365, 1925.
43. Britton, H. T. S., Gregg, S. J., and Winsor, G. W., *Trans. Faraday Soc.,* 48, 63, 1952.
44. Cremer, E., Allgauer, K., and Aschenbrenner, W., *Radex Rundschau,* 494, 1953.
45. Muraishi, K., *Yamgata Daigaku Kiyo, Shizen Kagaku,* 6, 187, 1964.
46. Bergstein, A. and Vintera, J., *Coll. Czech. Chem. Comm.*, 22, 884, 1957.
47. Dollimore, D. and Tonge, K. H., *Proc. 5th Symp. Reactivity of Solids 1964*, 497, Elsevier, Amsterdam, 1965.

48. Mahanty, S.R. and Ray, M.N., *Indian Chem. J.,* 3, 37, 1965.
49. Mah, A. D., *U.S. Bur. Mines Rept. Invest.,* 5600, 1960.
50. Coughlin, J. P., *Contributions to the Data of Theoretical Metallurgy XII. Heats and Free Energies of Formation of Inorganic Oxides*, U.S. Bur. Mines Bull. 584, 1954.
51. Anderson, C.T., *J. Am. Ceram. Soc.,* 56, 894, 1934.
52. Lamure, J. and Iltis, C., *Compt. Rend. Acad. Sci. Paris,* C 264, 583, 1967.
53. Mu, J. and Perlmutter, D. D., *Thermochim. Acta,* 49, 207, 1981.
54. Ehlert, T. C., *High Temp. Science*, 9, 237, 1977 .
55. Janz, G. J., Neuenschwander, E., and Kelly, F. J., *Trans. Faraday Soc.,* 59, 841, 1963 .
56. Erdey, L., Liptay, G., and Gal, S., *Talanta,* 12, 883, 1965.
57. Otto, E., *J. Electrochem. Soc.,* 113, 646, 1966.
58. Spencer, W. F. and Toley, B., *J. Chem. Soc.,* 2633, 1929.
59. Swanson, H.E., Morris, M. C., Stichfield, R. P., and Evans, E. H., *Standard X-Ray Diffraction Powder Patterns, Nat. Bur. Stds.,* Monograph 25, part 1, 1962.
60. Ginzburg, D. M., *J. Gen. Chem. USSR*, 26, 1107, 1956.
62. Jaffray, J. and Martin, P., *Compt. Rend.,* 236, 1755, 1953.
63. Khlapova, A. N., *Dokl. Akad. Nauk. SSSR,* 116, 979, 1957.
64. Motzfeld, K., *J. Phys. Chem.*, 59, 139, 1955.
65. Chase, M. W., Davis, C. A., Downey, J. R., Frurip, D. J., McDonald, R. A., and Syverud, A. N., JANAF Thermochemical Tables, *J. Phys. Chem. Ref. Data,* 14, Suppl.1, 1985.
66. Wanmaker, W. L. and Radielovic, D., *5th Int. Symp. Reactivity of Solids,* 529, 1964.
67. Batuecas, T., Rancano, A., and Ibarz, J., *An. Soc. Esp. Fis. Quim.*, 27, 290, 1929.
68. Baker, E. H., *J. Chem. Soc.,* 2525, 1962.
69. Fahim, R. B., Zaki, M.I., and Hussien, G. A. M., *Powder Technol.*, 33, 161, 1982.
70. Duval, C., Wadier, C., and Sevigne, C., *Anal. Chim. Acta*, 20, 263, 1959.
71. Gattow, G., *Pure Appl. Chem.,* 2, 121, 1961.
72. Rassa, M., *Compt. Rend.,* 236, 1673; *Ann. Chim.*, 8, 755, 1953.
73. Tranguard, A., Lacassagne, C., Boinon, M. J., and Capella, L. R., *Compt. Rend. Acad. Sci. Paris,* C264, 1111, 1967.
74. Vorlander, D., Hollatz, J., and Fischer, *J., Ber.*, 65, 536, 1932.
75. Huttig, G. F., Meller, A., and Lehman, E., *Z. Phys. Chem.,* B 19, 1, 1932.
76. Bretsnajder, S. and Cibor, E., *Bull. Acad. Polon. Sci. Chim.*, 13, 709, 765, 1965.
77. Rose, A. J., *Bull. Soc. Franc. Min.*, 71, 16, 1948.
78. Zawadzki, M. J. and Szamborska, W., *Bull. Intern. Acad. Polon. Sci. Classe Sci. Math. Nat. Ser. A (1940–46)* 27, 1948.
79. Gedalla, A. M., *Int. J. Kinetics,* 16, 1471, 1984.
80. Vickery, R. C., *Chemistry of the Lanthanons*, Academic Press, New York, 1953, 217–219.
81. Ambrozhii, M. N., Luchnikova, E. F., and Sidorova, M. I., *Zhur. Neorg. Khim.*, 5, 366, 1960.
82. Head, E. L. and Holly, C. E., *Proc. 3rd Conf. Rare Earth Research,* Gordon and Breach, New York, 1964, 51–63.
83. Glasner, A., Levy, E., Steinberg, M., and Badenheimer, W., *Talanta*, 11, 405, 1964.
84. Illner, E., *Z. Chem.*, 6, 190, 1966.
85. Pannetier, G., Nataf, J., and Dereigne, A., *Bull. Soc. Chim., France,* 318, 1965.
86. Jenkins, I. L. and Waterman, M. J., *J. Inorg. Nucl. Chem.,* 26, 131, 1964 .
87. Erdoes, E., *Coll. Czech. Comm.,* 27, 2152, 1962.
88. Brauer, G., *Handbuch der Praeparitiven Anorganischen Chemie*, Stuttgart, 1954, 828.
89. Head, E. L. and Holley, C.E., in *Proc. 3rd Conf. Rare Earth Research*, Gordon and Breach, New York, 1964, 51–63.

References used in tables:

Karapet 70: Karapetyants, M. Kh., Karapetyants, M. L., *Thermodynamic Constants of Inorganic and Organic Compounds*, Humphrey Science, Ann Arbor, 1970.

CRC 94: *Handbook of Chemistry and Physics*, 74th ed., CRC Press, Boca Raton, FL, 1994.

3 Sulfites

3.1 GENERAL

In contrast to the many sulfates (see Chapter 4), there are very few anhydrous sulfites about which much is known. (See Figure 3.1.) They are usually prepared by the dehydration of hydrates and thus may not be entirely water-free. This may affect the decomposition reactions, since in some instances water and SO_2 evolve simultaneously.

High temperature thermodynamic data are available only for three sulfites: calcium, potassium, and sodium. Most sulfites are fairly unstable, decomposing at relatively low temperatures. The decomposition reactions are not always exactly known, with diverse decomposition products, including sulfur, being reported. There are two major decomposition reactions: (1) decomposition to the oxide and SO_2, and (2) oxidation-reduction (disproportionation) to the sulfate and oxide and SO_2, i.e.,

$$2\ MSO_3 = MSO_4 + MO + SO_2$$

This reaction generally occurs at higher temperatures. Details of these two reactions are discussed under the individual elements.

The sulfite anion is pyramidal, with the three oxygens arranged around the sulfur apex. In Na_2SO_3 the O–S distance is 1.39 Å. Detailed bonding information has been discussed in a review of sulfite spectroscopy by Verma[1] which also classifies sulfites according to bonding: ionic or covalent.

In this chapter information on the various sulfites is summarized. Much of it comes from older literature which can be found in the relevant volumes of Gmelin.[2] This multivolume compilation which has been added to for many decades and is still a work "in progress," is a valuable source of references for all inorganic compounds, but has not been critically evaluated.

Thermodynamic data at 298.15 K were mostly taken from Wagman et al.[3] High temperature data for the above-mentioned three sulfites were taken from the HSC database referred to in Chapter 1.1. No thermodynamic data for oxides are included, since they were already given in Chapter 2, and sulfates, which may also be decomposition products, are described in Chapter 4. Therefore, when only minimal data are available, they are given in the text rather than in individual tables.

3.2 INDIVIDUAL SULFITES

3.2.1 Barium

The sulfite decomposes, beginning at 653 K.[2] Decomposition is complete at 1173 K. The decomposition reaction is most likely

$$2\ BaSO_3 = BaSO_4 + BaO + SO_2$$

$\Delta Hf°$ at 298.15 K = –1179.5 kJ mol^{-1}.[3]

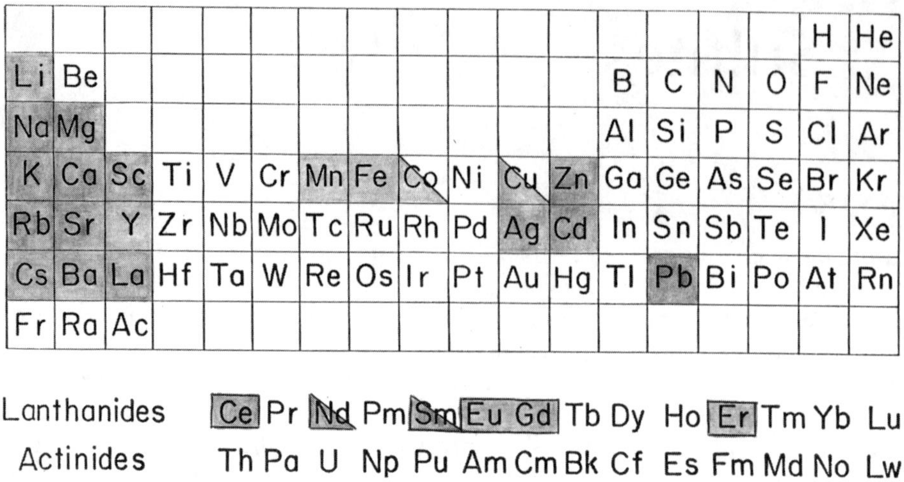

FIGURE 3.1

3.2.2 Cadmium

Lutz and El Suradi[4] found that $CdSO_3$ exists in two modifications, α and β, the latter resulting from the decomposition of the hydrate above 300°C. Above this temperature, decomposition of α-$CdSO_4$ forms CdO and 2 $CdO.CdSO_4$. Between 375 and 450°C SO_2 evolves.

3.2.3 Calcium

Decomposition of $CaSO_3$ has been studied by several workers, using a variety of methods, usually DTA, TGA, and X-ray diffraction of the products.[4–7] There is general agreement that there are two decomposition reactions:

$$CaSO_3 = CaO + SO_2$$

$$4\ CaSO_3 = 3\ CaSO_4 + CaS\ \text{(disproportionation)}$$

but the temperature range in which each occurs depends on the gas phase.

Cubicciotti et al.[5] used a torsion-effusion method to determine decomposition pressures in the range 723 to 767 K and found only SO_2 in the vapor. They surmised that the salt is in metastable equilibrium with respect to disproportionation below 775 K.

Thermodynamic data are given in Table 3.1.

3.2.4 Cesium

The preparation of the salt has been reported, but not its properties. ΔHf° at 298.15K = –1134.7 kJ mol^{-1}.[3]

3.2.5 Cobalt

The preparation of anhydrous $CoSO_3$ was tried in non-aqueous solution,[8] but a truly anhydrous compound may not exist.

3.2.6 Iron

The decomposition by TGA was studied.[9] In an inert atmosphere it occurs in two steps: an endothermic decomposition in the range 200 to 310°C,

$$16\ FeSO_3 = 5\ Fe_3O_4 + FeS_2 + 14\ SO_2$$

for which $\Delta H° = 1117$ kJ, and an exothermic decomposition above 320°C:

$$20\ FeSO_3 = 14\ FeSO_4 + 3\ FeS_2 + Fe_3O_4$$

for which $\Delta H° = -1192$ kJ.

Since $FeSO_4$ decomposes simultaneously, the stoichiometries of the decompositions are doubtful.

3.2.7 Lead

Depending on conditions, the dissociation between 300 and 500°C occurs either as[4]

$$PbSO_3 = PbO + SO_2$$

or

$$2\ PbSO_3 = PbO + PbSO_4 + SO_2$$

The first reaction was also studied by Vishnyakov et al.[10] who list dissociation pressures in the range 328 to 375°C. Above these temperatures they also find disproportionation occurs. $\Delta Hf°$ at 298.15K = –669.9 kJ mol^{-1}.[3]

3.2.8 Lithium

Decomposition of Li_2SO_3 has been studied in dry nitrogen.[11] It apparently occurs by disproportionation at 650 to 850°C, but the results are sketchy.

3.2.9 Magnesium

Decomposition has been reported above 400°C,[12,13] MgO and $MgSO_4$ being formed as products. At 298.15 K, $\Delta Hf° = -1008.3$ kJ mol^{-1}, $S° = 87.9$ $Jmol^{-1}deg^{-1}$.[3]

3.2.10 Manganese

$MnSO_3$.3 H_2O loses water in steps up to 260°. Above this temperature (280 to 540°C) the decomposition occurs in two steps. In both, $MnSO_4$ is formed, but the other products, MnO and MnS, depend on conditions.[13,14]

3.2.11 Potassium

Most information on decomposition predates 1900. Thermodynamic data are given in Table 3.2.

3.2.12 Rare Earths

The rare earth sulfites have potential applications in thermochemical cycles and as starting materials for the preparation of sulfides which can be formed by the reduction of sulfites, e.g., by CO.[15,16] One problem is that the anhydrous sulfites apparently cannot be synthesized directly, but only by the dehydration of hydrates. Thus, great care must be taken to avoid decomposing the salt while water is removed, frequently by removal of SO_2. The decomposition product is frequently an oxysulfate, or sulfite-sulfate, e.g., $Ce_2(SO_3)_2SO_4$ and $Ce_2SO_3(SO_4)_4$. Listed are several references which will provide the reader with entry into the field, many by workers at Helsinki University.[16–19]

3.2.13 Silver

Above 100°C decomposition in a vacuum or under nitrogen gas proceeds according to[20,21]

$$2\ AgSO_3 = 2\ Ag + Ag_2\ SO_4$$

The density is 5.4 g cm^{-3}.[22] Thermodynamic data at 298.15 K are: ΔHf° = –490.8 kJ mol^{-1}, S° = 158.2 J $mol^{-1}deg^{-1}$.[3]

3.2.14 Sodium

Decomposition has been studied in an inert atmosphere by Foerster and Kubel[23] and in air by Sack.[24] Both find disproportionation to be the predominant reaction above 700°C.

$$4\ Na_2SO_3 = 3Na_2SO_4 + Na_2S$$

Thermodynamic data for Na_2SO_3 are given in Table 3.3.

3.2.15 Strontium

Decomposition probably occurs according to Gmelin[2] as

$$4\ SrSO_3 = 3\ SrSO_4 + SrS$$

ΔHf° at 298.15 K = –1177.0 kJ.[3]

3.2.16 Zinc

Decomposition of $ZnSO_3$ has been studied, but only with a hydrate as the starting material.[25,26] Heating results in the simultaneous loss of water and SO_2; thus the results are suspect with respect to temperature range and the nature of the decomposition reaction.

TABLE 3.1
Thermodynamic Data of $CaSO_3$

T K	ΔHf° kJ mol^{-1}	S° J mol^{-1} deg^{-1}	ΔGf° kJ mol^{-1}	Reference
298.15	–1159.4	101.4	–1189.6	Barin 1993
400.00	–1149.8	128.9	–1201.4	Barin 1993
500.00	–1139.9	151.0	–1215.4	Barin 1993
600.00	–1129.6	169.9	–1231.5	Barin 1993
700.00	–1118.7	186.6	–1249.3	Barin 1993
800.00	–1107.4	201.7	–1268.8	Barin 1993
900.00	–1095.6	215.6	–1289.6	Barin 1993
1000.00	–1083.2	228.6	–1311.9	Barin 1993

TABLE 3.2
Thermodynamic Data of K_2SO_3

T K	ΔHf° kJ mol^{-1}	S° J mol^{-1} deg^{-1}	ΔGf° kJ mol^{-1}	Reference
298.15	–1126.8	171.5	–1177.9	Barin 1989
400.00	–1113.8	209.0	–1197.4	Barin 1989
500.00	–1100.2	239.3	–1219.8	Barin 1989
600.00	–1085.7	265.6	–1245.1	Barin 1989
700.00	–1070.5	289.1	–1272.8	Barin 1989
800.00	–1054.4	310.6	–1302.8	Barin 1989
900.00	–1031.5	330.5	–1334.9	Barin 1989
1000.00	–1019.7	349.2	–1368.9	Barin 1989

TABLE 3.3
Thermodynamic Data of Na_2SO_3

T K	ΔHf° kJ mol^{-1}	S° J mol^{-1} deg^{-1}	ΔGf° kJ mol^{-1}	Reference
298.15	–1089.4	146.0	–1133.0	Barin 1993
400.00	–1077.0	182.0	–1149.7	Barin 1993
500.00	–1064.3	210.3	–1169.4	Barin 1993
600.00	–1051.2	234.2	–1191.7	Barin 1993
700.00	–1037.6	255.1	–1216.1	Barin 1993
800.00	–1023.6	273.7	–1242.6	Barin 1993
900.00	–1009.2	290.7	–1270.8	Barin 1993
1000.00	–994.3	306.4	–1300.7	Barin 1993

REFERENCES

1. Verma, P., *Thermochim. Acta*, 89, 363, 1985.
2. Gmelin's *Handbook of Inorganic Chemistry*, Springer Verlag, Berlin-Heidelberg. This is a multivolume work that has been published for several decades and is continuously revised. It is currently in its 8th edition. Information on sulfites is given in the volumes on the individual metals (cations).
3. Wagman, D. D., Evans, W. H., Parker, V. B., Schumm, R. H., Halow, I., Bailey, S. M., Churney, K. L., and Nuttall, R.L., The NBS Tables of Chemical Thermodynamic Properties, *J. Phys. Chem. Ref. Data,* 11, Suppl. 2, 1982.
4. Lutz, H. D. and El Suradi, S., *Z. Anorg. Allgem. Chem.*, 425, 134, 1976.
5. Cubicciotti, D., Sanjuro, A., and Hildenbrand, D. L., *J. Electrochem. Soc.,* 124, 933, 1977.
6. Matsusaki, R., Mazumizu, H., and Murakami, N., *Bull. Chem. Soc. Japan*, 51, 121, 1978.
7. Wieczorek-Ciurowa, K., *J. Thermal Analysis*, 38, 523, 1992.
8. Maylor, R., Gill, J. B., and Godall, D. C., *J. Inorg. Nucl. Chem.*, 33, 1975, 1971.
9. Bugli, G., and Pannetier, G., *J. Thermal Anal.*, 16, 355, 1979.
10. Vishnyakov, A. V., Dubrovin, V. I., and Kovtunenko, P. V., Russ. *J. Inorg. Chem.*, 23, 1235, 1978.
11. Castellai-Bisi, C. and Garrini, E. R., *Gazz. Chim. Ital.*, 93, 1252, 1963, 1977.
12. Hartman, M. and Martinovsky, A., *Chem. Eng. Comm.*, 111, 149, 1992.
13. Butler, F. G. and Mitchell, A., *J. Thermal Anal.*, 10, 257, 1976.
14. Harrison, W.D., Gill, J. B., and Goodall, D. C., *Polyhedron,* 2, 153, 1983.
15. Leskela, M. and Niinisto, L., *Thermochim. Acta,* 37, 125, 1980.
16. Koskelinna, M. and Niinisto, L., *Finn. Chem. Lett.*, 83, 1975.
17. Leskela, T., Niinisto, L., and Lippmaa, M., *Thermochim. Acta,* 262, 195, 1995.
18. Leskela, T., Leskela, M., and Niinisto, L., *Thermochim. Acta,* 256, 67, 1995.
19. Koskelinna, M., Leskela, M., de Matos, J. E. K., and Niinisto, L., *Thermochim. Acta,* 95, 401, 1985.
20. Savel'yev, G. G., Budkov, V. A., Kabanov, A. A., Zakharov, Yu., *Izv. Tomsk. Politekh. Inst.*, 151, 36, 1967.
21. Savel'yev, G. G., Zakharov, Yu. A., Kabanov, A. A., Budkov, V.A., *Izv. Vyssh. Uchebn. Zaved., Khim. Tekhnol.*, 10, 735, 1967.
22. Larsson, L. O., *Acta Chem. Scand.*, 23, 2261, 1969.
23. Foerster, F. and Kubel, K., *Z. Anorg. Algem. Chem.*, 139, 261, 1925.
24. Sack, W., *Z. Anorg. Algem. Chem.*, 264, 272, 1951.
25. Pechkovskii, V. V. and Ketov, A. N., Russ. *J. Appl. Chem.*, 30, 1708, 1957.
26. Margulis, E. V. and Grishankina, N. S., *Russ. J. Inorg. Chem.*, 8, 1383, 1963.

4 Sulfates

4.1 GENERAL

The sulfates constitute one of the largest classes of inorganic oxyanion salts. (See Figure 4.1.) Because they are generally stable, they are used extensively in many areas of chemical technology. As can be seen in this chapter, many of their properties have been measured and their decomposition is well characterized.

The sulfate ion is tetrahedral, but both the S–O bond length and angle vary considerably from salt to salt, the S–O distance from 1.4 to 1.8Å, the O–S–O angle from 95 to 120°. A Hückel molecular calculation of these features has been published.[1]

The thermal stability of the sulfate anion varies with the cation, as was described for the carbonates, i.e., stability decreases with increasing polarizing power of the cation. Thus small cationic size and high charge promote decomposition. The primary decomposition is to the oxide and SO_3, accompanied by decomposition of that gas to SO_2 and O_2. Thus in a closed system there will also be two simultaneous equilibria:

$$MSO_4 = MO + SO_3 \tag{4.1}$$

$$SO_3 = SO_2 + 1/2\ O_2 \tag{4.2}$$

Equilibrium data for the second reaction are given in separate tables (see below).

As in the preceding chapters, data for densities and phase transitions are also provided. The discussion of these properties in Chapter 1 applies here also.

4.2 DECOMPOSITION EQUILIBRIA

The following data relevant to the thermal decomposition of each sulfate have been listed: ΔHf°, S°, and ΔGf° for each sulfate from 298.15 K to as high a temperature as is available; similar data for the corresponding oxides which were not given in Chapter 2, and separate tables for ΔH°, ΔS°, and ΔG° for the decomposition to (a) the oxide and SO_3, and (b) the oxide and SO_2 and O_2; and the equilibrium constant for these reactions. All of these parameters were calculated using a computer program.[2] The source of the data used in these calculations is given in the same manner as in Chapter 2.

Partial pressures of the three gases can be calculated if the initial conditions are known. For example, if the decomposition is carried out in air, $p(O_2) = 0.21$. If it is carried out in a closed, originally evacuated container in which the only source of the gases is the decomposing sulfate, partial pressures can be calculated from the simultaneous solution of Equations 4.3 and 4.4, corresponding to Equations 4.1 and 4.2

$$K_1 = p(SO_3) \tag{4.3}$$

$$K_2 = p(SO_2).p^{1/2}(O_2)/p(SO_3) \tag{4.4}$$

As a good approximation $p(SO_3) = K_1$, $p(O_2) \simeq 1/2\ p(SO_2)$ and the pressures of these two gases can be calculated since $p(SO_3) << p(SO_2)$. Exact calculations require the simultaneous solutions of 4.3 and 4.4, but this is somewhat complicated.

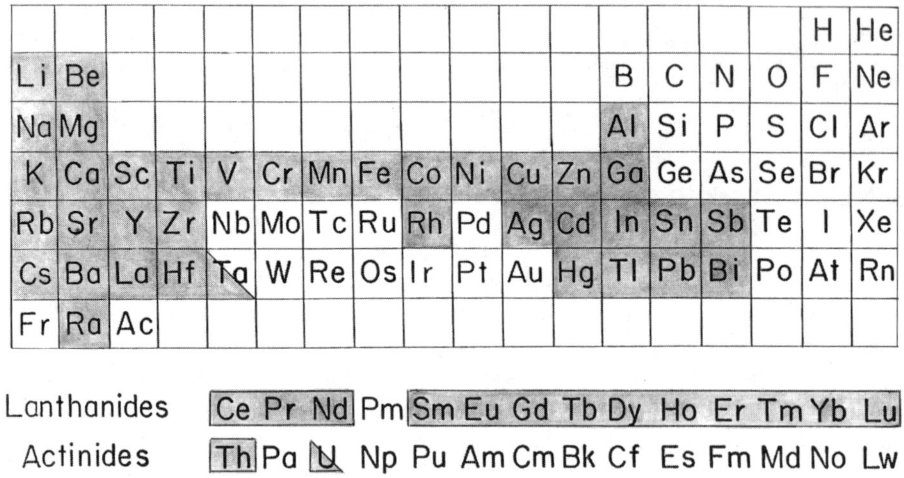

FIGURE 4.1 The known sulfates.

Thermodynamic data for SO_3, SO_2, and for the equilibrium 4.2 are given in Tables 4.1 to 4.3. As can be seen from Table 4.3, the equilibrium shifts to the right with increasing temperature.

4.3 INDIVIDUAL SULFATES

4.3.1 Aluminum

$Al_2(SO_4)$ decomposes directly to Al_2O_3 without the formation of an intermediate.[3] However, different researchers have obtained Al_2O_3 in different states. Knutsen and Searcy found it to be amorphous,[4] whereas Ingraham obtained γ-Al_2O_3.[5]

Since the volume of Al_2O_3 is much less than that of the sulfate, it is not surprising that the product is porous and that the kinetics follows the contracting sphere model.[6]

No solid–solid transitions are known and the solid decomposes without melting. Ambient temperature density is 2.71.[7] Thermodynamic data are given in Tables 4.4 to 4.7.

4.3.2 Antimony

On heating, $Sb_2(SO_4)_3$ decomposes with partial oxidation. Oxysulfates exist about which little is known. From the thermodynamic data of the sulfate one surmises that it is stable to 1000 K, but since the course of the decomposition is unknown, only the thermodynamic data of the sulfate and the oxides have been listed (Tables 4.8 to 4.12).

4.3.3 Barium

$BaSO_4$ melts without obvious decomposition. The density of c, II is given as 4.499 at 288 K.[8] Phase transitions are listed in Table 4.13, and thermodynamic data in Tables 4.14 to 4.16.

4.3.4 Beryllium

Thermodynamic calculations (Tables 4.18 to 4.21) indicate that $BeSO_4$ is unstable above 1000 K. Kinetics of the decomposition have been studied by Johnson and Gallagher.[11] The authors use their data to study various kinetic models and find the contracting sphere model most satisfactory.

The density is given in Table 4.17.

4.3.5 Bismuth

Thermodynamic calculations (Tables 4.22 to 4.25) indicate that $Bi_2(SO_4)_3$ should be stable to above 1200 K, but an experimental study[13] reports that decomposition proceeds through several oxysulfates, beginning at 700 K. One complication is the low melting point of Bi_2O_3, 1090 K.[10]

The density is 5.08 at 288 K.[14]

4.3.6 Cadmium

Solid-phase transitions and decomposition products of $CdSO_4$ have been studied by two groups of researchers,[15,16] both of whom review the confused earlier literature. On heating, the salt undergoes at least one, possibly two, transformations, but the transition temperature is not certain. The highest temperature form melts near 1300 K. However, slow decomposition to $2CdO.CdSO_4$ is observed well below this temperature. The final decomposition product is CdO. No thermodynamic data for the double salt are available, and therefore only data relevant for the decomposition to CdO are given (Tables 4.28 to 4.30). Data for CdO are given in Table 2.13.

There exist three high-temperature modifications of $CdSO_4$ whose structure was determined by X-ray diffraction.[18]

Densities are listed in Table 4.26, transitions in Table 4.27.

4.3.7 Calcium

$CaSO_4$ decomposes directly to CaO. The supposed existence of an intermediate oxysulfate and discrepant values for the melting point of $CaSO_4$ are explained[20] by a eutectic between CaO and $CaSO_4$ near 1638 K, with less than 20% CaO.

Kinetic studies of the decomposition are not strictly comparable, since one was carried out in a vacuum[21] and another in air,[22] but both agree that the rate-determining step occurs at the $CaSO_4$-CaO boundary.

The density is 4.691.[8] Transitions are in Table 4.31, and thermodynamic data in Tables 4.32 to 4.34. Data for CaO are in Table 2.19.

4.3.8 Cesium

Very careful work by Lau et al.[23] has established that the main mode of vaporization of Cs_2SO_4 is sublimation, by which 95% of the material reaches the vapor state, ΔH_{sub} at 298.15 K being 320.1 $\pm$ 2.0 $kJmol^{-1}$. Five percent is accounted for by the congruent decomposition

$$Cs_2SO_4(s,l) = 2\ Cs(g) + SO_2(g) + O_2(g)$$

A thermodynamic table for this reaction has been included, but it shows that it is minor compared to decomposition to the oxide:

$$Cs_2SO_4 = Cs_2O + SO_2(g) + 1/2\ O_2(g)$$

The density of the solid and liquid salt is given in Table 4.35, transitions in Table 4.36. Thermodynamic data are given in Tables 4.37 to 4.41. Data for Cs_2O are in Table 2.23.

4.3.9 Chromium

Cr(II). $CrSO_4$ is known only in solution and as hydrates, and double salts, all easily oxidized by air. The monohydrate yields Cr_2O_3 on heating.

Cr(III). $Cr_2(SO_4)_3$ is stable to 573 K, but decomposes not far above this temperature, ultimately to Cr_2O_3. The course of the decomposition is unknown, but we have included tables for the decomposition reactions

$$Cr_2(SO_4)_3 = Cr_2O_3 + 3\ SO_2(g) + 3/2\ O_2(g)$$

$$Cr_2(SO_4)_3 = Cr_2O_3 + 3\ SO_3(g)$$

which indicate that decomposition is slight up to 1000 K. Thermodynamic data are given in Tables 4.42 to 4.45.

The density is 2.706 at 298 K.[27]

4.3.10 Cobalt

The decomposition of $CoSO_4$ has been studied by Pechkovsky[28] and Pechkovsky, Zvedin, and Beresneva[29] in a dynamic system at a flow rate of 3 l/hr between 1023 and 1223 K. Increase of SO_2 in the carrier gas inhibited decomposition. The decomposition is linear in time with no evidence of an oxysulfate, i.e., $\alpha = kt$. A linear function of 1/T between 1113 and 1133 K is ln k, the slope yielding an activation energy of 319 kJ. This is also consistent with the work of Tagawa and Saijo.[30] The thermodynamic calculations (Tables 4.48 to 4.50) indicate that decomposition becomes appreciable above 1100 K. Data for CoO are in Table 2.27.

Densities are given in Table 4.46, and transitions in Table 4.47.

4.3.11 Copper

Cu(I). Cu_2SO_4 oxidizes in air.

Cu(II). There is general agreement that the first product of the decomposition of $CuSO_4$ is an oxysulfate, $CuO.CuSO_4$. In kinetic studies, the temperature range of this reaction has been reported as 1092 to 1133 K,[33] and 1053 to 1123 K,[29] possibly because of differences in the gas phase composition. The decomposition rate decreases with increasing oxygen content of the flow gas and increases with increasing SO_3 because these gases displace the $SO_3 = SO_3 + 1/s\ O_2$ equilibrium in opposite directions.

Decomposition of both sulfate and oxysulfate is zero-order. For the former, an Arrhenius activation energy of 212 kJ was obtained between 1053 and 1123 K, and for decomposition of the oxysulfate it was 346 kJ between 1133 and 1193 K. However, somewhat different values of 238 and 269 kJ for these reactions have also been reported,[34] and agree with those of Ingraham and Marier.[35]

Although the values for sulfate decomposition agree fairly well, those for the oxysulfates are quite different, possibly because the Russian researchers ran this decomposition at a lower temperature, 1033 to 1073 K.

Two sets of measurements[36,37] in fair agreement have been made of the equilibrium decomposition pressures, 948 to 1033 K, in the systems

$$2\ CuSO_4 = CuO.CuSO_4 + \text{gases}$$

and

$$2\ CuO.CuSO_4 = CuO + \text{gases}$$

where the pressures in the first system are higher than in the second. Because of lack of data on the oxysulfate it is not yet possible to calculate these pressures thermodynamically. We have, however, calculated equilibrium constants for the reactions

$$CuSO_4 = CuO + SO_3(g)$$

and

$$CuSO_4 = CuO + SO_2(g) + 1/2\ O_2$$

which indicate appreciable reaction in the range found experimentally. These data are in Tables 4.52 to 4.54. Densities are in Table 4.51, and transitions in Table 4.52. Data for CuO are in Table 2.30.

4.3.12 Gallium

Ga_2SO_4 decomposes without melting to Ga_2O_3. Marchal[17] measured equilibrium pressures over the range 833 to 973 K, from which it appears that the salt is less stable than $Al_2(SO_4)_3$. Equilibrium constants cannot yet be calculated thermodynamically.

The density is 3.229 at 296 K.[39]

4.3.13 Hafnium

Von Hevesy and Cremer[40] attempted to measure equilibrium pressures of $Hf(SO_4)_2$ in the range 823 to 923 K, but were not sure of obtaining true equilibrium. Decomposition was direct to HfO_2, not through an intermediate oxysulfate, and the sulfate appears to be more stable than $Zr(SO_4)_2$. No thermodynamic data for the salt are available.

4.3.14 Indium

Seward[41] measured equilibrium decomposition pressures of $In_2(SO_4)_3$ over the range 918 to 1093 K, from which it appeared to be more stable than $Al_2(SO_4)_3$. The product was In_2O_3 without an oxysulfate intermediate. Thermodynamic data for the decomposition are given in Tables 4.55 to 4.58. These indicate that the salt is stable to 1100 K.

A kinetic study[42] above 1073 K showed that the initial decomposition rate is nearly constant, but decreases with time. The degree of decomposition is given by

$$\alpha = 1 - \exp(-kt^n)$$

which can be written in logarithmic form as

$$\ln[-\ln(1 - \alpha)] = n \ln t + \ln k$$

From the linear plot of the left-hand side against ln t, the constants n and k are obtained; n increases linearly from 1.0 at 1123 K to 1.6 at 1273 K, while k increases logarithmically from 0.021 to 0.410 between these temperatures. The values of n indicate that decomposition proceeds from nuclei in two dimensions.

The density is 3.43 at 298 K.[8]

4.3.15 Iron

The sulfates of both Fe(II) and Fe(III) decompose without melting. Decomposition of $Fe_2(SO_4)_3$ to Fe_2O_3 is reversible, without the formation of an intermediate oxysulfate. $FeSO_4$ decomposes irreversibly, with oxidation, to Fe_2O_3, through intermediates that have not been identified with certainty, though FeO is often assumed.

A number of researchers have measured equilibrium decomposition pressures,[43–45] but these are no longer required for the calculation of thermodynamic functions that are now available and used for the calculations of equilibrium constants for the decomposition of both $FeSO_4$ and $Fe_2(SO_4)_3$ (Tables 4.59 to 4.62).

When $FeSO_4$ is heated above 823 K, the FeO formed is immediately oxidized by the oxygen or SO_3 liberated in the reaction.[46–48] The reactions occurring are

$$2\ FeSO_4 = 2\ FeO + 2\ SO_3$$

$$SO_3 = SO_2 + 1/2\ O_2$$

$$2\ FeSO_4 = Fe_2O_3 + SO_2 + SO_3$$

corresponding to the net reaction

$$2\ FeSO_4 = Fe_2O_3 + SO_2 + SO_3$$

If the SO_3 remains in the system the following reaction may occur:

$$Fe_2O_3 + 3\ SO_3 = Fe_2(SO_4)_3$$

However, the reaction is slow and may be suppressed by removing the SO_3. As expected, an increase of O_2 in the sweep gas inhibits the decomposition.

When the weight loss was measured in air[49] between 953 and 1003 K, decomposition followed the Roginsky equation

$$1 - (1 - \alpha)^{2/3} = \alpha = kt$$

From the linear temperature dependence of ln k on 1/T an Arrhenius activation energy of 253 kJ was obtained. Since, at any temperature in the above range, the decomposition pressure of $FeSO_4$ is greater than that of $Fe_2(SO_4)_3$; the rate-determining step for decomposition of the former is decomposition of the ferric salt. Decomposition appears to occur exclusively on the surface of the original material, rather than in the interior of the grains.

Decomposition of $Fe_2(SO_4)_3$ is zero-order between 973 and 1073 K. Activation energies of 89 kJ[50] and 154 kJ[51] for the reaction have been reported, the latter for vacuum, the former for a stream of dry N_2.

When the gas phase has a constant composition, the driving force for the reaction is the difference between the actual and equilibrium pressure of SO_3.

In air between 953 and 1053 K Roginsky's equation holds,[49] i.e., the degree of decomposition is a linear function of time. This implies that the decomposition is not autocatalytic, and that the new phase is instantaneously formed on the surface of the original material. During the later stages the rate is controlled by the diffusion of product gases through the solid oxide. The Arrhenius activation energy, calculated from the slope of ln k vs. 1/T is 289 kJ, quite different from the other values reported.

No explanation is evident for the wide range of reported activation energies reported by different authors, since one would not expect a radical change in mechanism to be produced by a change in experimental conditions.

Since the SO_2/SO_3 equilibrium is established relatively slowly, the rate- determining step is likely to involve the expulsion of an SO_3 molecule from the crystal lattice. All of the reported activation energies are much less than ΔH = 526 kJ, but ΔH per mole of SO_3 is 526/3 = 176 kJ which is at least in the right range.

Thermodynamic data for the decomposition reactions are given in Tables 4.59 to 4.64.

4.3.16 Lead

Because $PbSO_4$ decomposes noticeably below the melting point, this value is difficult to determine reliably. The ultimate decomposition product is PbO. Various oxysulfates have been prepared in ways other than by pyrolysis of $PbSO_4$. Some have congruent melting points and form eutectics with PbO or $PbSO_4$; for a summary and references, see Kelley[52]; Rossini et al.[9] lists $PbO.PbSO_4$, m.p. 1243 K, 2 $PbO.PbSO_4$, m.p. 1223 K. The former is the mineral lanarkite. Hoschek[53] found three intermediates before PbO in the pyrolysis of $PbSO_4$, the two oxysulfates above and 4 $PbO.PbSO_4$. Thermodynamic data are now available for all the above-mentioned oxysulfates, as well as 3 $PbO.PbSO_4$, and these have been listed, permitting the user to calculate equilibria. However, only equilibria for decomposition of $PbSO_4$ to PbO have been given (Tables 4.67 to 4.73). Densities are listed in Table 4.65 and transitions in Table 4.66. Data for PbO are in Table 2.37.

4.3.17 Lithium

A detailed study of Li_2SO_4 vaporization/decomposition[23] shows that decomposition occurs by the reaction

$$Li_2SO_4 = Li_2O + SO_2(g) + 1/2\ O_2$$

with Li_2O and Li_2SO_4 forming a solution. Sublimation of Li_2SO_4 accounts for about 0.3% of the pressure over the salt. The enthalpy of vaporization is 388.4 ± 4.1 kJ.

The density is given in Table 4.74, thermodynamic data in Tables 4.75 to 4.77. Data for Li_2O are in Table 2.41.

4.3.18 Magnesium

$MgSO_4$ begins to decompose below the melting point, directly to MgO without an oxysulfate intermediate. Melting points in the literature therefore vary widely. Thermodynamic calculations (Tables 4.80 and 4.81) indicate that decomposition probably becomes measurable above 1200 K. The initial decomposition temperature was observed to be near 1170 K, and was too rapid above 1450 K to be measurable.[56]

The decomposition kinetics of $MgSO_4$ have been studied in a flow system in the range 1223 to 1323 K.[57] When N_2 flows over the salt at 3 l/hr, decomposition is linear in time. Under these conditions the product gases are continuously removed, and the SO_2/SO_3 equilibrium is not established. If these gases and O_2 flow over the salt, the reaction rate is nearly independent of oxygen, but the SO_3/SO_2 ratio decreases with decreasing O_2 content. The activation energy in the range 1293 to 1333 K is 661 kJ.[49]

The density of $MgSO_4$ is given in Table 4.78. Transitions are given in Table 4.79, and thermodynamic data in Tables 4.80 to 4.82. Data for MgO are in Table 2.46.

4.3.19 Manganese

$MnSO_4$ decomposes before reaching its sometimes reported melting point of 973 K. The product usually observed on pyrolysis in air is Mn_3O_4,[58,59] though Mn_2O_3 may appear mixed with it in the lower temperature range, complicating the kinetic picture.[60] Since Mn_3O_4 is the expected product we have provided thermodynamic functions for it and for the decomposition reaction

$$3\ MnSO_4 = Mn_3O_4 + 3\ SO_2(g) + O_2(g)$$

These data show that decomposition becomes appreciable above 1000 K. Data are given in Tables 4.84 to 4.88. The density of $MnSO_4$ is in Table 4.83. Data for MnO are in Table 2.50.

4.3.20 Mercury

Hg(I). Pascal,[62] using very old references, describes Hg_2SO_4 as unstable to light, disproportionating to Hg° and$HgSO_4$, and melting with decomposition to Hg°, O_2, and SO_2.

Hg(II). According to a study by Hoschek,[15] $HgSO_4$ is unchanged after two hours at 673 K, but by 773 K decomposition occurs to 2 $HgO.HgSO_4$, identified through their X-ray spectra. All three compounds are volatile at elevated temperatures. The supposed next product, HgO, was not observed because it is even more volatile. If $HgSO_4$ is heated rapidly to 973 K, some Hg_2SO_4 appears along with the oxysulfates before volatilization is complete.

Since Hg_2SO_4 is so unstable, we have only listed thermodynamic functions for reactions involving $HgSO_4$. However, functions for Hg_2O and Hg_2SO_4 have been listed (Tables 4.89 to 4.92). Note that data for both the divalent sulfate and oxide were extrapolated from 800 K.

The density of $HgSO_4$ is 6.50 at 298 K.[63]

4.3.21 Nickel

$NiSO_4$ decomposes, without melting, to NiO. TGA curves show no evidence of an oxysulfate intermediate.[59,64–66] An experimental study[56] shows that the decomposition becomes noticeable at 900 K, and very rapid at 1223 K, consistent with thermodynamic data. These are listed in Tables 4.93 to 4.95. Data for NiO are given in Table 2.58.

A contracting sphere model applies, meaning that the reaction occurs at the oxide-sulfate boundary. Similar results were obtained by Tagawa and Saijo.[30]

The specific gravity is 3.543.[27]

4.3.22 Potassium

The vaporization of K_2SO_4 was studied by the torsion-effusion method, and by mass spectrometry in the range 1180 to 1280 K.[61] Sublimation contributes 63% of the flux, and decomposition

$$K_2SO_4(c) = 2\ K(g) + SO_2(g) + O_2(g)$$

the rest. According to the authors, vaporization is kinetically hindered. Comparison with decomposition to $K_2O + SO_2 + O_2$ shows that because of the exponents involved, congruent decomposition is actually more significant. Thermodynamic data for both kinds of decomposition are given (Tables 4.98 to 4.101).

Densities are listed in Table 4.96, and transitions in Table 4.97. Data for K_2O are in Table 2.63.

4.3.23 Radium

The density of $RaSO_4$ has been calculated as 5.77 gcm^{-3} from crystallographic data.[103] Thermodynamic data on $RaSO_4$ are given in Table 4.102.

RARE EARTHS

Since relatively little is known about the high-temperature behavior of the rare-earth sulfates, a summary covering all of them is given first, followed by information about the individual salts.

On heating, all of the anhydrous sulfates of the trivalent rare-earth elements and yttrium, type formula $R_2(SO_4)_3$ decompose, without first melting, to basic salts (oxysulfates) of the type $R_2O_3.SO_3$, then to an oxide. The oxide final product is R_2O_3 for all the elements except cerium, praesodymium, and terbium, which, in air at ambient pressure, form CeO_2, Pr_6O_{11}, and the somewhat variable oxide usually written as Tb_4O_7.

$$R_2(SO_4)_3 = R_2O_3.SO_3 + 2\ SO_3 \quad \text{(a)}$$

$$R_2O_3.SO_3 = R_2O_3 + SO_3 \quad \text{(b)}$$

There have been three comprehensive studies which, taken together, reveal the trend in stability of the sulfates and oxysulfates within the rare-earth series. Woehler and Gruenzweig[72,73] measured equilibrium decomposition pressures for reaction (a) for scandium, yttrium, and eight of the rare-earth elements in the range 973 to 1293 K. Nathans and Wenlandt[74] performed DTA and TGA, covering reactions (a) and (b) for yttrium and all the rare earths. Ivakin,[75] by a weight loss technique in a stream of dry N_2, working with yttrium and seven of the rare earths, investigated the kinetics of reaction (a) at 1223 and 1323 K and of reaction (b) at 1573 K. From these sources one may conclude that, with the exception of cerium, the stability of the sulfates in reaction (a) decreases from La to somewhere in the Sm Eu Gd group, and then increases to Lu, while the stability of the oxysulfates in reaction (b) decreases all the way from La to Lu. These conclusions hold for either kinetic or equilibrium definitions of "stability," and they ignore a few reversals of order which are probably attributable to experimental errors, in view of the close similarity of adjacent members of the rare-earth series. Among the more stable sulfates the order of decreasing stability in reaction (a) is La to Y to Lu (or Yb) according to kinetics,[75] equilibrium pressures,[76] and DTA,[74] but not by TGA.[74] In reaction (b) Y falls near Ho and Er,[74,75] in accordance with the crystal radii of R^{+3} and the lanthanide contraction.

By the equilibrium pressure criterion, $Sc_2(SO_4)_3$ is less stable than any of the rare-earth sulfates except cerium. However, unlike the rare earths, it seems to decompose directly to Sc_2O_3,[76] rather than through reactions (a) and (b).

$Ce_2(SO_4)_3$ is much less stable than the sulfates of the other rare earths,[72,74,76] and less stable than $Sc_2(SO_4)_3$.[76] The product is CeO_2, rather than Ce_2O_3, but the mechanism is unknown. Least stable by far is $Ce(SO_4)_3$ which also yields CeO_2 ultimately. Woehler, Plueddemann, and Woehler[45] measured decomposition pressures in the range 634 to 755 K in what they believed to be the system $2\ Ce(SO_4)_2 = Ce_2(SO_4)_3 + SO_2 + 1/2\ O_2$. Duval[77] found an intermediate $3\ CeO_2.4SO_3$ by TGA.

Ivakin's kinetic investigation[75] showed that the oxysulfates in reaction (b) decompose in accordance with a simple zero-order rate law, whereas reaction (a) is more complex kinetically. Thus a plot of the equation

$$1 - (1 - \alpha)^{1/3} = kt$$

is a line whose slope changes at one point from k_1 to k_2. This would suggest a compound intermediate between $R_2(SO_4)_3$ and $R_2O_3.SO_3$, but X-ray diffraction patterns were interpreted as showing that the only possible intermediate was a solid solution of variable composition.

The relative stabilities of the sulfates of Sc, Y, and the rare earths may be represented in the following order of decreasing stability:

La - Y - Lu - Gd - Sc - Ce(III) - Ce(IV)

This refers to decomposition of the normal anhydrous sulfate to its first decomposition product, oxysulfate or oxide as the case may be for each element. The dashed line preceding Gd represents the other rare earth elements. With increasing atomic number the stability first decreases in the order Pr - Nd - Sm - Eu and then increases in the order Tb - Dy - Ho - Er - Tm - Yb.

High-temperature behavior of the bivalent sulfates $SmSO_4$, $EuSO_4$, and $YbSO_4$ has not been investigated. They are expected to undergo self-oxidation, like $FeSO_4$, to ultimately yield R_2O_3.

4.3.24 Cerium

$Ce_2(SO_4)_3$ melts at 1143 K. Thermodynamic data are listed in Tables 4.103 to 4.106.

4.3.25 Dysprosium

TGA and DTA curves indicate that $Dy_2(SO_4)_3$ decomposes to DyO_2SO_4 in the range 1183 to 1413 K and then to Dy_2O_3 up to 1593 K.[78]

4.3.26 Erbium

TGA and DTA curves indicate that $Er_2(SO_4)_3$ decomposes to ErO_2SO_4 in the range 1213 to 1413 K, and then to Er_2O_3 from 1413 to 1543 K.[78]

4.3.27 Europium

The dry salt is quite stable in air. Decomposition proceeds according to

$$2\ EuSO_4 \rightarrow Eu_2(SO_4)_2 \rightarrow Eu_2O_2(SO_4)_2$$

and finally, Eu_2O_3 above 1173 K. Oxidation to Eu(III) occurs simultaneously.[79]

4.3.28 Gadolinium

Decomposition to $Gd_2O_2SO_4$ occurs at 1300 K and to Gd_2O_3 at 1543 K.[78]

4.3.29 Holmium

Decomposition of $Ho_2(SO_4)_3$ to $Ho_2O_2SO_4$ occurs from 1203 to 1443K, and to Ho_2O_3 from 1443 to 1583 K.[78]

4.3.30 Lanthanum

Thermodynamic functions of $La_2(SO_4)_3$ are given in Table 4.107. Decomposition to $La_2O_2SO_4$ begins at 973 K and becomes appreciable at 1173 K. At 1513 K decomposition proceeds to La_2O_3 and the reaction is complete at 1593 K. Both reactions are kinetically first order.[80]

4.3.31 Neodymium

$Nd_2(SO_4)_3$ decomposes to $Nd_2O_2SO_4$ at ~1300 K and then, up to 1630 K, to Nd_2O_3, consistent with the thermodynamic data in Tables 4.108 to 4.111.

4.3.32 Praesodymium

$Pr_2(SO_4)_3$ decomposes to $Pr_2O_2SO_4$ up to 1273 K. Between 1448 and 1658 K decomposition proceeds to Pr_2O_3.[78]

4.3.33 Samarium

$Sm_2(SO_4)_3$ is stable up to 1073 K and then decomposes to $Sm_2O_2SO_4$. At 1500 K decomposition proceeds to Sm_2O_3.

$$\Delta H°f(298) = -4473\ \text{kJ mol}^{-1}$$

$$\Delta G°f(298) = -3535\ \text{kJ mol}^{-1}$$

4.3.34 Terbium

$Tb_2(SO_4)_3$ decomposes to $Tb_2O_2SO_4$ in the range 1156 to 1363 K and to $Tb_2O_2SO_4$ at 1403 to 1590 K.[78]

4.3.35 Thulium

$Tm_2(SO_4)_3$ decomposes in the range 1213 to 1543 K, but the nature of the product is not certain.[78]

4.3.36 Ytterbium

$Yb_2(SO_4)_3$ decomposes in the range 1173 to 1523 K, but the nature of the product is not clear.[80]

4.3.37 Rhodium

The compound $Rh_2(SO_4)_3$ is difficult to prepare in pure form. The reported density of 3.44 at 291 K[81] needs confirmation.

4.3.38 Rubidium

The vaporization/decomposition of Rb_2SO_4 has been studied by mass spectrometry and the torsion-effusion method.[23] Molecular vaporization accounts for 75% of the total, with the decomposition

$$Rb_2SO_4 = 2\ Rb(g) + SO_2(g) + O_2(g)$$

accounting for the rest. The derived enthalpy of vaporization at 298 K is 340 ± 2.8 kJ mol^{-1}. Thermodynamic functions for this reaction are given in Tables 4.114 to 4.116. Data for Rb_2O are given in Table 2.66. Densities are in Table 4.112, phase changes in Table 4.113.

4.3.39 Scandium

$Sc_2(SO_4)_3$ is stable to 873 K and then decomposes directly to Sc_2O_3.[82] No high-temperature thermodynamic data for the salt are available, although they are known for the oxide.

4.3.40 Silver

The salt exists in two crystalline modifications: a lower temperature rhombic and a higher temperature hexagonal. The measured transition temperature seems to depend somewhat on the heating (or cooling) rate.

Since Ag_2O is only stable over a narrow temperature range above ambient, the decomposition of Ag_2SO_4 may be thought of as proceeding directly to the oxide. The tables given, 4.118–4.120, permit carrying out the desired calculations. Data for Ag_2O are in Tables 2.69 and 2.71. The density is 5.363 at 298 K,[83] and transitions are given in Table 4.117.

4.3.41 Sodium

Na_2SO_4 exists in several crystal modifications which are given in Table 4.122. The nomenclature for the crystal forms is that of Kracek and Gibson.[84] The transition enthalpy between III and V is 3.1 kJ mol^{-1} from 298 to 450 K, whence $\Delta H°f$ of III at 298.15 K is –1381.4 kJ mol^{-1}.

A mass spectrometric and torsion effusion study[23] found that sublimation accounted for 37% of the mass loss and the decomposition

$$Na_2SO_4 = 2\ Na(g) + SO_2(g) + O_2(g)$$

for the rest. The derived enthalpy of sublimation is 364.0 ± 1.3 kJ at 298.15 K. Other thermodynamic data are given in Tables 4.123 to 4.125. Data for Na_2O are given in Table 2.75. Densities are listed in Table 4.121, transitions in Table 4.122.

4.3.42 Strontium

$SrSO_4$ exists in two crystalline modifications. $SrSO_4$ is the mineral stelestite (stelestine). The ambient temperature density is 3.91.[86] Transition data are given in Table 4.126.

There are some discordant experimental decomposition studies [87], but the thermodynamic data (Tables 4.127 to 4.129) indicate stability to near 1500 K. Data for SrO are in Table 2.79.

4.3.43 Tantalum

The existence of anhydrous $Ta_2(SO_4)_3$ is uncertain.[88]

4.3.44 Thallium

Tl(I). According to most of the older literature,[89] Tl_2SO_4 may be heated as far as the melting point without noticeable decomposition or volatilization, after which it sublimes without decomposition. TGA studies have reported contradictory results. One has a plateau up to 628 K,[90] then a weight loss interpreted as formation of Tl_2O_3, not quite complete by 1153 K. The other[91] shows a plateau to 1000 K, followed by rapid volatilization.

Favorskaya and Ponomaryeva[92] measured the vapor pressure of Tl_2SO_4 in the range 973 to 1233 K by a dynamic method, finding no evidence of decomposition. Thermodynamic calculations suggest the decomposition begins above 1200 K. Densities are given in Tables 4.130, transitions in Table 4.131, and thermodynamic data in Tables 4.132 to 4.135.

Tl(III). Little is known about $Tl_2(SO_4)_3$. It has been reported[89] to decompose readily to Tl_2SO_4.

4.3.45 Thorium

$Th(SO_4)_2$ is stable to at least 673 K, the temperature commonly used to prepare the anhydrous salt from the hydrates. However, the thermodynamic Tables 4.136 to 4.139, indicate stability to near 1100 K. At higher temperatures, the salt decomposes to ThO_2 without melting or passing through an oxysulfate.

4.3.46 Tin

Sn(II). At room temperature $SnSO_4$ is stable in air for several months, longer in vacuum in the dark, but on heating it undergoes self-oxidation to SnO_2 and SO_2. Above 633 K in O_2-free N_2 complete conversion has been found at 651 K, but there is little decomposition below this temperature.[95] Various other studies[65,95] have obtained various contradictory results. Thermodynamic results (Tables 4.140 to 4.147) indicate stability to much higher temperatures, but these assume equilibrium rather than kinetic (vacuum or flowing gas) conditions.

Sn(IV). Anhydrous $Sn(SO_4)_2$ is very hygroscopic, and becomes thermally unstable not far above 373 K.[96] However, TGA studies indicate decomposition in a vacuum becomes noticeable near 573 K, with complete conversion to SnO_2 at about 860 K.[65]

4.3.47 Titanium

Ti(II) $TiSO_4$, existence uncertain.

Ti(III). $Ti_2(SO_4)_3$. Thermal properties unknown, except that it yields TiO_2 on ignition.

Ti(IV). Anhydrous $Ti(SO_4)_2$ has been prepared.[101,104] It decomposes in two steps,[101] beginning at 423 K, ultimately yielding TiO_2. Chuklantsev[104] heated the salt at 1.7°/min in dry argon by a procedure resembling TGA, and found decomposition beginning at 423 K to produce $TiOSO_4$; at 703 K this began decomposing to TiO_2, and was complete by approximately 873 K. The decomposition pressures reported by Woehler et al.[105] for $TiOSO_4$ at 741 to 853 K are not trustworthy.

The density of $TiSO_4$ has been reported to be 2.47 gcm^{-3} at 293 K.[104] No thermodynamic data for $TiSO_4$ are known and therefore decomposition pressures cannot be calculated.

4.3.48 URANIUM

U(III). Difficult to isolate from solution; it is readily oxidized by H_2O.

U(IV). Hydrates are well-known. Several methods for preparing anhydrous $U(SO_4)_2$ have been described.[97] Further heating produces UO_2SO_4, even in the absence of air. Thermodynamic data for the decomposition are in Tables 4.148 to 4.151.

U(VI). This occurs only in uranyl salts containing the UO_2^{2+} cation, which are outside the scope of this work.

4.3.49 VANADIUM

V(II). Only the hydrate $VSO_4.7H_2O$ is known.

V(III). Rivenq[98] described the preparation and pyrolysis of anhydrous $V_2(SO_4)_3$. On heating in a vacuum the compound was stable to 598 K, but decomposed above 653 K. According to the author's interpretation of weight loss curves, the decomposition is complicated, its course dependent on temperature. The final product below 673 K was V_2O_3, while above 681 K a mixture of V_2O_3 and V_2O_5 formed. Udupa[99] reported that the decomposition occurred in a single step

$$V_2(SO_4)_3 \rightarrow V_2O_5 + 3\ SO_2 + 1/2\ O_2$$

but in a higher temperature range, 723 to 893 K. Thermodynamic data are available for $VOSO_4$, but not for $V_2(SO_4)_3$.

V(IV). Anhydrous $VOSO_4$ decomposes, without melting, above 673 K, with oxidation to V_2O_5, SO_3, and SO_2. The reaction is reversible. Neumann and Sonntag[100] measured equilibrium decomposition pressures in the range 704 to 826 K.

V(V). Anhydrous $V_2O_5.2\ SO_3$, $V_2O_5.3\ SO_3$, and $V_2O_5.4\ SO_3$ are known. The preparation of the last of these has been described by Hayek and Engelbrecht,[101] who found it to be unstable above 373 to 473 K.

4.3.50 ZINC

The decomposition of $ZnSO_4$ proceeds through an intermediate oxysulfate $ZnO.2\ ZnSO_4$. The decomposition can thus be characterized as

$$3\ ZnSO_4 = ZnO.2\ ZnSO_4 + SO_3(g)\ (SO_2 + 1/2\ O_2)$$

For the oxysulfate, the corresponding reaction is

$$1/2\ ZnO.2\ ZnSO_4 = 3/2\ ZnO + SO_3(g)\ (SO_2 + 1/2\ O_2)$$

Both reactions were studied by Brittain, Lau, Knittel, and Hildenbrand[102] using a torsion-effusion method in the temperature range 800 to 900 K. They found the SO_3 pressure over both the sulfate and oxysulfate derived from extrapolation of steady-state pressures to the zero orifice area to be a factor of five lower than those calculated from established thermodynamic data. This indicates that the products are formed in a finely divided metastable state.

As for other sulfates, whether the product is SO_3 or $SO_2 + O_2$ seems to depend on experimental conditions which have somewhat different effects on different salts.

The kinetics was found to follow the contracting sphere model.[56]

Two groups[8,27] have reported different densities at 298 K: 3.74 and 3.546.

The $\alpha - \beta$ transition of $ZnSO_4$ was found at 1008 K,[9] a temperature at which decomposition also becomes noticeable.

Thermodynamic data are given in Tables 4.152 to 4.154. Thermodynamic data for ZnO are given in Table 2.84.

4.3.51 Zirconium

The decomposition of $Zr(SO_4)_2$ results in the formation of ZrO_2 without the formation of an intermediate. There is considerable disagreement about a "decomposition temperature," but is probably above 820 K.

Thermodynamic data are given in Tables 4.155 to 4.158.

TABLE 4.1
Thermodynamic Data of SO_3(g)

T K	ΔHf° kJ mol⁻¹	S° J mol⁻¹ deg⁻¹	ΔGf° kJ mol⁻¹	Reference
298.15	–395.77	256.77	–472.32	Barin 1989
400.00	–390.25	272.64	–499.30	Barin 1989
500.00	–384.20	286.11	–527.26	Barin 1989
600.00	–377.66	298.02	–556.47	Barin 1989
700.00	–370.75	308.67	–586.82	Barin 1989
800.00	–363.59	318.23	–618.17	Barin 1989
900.00	–356.22	326.91	–650.44	Barin 1989
1000.00	–348.70	334.83	–683.53	Barin 1989
1100.00	–341.05	342.12	–717.38	Barin 1989
1200.00	–333.31	348.85	–751.93	Barin 1989
1300.00	–325.49	355.11	–787.14	Barin 1989
1400.00	–317.60	360.96	–822.94	Barin 1989
1500.00	–309.65	366.45	–859.32	Barin 1989
1600.00	–301.65	371.60	–896.22	Barin 1989
1700.00	–293.62	376.47	–933.63	Barin 1989
1800.00	–285.66	381.08	–971.51	Barin 1989
1900.00	–277.47	385.46	–1009.84	Barin 1989
2000.00	–269.36	389.62	–1048.59	Barin 1989
2100.00	–261.23	393.58	–1087.75	Barin 1989
2200.00	–253.09	397.37	–1127.30	Barin 1989
2300.00	–244.93	400.99	–1167.22	Barin 1989
2400.00	–236.77	404.77	–1207.50	Barin 1989

TABLE 4.2
Thermodynamic Data of $SO_2(g)$

T K	ΔHf° kJ mol^{-1}	S° J mol^{-1} deg^{-1}	ΔGf° kJ mol^{-1}	Reference
298.15	–296.81	248.22	–370.82	Barin 1993
400.00	–292.57	260.42	–396.75	Barin 1993
500.00	–288.07	270.46	–423.30	Barin 1993
600.00	–283.27	279.20	–450.79	Barin 1993
700.00	–278.26	286.93	–479.11	Barin 1993
800.00	–273.08	293.84	–508.15	Barin 1993
900.00	–267.78	300.08	–537.85	Barin 1993
1000.00	–262.38	305.77	–568.15	Barin 1993
1100.00	–256.90	310.99	–598.99	Barin 1993
1200.00	–251.36	315.81	–630.34	Barin 1993
1300.00	–245.76	320.30	–662.14	Barin 1993
1400.00	–240.11	324.48	–694.38	Barin 1993
1500.00	–234.43	328.40	–727.03	Barin 1993
1600.00	–228.70	333.00	–760.06	Barin 1993
1700.00	–222.95	335.58	–793.44	Barin 1993
1800.00	–217.17	338.89	–827.17	Barin 1993
1900.00	–211.38	342.02	–861.21	Barin 1993
2000.00	–205.56	345.00	–895.57	Barin 1993
2100.00	–199.73	347.85	–930.21	Barin 1993
2200.00	–193.88	350.57	–965.13	Barin 1993
2300.00	–188.02	353.18	–1000.32	Barin 1993
2400.00	–182.14	355.66	–1035.76	Barin 1993

TABLE 4.3
$SO_3(g) = SO_2 + 1/2\ O_2(g)$

T K	ΔH° kJ	ΔS° J deg^{-1}	ΔG° kJ	K
298.15	98.95	94.02	70.92	3.752E – 013
400.00	99.20	94.76	61.39	9.880E – 009
500.00	99.24	94.85	51.81	3.861E – 006
600.00	99.11	94.61	42.34	2.060E – 004
700.00	98.86	94.24	32.90	3.509E – 003
800.00	98.55	93.82	23.49	2.924E – 002
900.00	98.18	93.38	14.13	1.513E – 001
1000.00	97.77	92.96	4.82	5.603E – 001
1100.00	97.35	92.55	–4.46	1.628E + 000
1200.00	96.90	92.17	–13.70	3.946E + 000
1300.00	96.45	91.81	–22.89	8.317E + 000
1400.00	96.00	91.47	–32.06	1.571E + 001
1500.00	95.54	91.16	–41.19	2.719E + 001
1600.00	95.09	90.86	–50.29	4.384E + 001
1700.00	94.65	90.59	–59.36	6.670E + 001
1800.00	94.21	90.34	–68.41	9.667E + 001
1900.00	93.79	90.11	–77.43	1.346E + 002
2000.00	93.37	89.90	–86.43	1.809E + 002
2100.00	92.97	89.71	–95.41	2.363E + 002
2200.00	92.58	89.52	–104.37	3.008E + 002
2300.00	92.20	89.36	–113.32	3.747E + 002
2400.00	91.84	89.20	–122.24	4.579E + 002

TABLE 4.4
Thermodynamic Data of $Al_2(SO_4)_3$

T K	ΔHf° kJ mol^{-1}	S° J mol^{-1} deg^{-1}	ΔGf° kJ mol^{-1}	Reference
298.15	–3440.75	239.30	–3512.10	Knacke et al. 1991
400.00	–3410.75	325.42	–3540.92	Knacke et al. 1991
500.00	–3376.89	400.86	–3577.32	Knacke et al. 1991
600.00	–3340.54	467.08	–3620.78	Knacke et al. 1991
700.00	–3302.49	525.69	–3670.48	Knacke et al. 1991
800.00	–3263.16	578.20	–3725.72	Knacke et al. 1991
900.00	–3222.76	625.77	–3785.96	Knacke et al. 1991
1000.00	–3181.42	669.32	–3850.74	Knacke et al. 1991
1100.00	–3139.24	709.52	–3919.71	Knacke et al. 1991
1200.00	–3096.35	746.92	–3992.55	Knacke et al. 1991
1300.00	–3052.51	781.92	–4069.01	Knacke et al. 1991
1400.00	–3008.04	814.87	–4148.87	Knacke et al. 1991
1500.00	–2962.87	846.04	–4231.93	Knacke et al. 1991

TABLE 4.5
Thermodynamic Data of Al_2O_3

T K	ΔHf° kJ mol⁻¹	S° J mol⁻¹ deg⁻¹	ΔGf° kJ mol⁻¹	Reference
298.15	−1675.27	50.94	−1690.46	Barin et al. 1977
400.00	−1666.25	76.86	−1696.99	Barin et al. 1977
500.00	−1656.13	99.39	−1705.83	Barin et al. 1977
600.00	−1645.27	119.17	−1716.78	Barin et al. 1977
700.00	−1633.88	136.73	−1729.59	Barin et al. 1977
800.00	−1622.04	152.53	−1744.06	Barin et al. 1977
900.00	−1609.08	166.85	−1760.04	Barin et al. 1977
1000.00	−1597.49	179.90	−1777.39	Barin et al. 1977
1100.00	−1584.91	191.89	−1795.99	Barin et al. 1977
1200.00	−1572.17	202.97	−1815.74	Barin et al. 1977
1300.00	−1559.28	213.29	−1836.56	Barin et al. 1977
1400.00	−1546.25	222.94	−1858.38	Barin et al. 1977
1500.00	−1533.15	232.02	−1881.13	Barin et al. 1977

TABLE 4.6
$Al_2(SO_4)_3 = Al_2O_3 + 3\ SO_3(g)$

T K	ΔH° kJ	ΔS° J deg⁻¹	ΔG° kJ	K
298.15	578.19	581.95	404.68	1.248E − 071
400.00	573.77	569.37	346.02	6.470E − 046
500.00	568.14	556.84	289.72	5.377E − 031
600.00	562.28	546.16	234.59	3.765E − 021
700.00	556.36	537.03	180.44	3.423E − 014
800.00	550.36	529.04	127.14	4.985E − 009
900.00	544.23	521.80	74.61	4.673E − 005
1000.00	537.85	515.08	22.77	6.467E − 002
1100.00	531.17	508.72	−28.42	2.237E + 001

TABLE 4.7
$Al_2(SO_4)_3 = Al_2O_3 + 3\ SO_2(g) + 3/2\ O_2(g)$

T K	ΔH° kJ	ΔS° J deg⁻¹	ΔG° kJ	K
298.15	875.04	864.02	617.43	6.594E − 109
400.00	871.37	853.66	529.91	6.240E − 070
500.00	865.85	841.38	445.16	3.095E − 047
600.00	859.60	829.99	361.60	3.290E − 032
700.00	852.95	819.75	279.12	1.479E − 021
800.00	846.00	810.47	197.62	1.246E − 013
900.00	838.76	801.95	117.00	1.617E − 007
1000.00	831.16	793.95	37.21	1.138E − 002
1100.00	823.20	786.37	−41.80	9.662E + 001

TABLE 4.8
Thermodynamic Data of $Sb_2(SO_4)_3$

T K	ΔHf° kJ mol⁻¹	S° J mol⁻¹ deg⁻¹	ΔGf° kJ mol⁻¹	Reference
298.15	–2402.50	291.21	–2489.32	Barin et al. 1977
400.00	–2373.48	374.80	–2523.40	Barin et al. 1977
500.00	–2343.11	442.49	–2564.36	Barin et al. 1977
600.00	–2310.89	501.19	–2611.60	Barin et al. 1977
700.00	–2276.80	553.69	–2664.39	Barin et al. 1977
800.00	–2240.86	601.66	–2722.19	Barin et al. 1977
900.00	–2203.07	646.15	–2784.60	Barin et al. 1977
1000.00	–2163.41	687.92	–2851.33	Barin et al. 1977
1100.00	–2121.90	727.47	–2922.11	Barin et al. 1977
1200.00	2078.52	765.20	–2996.76	Barin et al. 1977

TABLE 4.9
Thermodynamic Data of Sb_4O_6, Valentinite

T K	ΔHf° kJ mol⁻¹	S° J mol⁻¹ deg⁻¹	ΔGf° kJ mol⁻¹	Reference
298.15	–1417.54	246.02	–1490.89	Knacke et al. 1991
400.00	–1396.02	308.01	–1519.22	Knacke et al. 1991
500.00	–1373.81	357.53	–1552.58	Knacke et al. 1991
600.00	–1350.99	399.13	–1590.46	Knacke et al. 1991
700.00	–1327.74	434.96	–1632.21	Knacke et al. 1991
800.00	–1304.17	466.43	–1677.31	Knacke et al. 1991
900.00	–1280.33	494.51	–1725.38	Knacke et al. 1991
1000.00	–1256.24	519.88	–1776.12	Knacke et al. 1991
1100.00	–1231.94	543.04	–1829.29	Knacke et al. 1991
1200.00	–1207.42	564.37	–1884.67	Knacke et al. 1991

TABLE 4.10
Thermodynamic Data of Sb_2O_5

T K	ΔHf° kJ mol⁻¹	S° J mol⁻¹ deg⁻¹	ΔGf° kJ mol⁻¹	Reference
298.15	–971.90	125.10	–1009.2	Barin 1989
400.00	–959.36	161.23	–1023.85	Barin 1989
500.00	–946.40	190.13	–1041.46	Barin 1989
600.00	–933.14	214.29	–1061.72	Barin 1989
700.00	–919.73	234.96	–1084.20	Barin 1989
800.00	–906.24	252.98	–1108.62	Barin 1989
900.00	–892.70	268.92	–1134.73	Barin 1989
1000.00	–879.15	283.20	–1162.35	Barin 1989
1100.00	–865.59	296.13	–1191.33	Barin 1989
1200.00	–852.04	307.92	–1221.54	Barin 1989

TABLE 4.11
Thermodynamic Data of Sb_2O_4, Cervantite

T K	ΔHf° kJ mol^{-1}	S° J mol^{-1} deg^{-1}	ΔGf° kJ mol^{-1}	Reference
298.15	–907.51	126.98	–945.37	Barin 1993
400.00	–895.32	162.09	–960.16	Barin 1993
500.00	–882.72	190.19	–977.82	Barin 1993
600.00	–869.77	213.79	–998.05	Barin 1993
700.00	–856.60	234.08	–1020.46	Barin 1993
800.00	–843.27	251.89	–1044.78	Barin 1993
900.00	–829.80	267.75	–1070.78	Barin 1993
1000.00	–816.22	282.06	–1098.28	Barin 1993
1100.00	–802.54	296.00	–1127.15	Barin 1993
1200.00	–788.76	307.08	–1157.36	Barin 1993

TABLE 4.12
Thermodynamic Data of Sb_2O_3, Senarmontite

T K	ΔHf° kJ mol^{-1}	S° J mol^{-1} deg^{-1}	ΔGf° kJ mol^{-1}	Reference
298.15	–720.31	110.45	–753.24	Barin 1993
400.00	–708.58	144.23	–766.27	Barin 1993
500.00	–696.40	171.38	–782.09	Barin 1993
600.00	–683.56	194.78	–800.43	Barin 1993
700.00	–670.06	215.58	–820.96	Barin 1993
800.00	–655.90	234.48	–843.48	Barin 1993
900.00	–635.03	258.81	–867.96	Barin 1993
1000.00	–565.35	333.50	–898.85	Barin 1993
1100.00	–549.66	348.45	–932.96	Barin 1993
1200.00	–533.97	362.11	–968.50	Barin 1993

TABLE 4.13
Transitions of $BaSO_4$

Phase Change	T(K)	ΔH kJ/mol	ΔS J/(mol*K)	Reference
c,II → c,I	1422			9,10
c,I → l	1620	40.6	25	5,10

TABLE 4.14
Thermodynamic Data of $BaSO_4$, Barite

T K	ΔHf° kJ mol⁻¹	S° J mol⁻¹ deg⁻¹	ΔGf° kJ mol⁻¹	Reference
298.15	–1473.20	132.20	–1512.62	Barin 1989
400.00	–1461.81	164.94	–1527.79	Barin 1989
500.00	–1449.43	192.53	–1545.70	Barin 1989
600.00	–1436.46	216.16	–1566.16	Barin 1989
700.00	–1423.16	236.66	–1588.82	Barin 1989
800.00	–1409.65	254.70	–1613.41	Barin 1989
900.00	–1396.00	270.78	–1639.70	Barin 1989
1000.00	–1382.25	285.27	–1667.51	Barin 1989
1100.00	–1368.43	298.44	–1696.71	Barin 1989
1200.00	–1354.55	310.51	–1727.16	Barin 1989

TABLE 4.15
$BaSO_4 = BaO + SO_3(g)$

T K	ΔH° kJ	ΔS° J deg⁻¹	ΔG° kJ	K
298.15	523.89	194.86	465.79	2.444E – 082
400.00	522.99	192.31	446.06	5.562E – 059
500.00	521.76	189.57	426.97	2.459E – 045
600.00	520.59	187.44	408.13	2.926E – 036
700.00	519.58	185.87	389.47	8.615E – 030
800.00	518.71	184.72	370.94	5.999E – 025
900.00	518.01	183.88	352.51	3.459E – 021
1000.00	517.45	183.29	334.16	3.500E – 018
1100.00	517.02	182.88	315.85	1.001E – 015
1200.00	516.72	182.62	297.57	1.111E – 013

TABLE 4.16
$BaSO_4 = BaO + SO_2(g) + 1/2\ O_2(g)$

T K	ΔH° kJ	ΔS° J deg⁻¹	ΔG° kJ	K
298.15	622.84	288.88	536.71	9.171E – 095
400.00	622.19	287.08	507.36	5.495E – 067
500.00	621.00	284.42	478.79	9.492E – 051
600.00	619.70	282.05	450.47	6.028E – 040
700.00	618.44	280.11	422.36	3.023E – 032
800.00	617.26	278.53	394.53	1.754E – 026
900.00	616.18	277.26	366.65	5.233E – 022
1000.00	615.22	276.25	338.97	1.961E – 018
1100.00	614.36	275.43	311.39	1.630E – 015
1200.00	613.62	274.78	283.88	4.386E – 013

TABLE 4.17
Density of $BeSO_4$

Phase	T(K)	d	Reference
c	ambient	2.54 gcm^{-3}, crystallographic	12
c	298	2.44	8

TABLE 4.18
Thermodynamic Data of $BeSO_4$

T K	ΔHf° kJ mol^{-1}	S° J mol^{-1} deg^{-1}	ΔGf° kJ mol^{-1}	Reference
298.15	–1200.81	77.97	–1224.05	Barin 1989
400.00	–1191.13	105.77	–1233.44	Barin 1989
500.00	–1180.15	130.21	–1245.26	Barin 1989
600.00	–1167.95	152.43	–1259.41	Barin 1989
700.00	–1154.62	172.95	–1275.69	Barin 1989
800.00	–1140.21	192.18	–1293.95	Barin 1989
900.00	–1123.56	211.76	–1314.14	Barin 1989
1000.00	–1087.14	251.05	–1338.19	Barin 1989
1100.00	–1069.24	268.11	–1364.15	Barin 1989
1200.00	–1050.60	284.32	–1391.78	Barin 1989

TABLE 4.19
Thermodynamic Data of BeO, Bromellite

T K	ΔHf° kJ mol^{-1}	S° J mol^{-1} deg^{-1}	ΔGf° kJ mol^{-1}	Reference
298.15	–608.35	13.77	–612.46	Barin 1989
400.00	–605.31	22.51	–614.31	Barin 1989
500.00	–601.68	30.59	–616.97	Barin 1989
600.00	–597.61	37.98	–620.40	Barin 1989
700.00	–593.24	44.72	–624.54	Barin 1989
800.00	–588.62	50.88	–629.33	Barin 1989
900.00	–583.85	56.50	–634.70	Barin 1989
1000.00	–578.97	61.64	–640.61	Barin 1989
1100.00	–574.03	66.35	–647.01	Barin 1989
1200.00	–569.02	70.71	–653.87	Barin 1989

TABLE 4.20
$BeSO_4 = BeO + SO_3(g)$

T K	ΔH° kJ	ΔS° J deg⁻¹	ΔG° kJ	K
298.15	196.69	192.57	139.27	3.960E – 025
400.00	195.58	189.38	119.83	2.244E – 016
500.00	194.28	186.48	101.04	2.780E – 011
600.00	192.68	183.58	82.53	6.523E – 008
700.00	190.63	180.44	64.33	1.583E – 005
800.00	188.00	176.93	46.46	9.258E – 004
900.00	183.49	171.65	29.00	2.073E – 002
1000.00	159.47	145.42	14.05	1.845E – 001
1100.00	154.15	140.36	-0.241	1.027E + 000
1200.00	148.27	135.25	–14.020	4.077E + 000

TABLE 4.21
$BeSO_4 = BeO + SO_2(g) + 1/2\ O_2(g)$

T K	ΔH° kJ	ΔS° J deg⁻¹	ΔG° kJ	K
298.15	295.64	286.60	210.19	1.486E – 037
400.00	294.78	284.14	181.12	2.217E – 024
500.00	293.51	281.33	152.85	1.073E – 016
600.00	291.78	278.19	124.87	1.344E – 011
700.00	289.50	274.68	97.22	5.554E – 011
800.00	286.55	270.75	69.95	2.707E – 005
900.00	281.67	265.03	43.13	3.136E – 003
1000.00	257.25	238.38	18.87	1.034E – 001
1100.00	251.50	232.91	–4.70	1.672E + 000
1200.00	245.18	227.41	–27.71	1.609E + 001

TABLE 4.22
Thermodynamic Data of $Bi_2(SO_4)_3$

T K	ΔHf° kJ mol⁻¹	S° J mol⁻¹ deg⁻¹	ΔGf° kJ mol⁻¹	Reference
298.15	–2544.30	312.55	–2637.48	Barin et al. 1977
400.00	–2512.02	396.89	–2673.78	Barin et al. 1977
500.00	–2484.57	464.77	–2716.96	Barin et al. 1977
600.00	–2452.43	523.33	–2766.43	Barin et al. 1977
700.00	–2418.60	575.45	–2821.41	Barin et al. 1977
800.00	–2383.08	622.85	–2881.36	Barin et al. 1977
900.00	–2345.86	666.66	–2945.86	Barin et al. 1977
1000.00	–2396.96	707.64	–3014.60	Barin et al. 1977
1100.00	–2266.37	746.31	–3087.31	Barin et al. 1977
1200.00	–2224.08	783.10	–3163.80	Barin et al. 1977

TABLE 4.23
Thermodynamic Data of Bi_2O_3

T K	ΔHf° kJ mol⁻¹	S° J mol⁻¹ deg⁻¹	ΔGf° kJ mol⁻¹	Reference
298.15	–573.88	151.50	–619.05	Barin 1993
400.00	–562.15	185.33	–636.28	Barin 1993
500.00	–550.29	211.77	–656.17	Barin 1993
600.00	–538.10	233.99	–678.49	Barin 1993
700.00	–525.57	253.30	–702.88	Barin 1993
800.00	–512.71	270.46	–729.08	Barin 1993
900.00	–499.51	286.00	–756.92	Barin 1993
1000.00	–485.98	300.26	–786.24	Barin 1993
1100.00	–423.73	360.26	–820.02	Barin 1993
1200.00	–405.74	375.91	–856.84	Barin 1993

TABLE 4.24
$Bi_2(SO_4)_3 = Bi_2O_3 + 3\ SO_3(g)$

T K	ΔH° kJ	ΔS° J deg⁻¹	ΔG° kJ	K
298.15	783.13	609.28	601.56	3.526E – 106
400.00	782.13	606.36	539.59	3.398E – 071
500.00	781.67	605.31	479.01	8.990E – 051
600.00	781.35	604.73	418.51	3.651E – 037
700.00	780.77	603.84	358.08	1.896E – 027
800.00	779.61	602.30	297.76	3.601E – 020
900.00	777.70	600.06	237.64	1.609E – 014
1000.00	774.89	597.12	177.78	5.167E – 10
1100.00	819.48	640.30	115.16	3.398E – 006
1200.00	818.42	639.38	51.16	5.925E – 003

TABLE 4.25
$Bi_2(SO_4)_3 = Bi_2O_3 + 3\ SO_2(g) + 3/2\ O_2(g)$

T K	ΔH° kJ	ΔS° J deg⁻¹	ΔG° kJ	K
298.15	1079.98	891.34	814.23	2.183E – 143
400.00	1079.74	890.65	723.48	3.277E – 095
500.00	1079.38	889.85	634.45	5.174E – 067
600.00	1078.67	888.57	545.53	3.191E – 048
700.00	1077.35	886.56	456.76	8.188E – 035
800.00	1075.24	883.75	368.24	9.003E – 025
900.00	1072.22	880.21	280.04	5.569E – 017
1000.00	1068.21	875.98	192.22	9.090E – 011
1100.00	1111.00	917.94	101.78	1.467E – 005
1200.00	1109.13	915.87	10.08	3.641E – 001

TABLE 4.26
Density of $CdSO_4$

Phase	T(K)	d	Reference
c	293	4.692 (20/4)	19
c	297	4.691	8

TABLE 4.27
Transitions of $CdSO_4$

Phase Change	T(K)	ΔH kJ/mol	ΔS J/(mol*K)	Reference
c → c	—			
c → l	1273			

TABLE 4.28
Thermodynamic Data of $CdSO_4$

T K	ΔHf° kJ mol⁻¹	S° J mol⁻¹ deg⁻¹	ΔGf° kJ mol⁻¹	Reference
298.15	–933.28	123.04	–969.96	Barin 1993
400.00	–922.73	153.42	–984.10	Barin 1993
500.00	–911.59	178.24	–1000.71	Barin 1993
600.00	–899.68	199.93	–1019.64	Barin 1993
700.00	–887.00	219.47	–1040.63	Barin 1993
800.00	–873.54	237.43	–1063.48	Barin 1993
900.00	–859.30	254.19	–1088.07	Barin 1993
1000.00	–844.30	269.99	–1114.29	Barin 1993
1100.00	–822.28	290.88	–1142.25	Barin 1993
1200.00	–795.90	313.97	–1172.66	Barin 1993

TABLE 4.29
$CdSO_4 = CdO + SO_3(g)$

T K	ΔH° kJ	ΔS° J deg⁻¹	ΔG° kJ	K
298.15	278.52	188.54	222.31	1.119E – 039
400.00	278.12	187.35	203.18	2.922E – 027
500.00	277.79	186.62	184.49	5.322E – 020
600.00	277.33	185.80	165.86	3.629E – 015
700.00	276.58	184.64	147.33	1.012E – 011
800.00	275.40	183.07	128.94	3.804E – 009
900.00	273.73	181.12	110.73	3.740E – 007
1000.00	271.53	178.79	92.73	1.432E – 005
1100.00	262.50	170.29	75.19	2.688E – 004
1200.00	249.29	158.66	58.90	2.728E – 003

TABLE 4.30
$CdSO_4 = CdO + SO_2(g) + 1/2O_2$

T K	ΔH° kJ	ΔS° J deg⁻¹	ΔG° kJ	K
298.15	377.48	282.57	293.23	4.200E – 052
400.00	377.32	282.11	264.47	2.887E – 035
500.00	377.02	281.47	236.29	2.055E – 025
600.00	376.44	280.41	208.19	7.475E – 019
700.00	375.44	278.88	180.23	3.550E – 014
800.00	373.95	276.89	152.43	1.112E – 010
900.00	371.91	274.50	124.86	5.658E – 008
1000.00	369.30	271.75	97.55	8.022E – 006
1100.00	359.85	262.84	70.73	4.377E – 004
1200.00	346.19	250.82	45.21	1.077E – 002

TABLE 4.31
Transitions of $CaSO_4$

Phase Change	T(K)	ΔH kJ/mol	ΔS J/(mol·K)	Reference
c,II → c,I	1453	21	$14._2$	10
c,I → l	See text			

Note: c,II or α is the mineral anhydrite.

TABLE 4.32
Thermodynamic Data of $CaSO_4$

T K	ΔHf° kJ mol⁻¹	S° J mol⁻¹ deg⁻¹	ΔGf° kJ mol⁻¹	Reference
298.15	–1434.11	106.69	–1465.92	Barin et al. 1977
400.00	–1423.45	137.38	–1478.40	Barin et al. 1977
500.00	–1411.98	162.92	–1493.44	Barin et al. 1977
600.00	–1399.53	185.60	–1510.89	Barin et al. 1977
700.00	–1386.09	206.29	–1530.50	Barin et al. 1977
800.00	–1371.67	225.54	–1552.10	Barin et al. 1977
900.00	–1356.25	243.69	–1575.57	Barin et al. 1977
1000.00	–1339.85	260.96	–1600.81	Barin et al. 1977
1100.00	–1322.46	277.52	–1627.74	Barin et al. 1977
1200.00	–1304.09	293.51	–1656.29	Barin et al. 1977
1300.00	–1286.09	307.91	–1686.37	Barin 1993
1400.00	–1268.10	321.24	–1717.84	Barin 1993
1500.00	–1250.11	333.65	–1750.59	Barin 1993
1600.00	–1232.12	345.26	–1784.54	Barin 1993

TABLE 4.33
$CaSO_4 = CaO + SO_3(g)$

T K	ΔH° kJ	ΔS° J deg⁻¹	ΔG° kJ	K
298.15	403.25	188.15	347.16	1.495E – 061
400.00	402.73	186.64	328.08	1.427E – 043
500.00	402.13	185.31	309.48	4.642E – 033
600.00	401.20	183.63	291.03	4.592E – 026
700.00	399.76	181.42	272.77	4.407E – 021
800.00	397.68	178.64	254.76	2.315E – 017
900.00	394.88	175.36	237.06	1.740E – 014
1000.00	391.32	171.61	219.70	3.333E – 012
1100.00	386.95	167.45	202.75	2.352E – 010
1200.00	381.74	162.93	186.23	7.819E – 009
1300.00	377.06	159.18	170.12	1.458E – 007
1400.00	372.49	155.79	154.38	1.736E – 006
1500.00	368.03	152.72	138.96	1.448E – 005
1600.00	363.67	149.90	123.83	9.061E – 005

TABLE 4.34
$CaSO_4 = CaO + SO_2 + 1/2\ O_2$

T K	ΔH° kJ	ΔS° J deg⁻¹	ΔG° kJ	K
298.15	502.21	282.18	418.08	5.609E – 074
400.00	501.94	281.41	389.37	1.409E – 051
500.00	501.36	280.15	361.29	1.792E – 038
600.00	500.31	278.24	333.36	9.455E – 030
700.00	498.62	275.65	305.66	1.546E – 023
800.00	496.22	272.46	278.25	6.767E – 019
900.00	493.06	268.74	251.19	2.631E – 015
1000.00	489.09	264.57	224.52	1.868E – 012
1100.00	484.29	260.00	198.29	3.831E – 010
1200.00	478.64	255.09	172.53	3.085E – 008
1300.00	473.51	250.98	147.23	1.213E – 006
1400.00	468.49	247.26	122.32	2.727E – 005
1500.00	463.58	243.87	97.77	3.937E – 004
1600.00	458.76	240.76	73.54	3.972E – 003

TABLE 4.35
Density of Cs_2SO_4

Phase	T(K)	d	Reference
Solid	293	4.246	24
	298	4.243	8
	333	4.222	25
Liquid	1309	3.037	26
	1336	3.018	26
	1378	2.988	26
	1438	2.937	26
	1494	2.889	26
	1547	2.841	26
	1604	2.787	26
	1645	2.743	26
	1696	2.690	26
	1743	2.636	26
	1803	2.566	26

TABLE 4.36
Transitions of Cs_2SO_4

Phase Change	T(K)	ΔH kJ/mol	ΔS J/(mol*K)	Reference
c,II → c,I	933			10
c,I → l	1277	$40._2$	$31._4$	10

TABLE 4.37
Thermodynamic Data of Cs_2SO_4

T K	ΔHf° kJ mol⁻¹	S° J mol⁻¹ deg⁻¹	ΔGf° kJ mol⁻¹	Reference
298.15	–1443.02	211.92	–1506.20	Barin 1993
400.00	–1428.50	253.73	–1529.99	Barin 1993
500.00	–1412.86	288.56	–1557.14	Barin 1993
600.00	–1395.80	319.62	–1587.57	Barin 1993
700.00	–1377.22	348.24	–1620.98	Barin 1993
800.00	–1356.98	375.24	–1657.17	Barin 1993
900.00	–1334.95	401.16	–1695.99	Barin 1993
1000.00	–1309.93	427.62	–1737.55	Barin 1993
1100.00	–1289.74	446.85	–1781.28	Barin 1993
1200.00	–1268.00	465.76	–1826.91	Barin 1993
1300.00	–1209.73	511.77	–1875.03	Barin 1993
1400.00	–1189.06	527.09	–1926.98	Barin 1993
1500.00	–1168.39	541.35	–1980.41	Barin 1993
1600.00	–1147.72	554.69	–2035.22	Barin 1993
1700.00	–1127.05	567.22	–2091.32	Barin 1993
1800.00	–1106.38	579.03	–2148.64	Barin 1993
1900.00	–1085.72	590.21	–2207.19	Barin 1993
2000.00	–1065.05	600.81	–2266.66	Barin 1993
2100.00				
2200.00				
2300.00				
2400.00				

TABLE 4.38
Thermodynamic Data of Cs_2SO_4 (g)

T K	ΔHf° kJ mol⁻¹	S° J mol⁻¹ deg⁻¹	ΔGf° kJ mol⁻¹	Reference
298.15	–1122.60	406.50	–1243.80	JANAF 1985
400.00	–1110.69	440.77	–1287.00	JANAF 1985
500.00	–1097.90	469.28	–1332.54	JANAF 1985
600.00	–1084.36	493.94	–1380.73	JANAF 1985
700.00	–1070.32	515.57	–1431.22	JANAF 1985
800.00	–1055.91	534.81	–1483.76	JANAF 1985
900.00	–1041.22	552.12	–1538.12	JANAF 1985
1000.00	–1026.32	567.81	–1594.13	JANAF 1985
1100.00	–1011.28	582.15	–1651.64	JANAF 1985
1200.00	–996.12	595.33	–1710.52	JANAF 1985
1300.00	–980.87	607.54	–1770.67	JANAF 1985
1400.00	–965.54	618.90	–1832.00	JANAF 1985
1500.00	–950.15	629.52	–1894.43	JANAF 1985

TABLE 4.39
$Cs_2SO_4 = Cs_2O + SO_3$ (g)

T K	ΔH° kJ	ΔS° J deg⁻¹	ΔG° kJ	K
298.15	701.28	191.71	644.12	1.390E – 113
400.00	700.19	188.59	624.75	2.563E – 082
500.00	698.71	185.30	606.06	4.791E – 064
600.00	696.63	181.54	587.71	6.777E – 052
700.00	693.73	177.09	569.77	3.017E – 043
800.00	689.77	171.81	552.32	8.593E – 037
900.00	684.55	165.67	535.44	8.343E – 032
1000.00	676.81	157.42	519.39	7.372E – 028
1100.00	674.36	155.09	503.76	1.193E – 024
1200.00	670.79	151.99	488.40	5.481E – 022
1300.00	631.10	120.85	474.00	8.981E – 020
1400.00	629.41	119.60	461.97	5.785E – 018
1500.00	628.10	118.70	450.06	2.120E – 016
1600.00	627.17	118.09	438.22	4.935E – 015

TABLE 4.40
$Cs_2SO_4 = Cs_2O + SO_2(g) + 1/2\ O_2$ (g)

T K	ΔH° kJ	ΔS° J deg⁻¹	ΔG° kJ	K
298.15	800.23	285.73	715.04	5.217E – 126
400.00	799.39	283.35	686.05	2.532E – 090
500.00	797.94	280.15	657.87	1.850E – 069
600.00	795.74	276.15	630.05	1.396E – 055
700.00	792.59	271.32	602.67	1.058E – 045
800.00	788.31	265.62	575.81	2.513E – 038
900.00	782.72	259.06	549.57	1.262E – 032
1000.00	774.58	250.38	524.21	4.131E – 028
1100.00	771.79	247.64	499.30	1.943E – 024
1200.00	767.69	244.16	474.70	2.163E – 021
1300.00	727.55	212.66	451.10	7.469E – 019
1400.00	725.40	211.07	429.91	9.088E – 017
1500.00	723.64	209.85	408.87	5.764E – 015
1600.00	722.26	208.96	387.93	2.159E – 013

TABLE 4.41
Cs_2SO_4 = Cs(g) + SO_2(g) + O_2(g)

T K	ΔH° kJ	ΔS° J deg⁻¹	ΔG° kJ	K
298.15	1299.21	592.64	1122.51	2.111E – 197
400.00	1296.22	584.08	1062.59	1.694E – 139
500.00	1292.39	575.57	1004.60	1.100E – 105
600.00	1287.51	566.72	947.48	3.216E – 083
700.00	1281.40	557.32	891.28	3.066E – 067
800.00	1273.84	547.25	836.04	2.555E – 055
900.00	1264.67	536.47	781.85	4.157E – 046
1000.00	1252.65	523.70	728.95	8.327E – 039
1100.00	1245.57	516.96	676.91	7.138E – 033
1200.00	1237.04	509.55	625.57	5.848E – 028
1300.00	1192.06	474.18	575.63	7.398E – 024
1400.00	1184.78	468.78	528.48	1.907E – 020
1500.00	1177.57	463.81	481.86	1.655E – 017

TABLE 4.42
Thermodynamic Data of $Cr_2(SO_4)_3$

T K	ΔHf° kJ mol⁻¹	S° J mol⁻¹ deg⁻¹	ΔGf° kJ mol⁻¹	Reference
298.15	–2942.72	258.78	–3026.88	Barin 1993
400.00	–2919.24	346.44	–3057.82	Barin 1993
500.00	–2884.91	422.85	–3096.34	Barin 1993
600.00	–2846.39	493.15	–3142.19	Barin 1993
700.00	–2804.41	557.67	–3194.78	Barin 1993
800.00	–2760.58	616.17	–3253.52	Barin 1993
900.00	–2716.32	668.31	–3386.96	Barin 1993
1000.00	–2632.84	713.76	–3386.96	Barin 1993

TABLE 4.43
Thermodynamic Data of Cr_2O_3

T K	ΔHf° kJ mol⁻¹	S° J mol⁻¹ deg⁻¹	ΔGf° kJ mol⁻¹	Reference
298.15	–1139.70	81.20	–1163.91	Barin 1989
400.00	–1127.61	116.10	–1174.05	Barin 1989
500.00	–1115.78	142.51	–1187.03	Barin 1989
600.00	–1103.86	164.22	–1202.40	Barin 1989
700.00	–1091.82	182.78	–1219.77	Barin 1989
800.00	–1079.64	199.05	–1238.88	Barin 1989
900.00	–1067.30	213.58	–1259.52	Barin 1989
1000.00	–1054.81	226.74	–1281.55	Barin 1989

TABLE 4.44
$Cr_2(SO_4)_3 = Cr_2O_3 + 3\ SO_3$

T K	ΔH° kJ	ΔS° J deg⁻¹	ΔG° kJ	K
298.15	622.73	592.74	446.00	7.176E – 079
400.00	620.90	587.60	385.86	4.056E – 051
500.00	616.53	577.97	327.54	6.010E – 035
600.00	609.45	565.14	270.37	2.885E – 024
700.00	600.33	551.10	214.56	9.734E – 017
800.00	590.19	537.57	160.13	3.495E – 011
900.00	580.37	525.99	106.98	6.176E – 007
1000.00	572.31	517.48	54.83	1.366E – 003

TABLE 4.45
$Cr2(SO_4)_3 = Cr_2O_3 + 3\ SO_2 + 3/2\ O_2$

T K	ΔH° kJ	ΔS° J deg⁻¹	ΔG° kJ	K
298.15	622.73	592.74	446.00	7.176E – 079
400.00	620.90	587.60	385.86	4.056E – 051
500.00	616.53	577.97	327.54	6.010E – 035
600.00	609.45	565.14	270.37	2.885E – 024
700.00	600.33	551.10	214.56	9.734E – 017
800.00	590.19	537.57	160.13	3.495E – 011
900.00	580.37	525.99	106.98	6.176E – 007
1000.00	572.31	517.48	54.83	1.366E – 003

TABLE 4.46
Density of $CoSO_4$

Phase	T(K)	d	References
α	298	3.86 gcm^{-3}	31
β	298	3.770 gcm^{-3}	32
Materials used may be impure		Densities are crystallographic	

TABLE 4.47
Transitions of $CoSO_4$

Phase Change	T(K)	ΔH kJ/mol	ΔS J/(mol*K)	Reference
α → β	705 ± 8	6.8 ± 0.6	9.6 ± 0.8	32

TABLE 4.48
Thermodynamic Data of $CoSO_4$

T K	ΔHf° kJ mol⁻¹	S° J mol⁻¹ deg⁻¹	ΔGf° kJ mol⁻¹	Reference
298.15	–888.26	117.36	–923.25	Knacke et al. 1991
400.00	–876.88	150.12	–936.32	Knacke et al. 1991
500.00	–864.31	178.10	–953.36	Knacke et al. 1991
600.00	–850.69	202.90	–972.43	Knacke et al. 1991
700.00	–836.28	225.10	–993.85	Knacke et al. 1991
800.00	–821.28	245.12	–1017.38	Knacke et al. 1991
900.00	–805.90	263.23	–1042.81	Knacke et al. 1991
1000.00	–788.14	281.91	–1070.05	Knacke et al. 1991
1100.00	–772.21	297.09	–1099.01	Knacke et al. 1991
1200.00	–756.07	311.13	–1129.43	Knacke et al. 1991

TABLE 4.49
$CoSO_4 = CoO + SO_3(g)$

T K	ΔH° kJ	ΔS° J deg⁻¹	ΔG° kJ	K
298.15	254.55	192.38	197.20	2.813E – 035
400.00	254.21	191.47	177.63	6.348E – 024
500.00	253.06	188.92	158.60	2.692E – 017
600.00	251.36	185.84	139.86	6.658E – 013
700.00	249.26	182.69	121.44	8.663E – 010
800.00	246.87	179.42	103.34	1.788E – 007
900.00	244.37	176.48	85.54	1.083E – 005
1000.00	239.71	171.61	68.11	2.767E – 004
1100.00	237.09	169.11	51.08	3.754E – 003
1200.00	234.45	166.81	34.28	3.219E – 002

TABLE 4.50
$CoSO_4 = CoO + SO_2(g) + 1/2\ O_2(g)$

T K	ΔH° kJ	ΔS° J deg⁻¹	ΔG° kJ	K
298.15	353.51	286.40	268.12	1.055E – 047
400.00	353.42	286.24	238.92	6.272E – 032
500.00	352.30	283.77	210.41	1.039E – 022
600.00	350.47	280.45	182.20	1.371E – 016
700.00	348.12	276.84	154.33	3.039E – 012
800.00	345.42	273.24	126.83	5.228E – 009
900.00	342.55	269.86	99.68	1.639E – 006
1000.00	337.49	264.56	72.93	1.551E – 004
1100.00	334.44	261.66	46.62	6.113E – 003
1200.00	331.35	258.97	20.59	1.270E – 001

TABLE 4.51
Density of $CuSO_4$

Phase	T(K)	d	Reference
?	298	3.542 (pycnometric)	27
α	298	4.113 gcm^{-3}	38
β	453	3.923	38

TABLE 4.52
Thermodynamic Data of $CuSO_4$

T K	ΔHf° kJ mol^{-1}	S° J mol^{-1} deg^{-1}	ΔGf° kJ mol^{-1}	Reference
298.15	–771.36	109.00	–803.86	Barin 1993
400.00	–760.39	140.52	–816.60	Barin 1993
500.00	–748.25	167.58	–832.04	Barin 1993
600.00	–735.08	191.56	–850.02	Barin 1993
700.00	–721.13	213.05	–870.26	Barin 1993
800.00	–706.58	232.48	–892.56	Barin 1993
900.00	–691.59	250.12	–916.70	Barin 1993
1000.00	–676.33	266.20	–942.53	Barin 1993
1100.00	–660.95	280.85	–969.89	Barin 1993
1200.00	–645.60	294.21	–998.66	Barin 1993

TABLE 4.53
$CuSO_4 = CuO + SO_3(g)$

T K	ΔH° kJ	ΔS° J deg^{-1}	ΔG° kJ	K
298.15	219.53	190.36	162.77	3.022E – 029
400.00	218.65	187.85	143.51	1.813E – 019
500.00	217.36	184.98	124.86	9.005E – 014
600.00	215.75	182.06	106.51	5.328E – 010
700.00	213.86	179.16	88.45	2.507E – 007
800.00	211.76	176.35	70.68	2.426E – 005
900.00	209.53	173.73	53.18	8.195E – 004
1000.00	207.27	171.35	35.92	1.329E – 002
1100.00	205.11	169.28	18.90	1.267E – 001
1200.00	203.15	167.58	2.05	8.139E – 001

TABLE 4.54
$CuSO_4 = CuO + SO_2(g) + 1/2\ O_2(g)$

T K	ΔH° kJ	ΔS° J deg⁻¹	ΔG° kJ	K
298.15	318.48	284.39	233.69	1.134E – 041
400.00	317.85	282.61	204.80	1.791E – 027
500.00	316.59	279.83	176.68	3.477E – 019
600.00	314.85	276.67	148.85	1.097E – 013
700.00	312.73	273.40	121.35	8.796E – 010
800.00	310.30	270.17	94.17	7.094E – 007
900.00	307.70	267.11	67.31	1.240E – 004
1000.00	305.04	264.31	40.74	7.445E – 003
1100.00	302.45	261.83	14.44	2.063E – 001
1200.00	300.06	259.75	–11.64	3.213E + 000

TABLE 4.55
Thermodynamic Data of $In_2(SO_4)_3$

T K	ΔHf° kJ mol⁻¹	S° J mol⁻¹ deg⁻¹	ΔGf° kJ mol⁻¹	Reference
298.15	–2787.00	272.00	–2868.10	Barin 1993
400.00	–2757.18	357.86	–2900.32	Barin 1993
500.00	–330.68	428.74	–2939.74	Barin 1993
600.00	–355.78	491.25	–2985.79	Barin 1993
700.00	–2654.21	547.98	–3037.80	Barin 1993
800.00	–2614.87	600.47	–3095.25	Barin 1993
900.00	–2573.02	649.74	–3157.78	Barin 1993
1000.00	–2528.65	696.46	–3225.11	Barin 1993
1100.00	–2481.78	741.12	–3297.01	Barin 1993
1200.00	–2432.39	784.07	–3373.28	Barin 1993

TABLE 4.56
Thermodynamic Data of In_2O_3

T K	ΔHf° kJ mol⁻¹	S° J mol⁻¹ deg⁻¹	ΔGf° kJ mol⁻¹	Reference
298.15	–925.79	104.20	–956.86	Knacke et al. 1991
400.00	–914.88	135.59	–969.12	Knacke et al. 1991
500.00	–903.34	161.31	–984.00	Knacke et al. 1991
600.00	–891.36	183.15	–1001.25	Knacke et al. 1991
700.00	–879.09	202.06	–1020.53	Knacke et al. 1991
800.00	–866.60	218.73	–1041.59	Knacke et al. 1991
900.00	–853.95	233.63	–1064.22	Knacke et al. 1991
1000.00	–128.59	247.11	–1088.27	Knacke et al. 1991
1100.00	–828.24	259.42	–1113.60	Knacke et al. 1991
1200.00	–815.20	270.76	–1140.12	Knacke et al. 1991

TABLE 4.57
$In_2(SO_4)_3 = In_2O_3 + 3SO_3(g)$

T K	ΔH° kJ	ΔS° J deg⁻¹	ΔG° kJ	K
298.15	673.92	602.51	494.28	2.496E – 087
400.00	671.56	595.66	433.39	2.587E – 057
500.00	669.42	590.88	373.97	8.475E – 040
600.00	666.70	585.97	315.12	3.662E – 028
700.00	662.87	580.08	256.81	6.837E – 020
800.00	657.51	572.95	199.15	9.902E – 014
900.00	650.41	564.61	142.26	5.528E – 009
1000.00	641.41	555.15	86.27	3.116E – 005
1100.00	630.39	544.65	31.27	3.274E – 002
1200.00	617.26	533.25	–22.63	9.668E + 000

TABLE 4.58
$In_2(SO_4)_3 = In_2O_3 + 3\ SO_2(g) + 3/2\ O_2(g)$

T K	ΔH° kJ	ΔS° J deg⁻¹	ΔG° kJ	K
298.15	970.77	884.58	707.04	1.319E – 124
400.00	969.17	879.95	617.19	2.497E – 081
500.00	967.12	875.42	529.41	4.880E – 056
600.00	964.02	869.80	442.14	3.202E – 039
700.00	959.45	862.79	355.50	2.954E – 027
800.00	953.14	854.40	269.63	2.476E – 018
900.00	944.94	844.76	184.66	1.913E – 011
1000.00	934.73	834.01	100.71	5.482E – 006
1100.00	922.42	822.30	17.89	1.414E – 001
1200.00	907.97	809.74	–63.72	5.941E + 002

TABLE 4.59
Thermodynamic Data of $FeSO_4$

T K	ΔHf° kJ mol⁻¹	S° J mol⁻¹ deg⁻¹	ΔGf° kJ mol⁻¹	Reference
298.15	–928.85	120.96	–964.91	Knacke et al. 1991
400.00	–917.69	153.04	–978.90	Knacke et al. 1991
500.00	–905.34	180.56	–995.61	Knacke et al. 1991
600.00	–891.99	204.86	–1014.91	Knacke et al. 1991
700.00	–877.89	226.58	–1036.50	Knacke et al. 1991
800.00	–863.21	246.18	–1060.15	Knacke et al. 1991
900.00	–848.09	263.99	–1085.68	Knacke et al. 1991
1000.00	–832.68	280.22	–1112.90	Knacke et al. 1991
1100.00	–817.10	295.07	–1141.67	Knacke et al. 1991
1200.00	–801.49	308.65	–1171.87	Knacke et al. 1991

TABLE 4.60
Thermodynamic Data of $Fe_2(SO_4)_3$

T K	ΔHf° kJ mol⁻¹	S° J mol⁻¹ deg⁻¹	ΔGf° kJ mol⁻¹	Reference
298.15	–2582.99	307.52	–2674.68	Barin 1989
400.00	–2553.49	392.30	–2710.41	Barin 1989
500.00	–2520.77	465.20	–2753.37	Barin 1989
600.00	–2485.61	529.24	–2803.15	Barin 1989
700.00	–2448.64	586.20	–2858.98	Barin 1989
800.00	–2410.22	636.48	–2920.20	Barin 1989
900.00	–2370.61	684.12	–2986.32	Barin 1989
1000.00	–2330.03	726.86	–3058.90	Barin 1989
1100.00	–2288.66	766.29	–3131.58	Barin 1989
1200.00	–2246.65	802.84	–3210.06	Barin 1989

TABLE 4.61
Thermodynamic Data of FeO (wuestite)

T K	ΔHf° kJ mol⁻¹	S° J mol⁻¹ deg⁻¹	ΔGf° kJ mol⁻¹	Reference
298.15	–272.04	60.75	–290.16	Barin 1993
400.00	–266.85	75.73	–297.14	Barin 1993
500.00	–261.54	87.56	–305.32	Barin 1993
600.00	–256.10	97.48	–314.59	Barin 1993
700.00	–250.54	106.05	–324.77	Barin 1993
800.00	–244.87	113.62	–335.76	Barin 1993
900.00	–239.11	120.41	–347.47	Barin 1993
1000.00	–233.24	126.58	–359.82	Barin 1993
1100.00	–227.29	132.26	–372.77	Barin 1993
1200.00	–221.24	137.52	–386.26	Barin 1993

TABLE 4.62
Thermodynamic Data of Fe_2O_3 (hematite)

T K	ΔHf° kJ mol⁻¹	S° J mol⁻¹ deg⁻¹	ΔGf° kJ mol⁻¹	Reference
298.15	–824.25	87.40	–850.31	Barin 1993
400.00	–812.74	120.50	–860.94	Barin 1993
500.00	–800.15	148.54	–874.42	Barin 1993
600.00	–786.54	173.34	–890.54	Barin 1993
700.00	–772.01	195.72	–909.01	Barin 1993
800.00	–756.61	216.27	–929.62	Barin 1993
900.00	–740.37	235.39	–952.22	Barin 1993
1000.00	–723.85	252.81	–976.65	Barin 1993
1100.00	–709.05	266.92	–1002.66	Barin 1993
1200.00	–695.06	279.10	–1029.97	Barin 1993

TABLE 4.63
2 $FeSO_4 = Fe_2O_3 + SO_2(g) + SO_3(g)$

T K	ΔH° kJ	ΔS° J deg^{-1}	ΔG° kJ	K
298.15	340.87	350.48	236.38	3.842E – 042
400.00	339.81	347.48	200.82	5.938E – 027
500.00	338.25	344.00	166.25	4.276E – 018
600.00	336.51	340.84	132.01	3.212E – 012
700.00	334.77	338.15	98.06	4.809E – 008
800.00	331.14	335.97	64.36	6.272E – 002
900.00	331.81	334.40	30.85	1.620E – 002
1000.00	330.43	332.96	–2.53	1.356E + 000
1100.00	327.19	329.89	–35.68	4.950E + 001
1200.00	323.25	326.46	–68.50	9.594E + 002

TABLE 4.64
$Fe_2(SO_4)_3 = Fe_2O_3 + 3\ SO_3\ (g)$

T K	ΔHf° kJ mol^{-1}	S° J mol^{-1} deg^{-1}	ΔGf° kJ mol^{-1}	K
298.15	571.45	550.20	407.41	4.146E – 072
400.00	570.01	546.13	351.56	1.224E – 046
500.00	568.01	541.66	297.17	8.951E – 032
600.00	566.09	538.17	243.19	6.709E – 022
700.00	564.37	535.52	189.51	7.199E – 015
800.00	562.85	533.48	136.07	1.303E – 009
900.00	561.59	531.99	82.80	1.564E – 005
1000.00	560.10	530.44	29.66	2.821E – 002
1100.00	556.45	526.98	–23.22	1.267E + 001
1200.00	551.66	522.81	–75.71	1.976E + 003

TABLE 4.65
Density of $PbSO_4$

Phase	T(K)	d	Reference
c,II	288	6.2907	54
	298	6.2866	
	308	6.2824	
	318	6.2780	

TABLE 4.66
Transitions of $PbSO_4$

Phase Change	T(K)	ΔH kJ/mol	ΔS J/(mol*K)	Reference
c,II → c,I	1139	17.0	14.9	9.10
c,I → l	1360	40	29.7	

TABLE 4.67
Thermodynamic Data of $PbSO_4$ (anglesite)

T K	ΔHf° kJ mol⁻¹	S° J mol⁻¹ deg⁻¹	ΔGf° kJ mol⁻¹	Reference
298.15	–923.14	148.49	–967.41	Barin 1993
400.00	–912.36	179.54	–984.18	Barin 1993
500.00	–901.07	204.71	–1003.42	Barin 1993
600.00	–888.77	227.10	–1025.03	Barin 1993
700.00	–875.35	247.77	–1048.79	Barin 1993
800.00	–860.73	267.27	–1074.55	Barin 1993
900.00	–844.88	285.92	–1102.21	Barin 1993
1000.00	–827.79	303.92	–1131.71	Barin 1993
1100.00	–809.43	321.41	–1162.98	Barin 1993
1200.00	–773.11	353.15	–1196.88	Barin 1993
1300.00	–753.71	368.67	–1232.98	Barin 1993
1400.00	–734.31	383.05	–1270.58	Barin 1993
1500.00	–675.20	423.96	–1311.14	Barin 1993

TABLE 4.68
Thermodynamic Data of $PbO.PbSO_4$

T K	ΔHf° kJ mol⁻¹	S° J mol⁻¹ deg⁻¹	ΔGf° kJ mol⁻¹	Reference
298.15	–1157.45	223.32	–1224.04	Barin 1993
400.00	–1141.89	268.16	–1249.16	Barin 1993
500.00	–1125.68	304.30	–1277.83	Barin 1993
600.00	–1108.29	335.98	–1309.87	Barin 1993
700.00	–1089.62	364.73	–1344.93	Barin 1993
800.00	–1069.62	391.41	–1382.75	Barin 1993
900.00	–1048.26	416.55	–1423.16	Barin 1993
1000.00	–1025.52	440.49	–1466.02	Barin 1993
1100.00	–1001.40	463.48	–1511.22	Barin 1993
1200.00	–975.87	485.68	–1558.69	Barin 1993
1300.00	–948.95	597.22	–1608.34	Barin 1993
1400.00	–920.61	528.22	–1660.11	Barin 1993
1500.00	–890.86	548.74	–1713.97	Barin 1993

TABLE 4.69
Thermodynamic Data of 2 $PbO.PbSO_4$

T K	ΔHf° kJ mol⁻¹	S° J mol⁻¹ deg⁻¹	ΔGf° kJ mol⁻¹	Reference
298.15	–1373.06	294.24	–1460.79	Barin 1989
400.00	–1351.32	356.88	–1494.08	Barin 1989
500.00	–1328.74	407.22	–1532.35	Barin 1989
600.00	–1304.72	450.98	–1575.30	Barin 1989
700.00	–1279.17	490.33	–1622.40	Barin 1989
800.00	–1252.05	526.52	–1673.26	Barin 1989
900.00	–1223.34	560.32	–1727.62	Barin 1989
1000.00	–1193.02	592.25	–1785.26	Barin 1989
1100.00	–1161.07	622.68	–1848.02	Barin 1989
1200.00	–1157.51	651.88	–1906.76	Barin 1989
1300.00	–1092.31	680.04	–1976.36	Barin 1989
1400.00	–1055.49	707.32	–2045.74	Barin 1989
1500.00	–1017.02	733.85	–2117.80	Barin 1989

TABLE 4.70
Thermodynamic Data of 3 $PbO.PbSO_4$

T K	ΔHf° kJ mol⁻¹	S° J mol⁻¹ deg⁻¹	ΔGf° kJ mol⁻¹	Reference
298.15	–1623.74	340.58	–1728.28	Barin 1993
400.00	–1601.25	413.99	–1766.55	Barin 1993
500.00	–1574.49	473.64	–1811.31	Barin 1993
600.00	–1546.16	525.25	–1861.31	Barin 1993
700.00	–1516.32	571.22	–1916.17	Barin 1993
800.00	–1485.01	613.01	–1975.41	Barin 1993
900.00	–1452.22	651.60	–2038.67	Barin 1993
1000.00	–1418.00	687.65	–2105.65	Barin 1993
1100.00	–1382.33	721.64	–2176.13	Barin 1993
1200.00	–1345.22	753.92	–2249.92	Barin 1993
1300.00	–1306.67	784.76	–2326.86	Barin 1993
1400.00	–1266.69	814.39	–2406.83	Barin 1993
1500.00	–1225.28	842.95	–2489.70	Barin 1993

TABLE 4.71
Thermodynamic Data of 4 PbO.$PbSO_4$

T K	ΔHf° kJ mol^{-1}	S° J mol^{-1} deg^{-1}	ΔGf° kJ mol^{-1}	Reference
298.15	–1828.94	424.59	–1955.54	Barin 1993
400.00	–1798.24	513.07	–2003.47	Barin 1993
500.00	–1766.44	583.97	–2058.42	Barin 1993
600.00	–1732.90	645.06	–2119.94	Barin 1993
700.00	–1697.59	699.46	–2187.21	Barin 1993
800.00	–1660.48	748.98	–2259.67	Barin 1993
900.00	–1621.57	794.79	–2336.88	Barin 1993
1000.00	–1580.85	837.68	–2418.53	Barin 1993
1100.00	–1538.31	878.21	–2504.34	Barin 1993
1200.00	–1493.94	916.80	–2594.11	Barin 1993
1300.00	–1447.76	953.76	–2687.65	Barin 1993
1400.00	–1399.75	989.33	–2784.81	Barin 1993
1500.00	–1349.92	1023.70	–2885.47	Barin 1993

TABLE 4.72
$PbSO_4$ = PbO + SO_3(g)

T K	ΔH° kJ	ΔS° J deg^{-1}	ΔG° kJ	K
298.15	309.31	176.98	256.54	1.124E – 045
400.00	308.87	175.66	238.60	6.907E – 032
500.00	308.57	175.01	221.07	8.007E – 024
600.00	307.95	173.89	203.61	1.872E – 018
700.00	306.71	172.00	186.31	1.247E – 014
800.00	304.69	169.32	169.24	8.886E – 012
900.00	301.79	165.90	152.48	1.412E – 009
1000.00	297.93	161.84	136.08	7.783E – 008
1100.00	293.06	157.21	120.13	1.973E – 006
1200.00	296.17	159.60	104.66	2.780E – 005
1300.00	291.10	155.53	88.91	2.676E – 004
1400.00	286.09	151.82	73.54	1.803E – 003
1500.00	241.42	120.88	60.11	8.065E – 002

TABLE 4.73
$PbSO_4 = PbO + SO_2(g) + 1/2\ O_2(g)$

T K	ΔH° kJ	ΔS° J deg^{-1}	ΔG° kJ	K
298.15	408.26	271.00	327.46	4.219E – 058
400.00	408.07	270.43	299.90	6.825E – 040
500.00	407.81	269.86	272.88	3.092E – 029
600.00	407.05	268.50	245.95	3.855E – 022
700.00	405.57	266.24	219.21	4.376E – 017
800.00	403.24	263.13	192.73	2.598E – 013
900.00	399.96	259.29	166.61	2.136E – 010
1000.00	395.70	254.80	140.90	4.361E – 010
1100.00	390.40	249.76	115.67	3.214E – 006
1200.00	393.08	251.76	90.96	1.097E – 004
1300.00	387.55	247.34	66.01	2.225E – 003
1400.00	382.09	243.29	41.48	2.832E – 002
1500.00	336.97	212.03	18.92	2.193E – 001

TABLE 4.74
Density of Li_2SO_4

Phase	T(K)	d	Reference
c,II	298	2.221	8
c,I	883	2.07	55
liquid	1133	2.004	26
	1147	1.999	
	1170	1.989	
	1196	1.978	
	1236	1.962	
	1250	1.956	
	1274	1.947	
	1311	1.932	
	1330	1.924	
	1363	1.911	
	1385	1.901	
	1430	1.884	
	1457	1.873	
	1465	1.869	
	1487	1.860	

TABLE 4.75
Thermodynamic Data of Li_2SO_4

T K	ΔHf° kJ mol⁻¹	S° J mol⁻¹ deg⁻¹	ΔGf° kJ mol⁻¹	Reference
298.15	–1436.49	115.10	–1470.81	Barin 1993
400.00	–1423.39	152.78	–1484.50	Barin 1993
500.00	–1409.03	184.78	–1501.41	Barin 1993
600.00	–1393.16	213.66	–1521.36	Barin 1993
700.00	–1375.80	240.39	–1544.07	Barin 1003
800.00	–1356.95	265.54	–1569.38	Barin 1993
900.00	–1307.82	323.42	–1598.90	Barin 1993
1000.00	–1286.48	345.90	–1632.38	Barin 1993
1100.00	–1265.14	366.24	–1668.01	Barin 1993
1200.00	–1235.80	391.90	–1706.07	Barin 1993
1300.00	–1215.30	408.31	–1746.09	Barin 1993
1400.00	–1194.79	423.50	–1787.69	Barin 1993
1500.00	–1174.29	437.64	–1830.76	Barin 1993

TABLE 4.76
$Li_2SO_4 = Li_2O + SO_3(g)$

T K	ΔH° kJ	ΔS° J deg⁻¹	ΔG° kJ	K
298.15	441.99	179.56	388.46	8.670E – 069
400.00	440.51	175.26	370.41	4.224E – 049
500.00	439.00	171.90	353.05	1.301E – 037
600.00	436.98	168.24	336.04	5.531E – 030
700.00	434.20	163.96	319.42	1.454E – 024
800.00	430.47	159.00	303.27	1.574E – 020
900.00	396.92	119.46	289.40	1.593E – 017
1000.00	391.54	113.80	277.75	3.096E – 015
1100.00	386.51	109.00	266.61	2.181E – 013
1200.00	373.77	97.79	256.42	6.875E – 012
1300.00	370.16	94.90	246.79	1.210E – 010
1400.00	366.82	92.42	237.43	1.383E – 009
1500.00	363.72	90.28	228.30	1.120E – 008

TABLE 4.77
$Li_2SO_4 = Li_2O + SO_2(g) + 1/2\ O_2(g)$

T K	ΔH° kJ	ΔS° J deg⁻¹	ΔG° kJ	K
298.15	540.95	273.58	459.18	3.253E – 081
400.00	539.71	270.03	431.70	4.174E – 057
500.00	538.23	266.74	404.86	5.021E – 043
600.00	536.09	262.85	378.38	1.139E – 033
700.00	533.06	258.20	352.32	5.100E – 027
800.00	529.01	252.81	326.76	4.602E – 022
900.00	495.09	212.84	303.53	2.409E – 018
1000.00	489.32	206.75	282.56	1.735E – 015
1100.00	483.85	201.55	262.15	3.551E – 013
1200.00	470.68	189.96	242.73	2.713E – 011
1300.00	466.62	186.71	223.90	1.007E – 009
1400.00	462.81	183.89	205.37	2.172E – 008
1500.00	459.26	181.44	187.11	3.046E – 007

TABLE 4.78
Density of $MgSO_4$

Phase	T(K)	d	Reference
c,II	288	2.71	57a
c,II	298	2.66	8

TABLE 4.79
Transitions of $MgSO_4$

Phase Change	T(K)	ΔH kJ/mol	ΔS J/(mol*K)	Reference
c,II → c,I	1283	—	—	5
c,I → liq.	1400	14.6	10.4	10

TABLE 4.80
Thermodynamic Data of $MgSO_4$

T K	ΔHf° kJ mol^{-1}	S° J mol^{-1} deg^{-1}	ΔGf° kJ mol^{-1}	Reference
298.15	–1284.86	91.60	–1312.18	Barin 1993
400.00	–1274.26	122.10	–1323.10	Barin 1993
500.00	–1262.66	147.94	–1336.63	Barin 1993
600.00	–1250.25	170.56	–1352.58	Barin 1993
700.00	–1237.16	190.71	–1370.66	Barin 1993
800.00	–1223.49	208.97	–1390.66	Barin 1993
900.00	–1209.26	225.72	–1412.41	Barin 1993
1000.00	–1179.24	241.26	–1435.76	Barin 1993
1100.00	–1179.24	255.81	–1460.62	Barin 1993
1200.00	–1163.47	269.52	–1486.90	Barin 1993
1300.00	–1147.21	282.54	–1514.51	Barin 1993
1400.00	–1130.46	294.95	–1543.39	Barin 1993
1500.00	–1099.91	316.38	–1574.48	Barin 1993

TABLE 4.81
$MgSO_4 = MgO + SO_3(g)$

T K	ΔH° kJ	ΔS° J deg^{-1}	ΔG° kJ	K
298.15	287.86	192.10	230.59	3.966E – 041
400.00	286.88	189.27	211.17	2.642E – 028
500.00	285.77	186.80	192.37	7.975E – 021
600.00	284.57	184.61	173.80	7.382E – 016
700.00	283.21	182.52	155.44	2.510E – 012
800.00	281.62	180.40	137.30	1.083E – 009
900.00	279.76	178.22	119.37	1.179E – 007
1000.00	277.61	175.95	101.66	4.893E – 006
1100.00	275.12	173.59	84.18	1.006E – 004
1200.00	272.29	171.13	66.94	1.219E – 003
1300.00	269.10	168.58	49.96	9.831E – 003
1400.00	265.54	165.94	33.23	5.756E – 002
1500.00	248.29	153.68	17.77	2.405E – 001

TABLE 4.82
$MgSO_4 = MgO + SO_2(g) + 1/2\ O_2(g)$

T K	ΔH° kJ	ΔS° J deg^{-1}	ΔG° kJ	K
298.15	386.81	286.12	301.51	1.488E – 053
400.00	386.08	284.04	272.47	2.610E – 036
500.00	385.00	281.65	244.18	3.079E – 026
600.00	383.67	279.22	216.14	1.520E – 019
700.00	382.07	276.76	188.34	8.808E – 015
800.00	380.16	274.22	160.79	3.167E – 011
900.00	377.94	271.60	133.50	1.784E – 008
1000.00	375.38	268.91	106.47	2.742E – 006
1100.00	372.47	266.14	79.72	1.637E – 004
1200.00	369.20	263.29	53.29	4.809E – 003
1300.00	365.56	260.38	27.06	6.176E – 002
1400.00	361.54	257.40	1.17	9.042E – 001
1500.00	343.83	244.83	–23.41	6.538E + 000

TABLE 4.83
Density of Manganese Sulfates

Phase	T(K)	d	Reference
$MnSO_4$(c)	298	3.181 (25/4)	27
$Mn_2(SO_4)_3$	288	3.24 (g/cm^3)	61

TABLE 4.84
Thermodynamic Data of $MnSO_4$

T K	ΔHf° kJ mol^{-1}	S° J mol^{-1} deg^{-1}	ΔGf° kJ mol^{-1}	Reference
298.15	–1065.25	112.10	–1098.67	Knacke et al. 1991
400.00	–1054.00	144.44	–1111.77	Knacke et al. 1991
500.00	–1041.57	172.12	–1127.63	Knacke et al. 1991
600.00	–1028.28	196.33	–1146.08	Knacke et al. 1991
700.00	–1014.34	217.81	–1166.81	Knacke et al. 1991
800.00	–999.85	237.15	–1189.57	Knacke et al. 1991
900.00	–984.87	254.79	–1214.18	Knacke et al. 1991
1000.00	–969.44	271.05	–1240.48	Knacke et al. 1991
1100.00	–953.57	286.17	–1268.35	Knacke et al. 1991
1200.00	–937.28	300.33	–1297.68	Knacke et al. 1991
1300.00	–920.59	313.69	–1328.39	Knacke et al. 1991
1400.00	–903.50	326.36	–1360.40	Knacke et al. 1991
1500.00	–886.01	338.42	–1393.64	Knacke et al. 1991

TABLE 4.85
Thermodynamic Data of Mn_3O_4

T K	ΔHf° kJ mol⁻¹	S° J mol⁻¹ deg⁻¹	ΔGf° kJ mol⁻¹	Reference
298.15	–1387.80	155.60	–1434.19	Barin 1993
400.00	–1372.70	199.06	–1452.33	Barin 1993
500.00	–1356.77	234.57	–1474.06	Barin 1993
600.00	–1340.05	265.04	–1499.07	Barin 1993
700.00	–1322.67	291.82	–1526.94	Barin 1993
800.00	–1304.70	315.81	–1557.34	Barin 1993
900.00	–1286.16	337.63	–1590.03	Barin 1993
1000.00	–1267.10	357.71	–1624.81	Barin 1003
1100.00	–1247.51	376.38	–1661.53	Barin 1993
1200.00	–1227.41	393.86	–1700.05	Barin 1993
1300.00	–1206.80	410.36	–1740.27	Barin 1993
1400.00	–1185.69	426.00	–1782.09	Barin 1993
1500.00	–1146.36	453.18	–1826.13	Barin 1993

TABLE 4.86
$MnSO_4 = MnO + SO_3(g)$

T K	ΔH° kJ	ΔS° J deg⁻¹	ΔG° kJ	K
298.15	284.27	204.38	223.33	7.422E – 040
400.00	283.24	201.48	202.65	3.428E – 027
500.00	281.69	198.03	182.67	8.223E – 020
600.00	279.91	194.81	163.03	6.394E – 015
700.00	277.97	191.81	143.70	1.888E – 011
800.00	275.84	188.96	124.66	7.238E – 009
900.00	273.51	186.23	105.91	7.127E – 007
1000.00	270.98	183.56	87.42	2.713E – 005
1100.00	268.22	180.93	69.19	5.177E – 004
1200.00	265.23	178.33	51.23	5.887E – 003
1300.00	262.00	175.75	33.52	4.496E – 002
1400.00	258.52	173.17	16.08	2.512E – 001
1500.00	254.79	170.60	–1.11	1.093E + 000

TABLE 4.87
$MnSO_4 = MnO + SO_2(g) + 1/2\ O_2(g)$

T K	ΔH° kJ	ΔS° J deg⁻¹	ΔG° kJ	K
298.15	383.22	298.41	294.25	2.785E – 052
400.00	382.44	296.24	263.94	3.387E – 035
500.00	380.92	292.88	234.48	3.175E – 025
600.00	379.02	289.42	205.37	1.317E – 018
700.00	376.83	286.05	176.60	6.624E – 014
800.00	374.38	282.78	148.16	2.116E – 010
900.00	371.69	279.61	120.04	1.078E – 007
1000.00	368.75	276.52	92.23	1.520E – 005
1100.00	365.56	273.48	64.73	8.431E – 004
1200.00	362.13	270.50	37.53	2.323E – 002
1300.00	358.45	267.55	10.63	3.740E – 001
1400.00	354.52	264.64	–15.98	3.947E + 000
1500.00	350.33	261.75	–42.30	2.972E + 001

TABLE 4.88
$3\ MnSO_4 = Mn_3O_4 + 3\ SO_2(g) + 1/2\ O_2(g)$

T K	ΔH° kJ	ΔS° J deg⁻¹	ΔG° kJ	K
298.15	917.52	769.12	688.20	2.628E – 121
400.00	914.62	760.98	610.23	2.019E – 080
500.00	909.94	750.58	534.65	1.382E – 056
600.00	904.42	740.53	460.10	8.733E – 041
700.00	898.32	731.13	386.53	1.428E – 029
800.00	891.69	722.30	313.86	3.202E – 021
900.00	884.58	713.93	242.05	8.925E – 015
1000.00	876.98	705.92	171.06	1.159E – 009
1100.00	868.88	698.20	100.86	1.623E – 005
1200.00	860.27	690.71	31.41	4.291E – 002

TABLE 4.89
Thermodynamic Data of Hg_2SO_4

T K	ΔHf° kJ mol⁻¹	S° J mol⁻¹ deg⁻¹	ΔGf° kJ mol⁻¹	Reference
298.15	–743.12	200.83	–803.00	Barin et al. 1977
400.00	–728.94	241.64	–825.60	Barin et al. 1977
500.00	–713.59	275.84	–851.51	Barin et al. 1977
600.00	–696.81	306.39	–880.64	Barin et al. 1977
700.00	–678.61	334.41	–912.70	Barin et al. 1977
800.00	–658.99	360.59	–947.47	Barin et al. 1977
900.00	–637.95	385.36	–984.77	Barin et al. 1977
1000.00	–615.49	409.02	–1024.50	Barin et al. 1977

TABLE 4.90
Thermodynamic Data of $HgSO_4$

T K	ΔHf° kJ mol⁻¹	S° J mol⁻¹ deg⁻¹	ΔGf° kJ mol⁻¹	Reference
298.15	–707.50	140.16	–749.29	Barin 1989
400.00	–696.33	172.29	–765.25	Barin 1989
500.00	–683.88	200.01	–783.89	Barin 1989
600.00	–669.97	225.33	–805.17	Barin 1989
700.00	–654.59	249.01	–828.90	Barin 1989
800.00	–637.75	271.47	–854.93	Barin 1989
900.00	–619.45	293.01	–883.16	Barin 1989
1000.00	–599.68	313.83	–913.51	Barin 1989

TABLE 4.91
$HgSO_4 = HgO + SO_3(g)$

T K	ΔH° kJ	ΔS° J deg⁻¹	ΔG° kJ	K
298.15	220.95	186.88	165.23	1.122E – 029
400.00	220.01	184.20	146.33	7.754E – 020
500.00	218.61	181.10	128.06	4.173E – 014
600.00	216.53	177.33	110.13	2.578E – 010
700.00	213.58	172.80	92.62	1.225E – 007
800.00	209.59	167.49	75.60	1.157E – 005
900.00	204.45	161.45	59.15	3.689E – 004
1000.00	198.07	154.73	43.33	5.450E – 003

TABLE 4.92
$HgSO_4 = HgO + SO_2(g) + 1/2\ O_2(g)$

T K	ΔH° kJ	ΔS° J deg⁻¹	ΔG° kJ	K
298.15	319.90	280.90	236.15	4.209E – 042
400.00	319.21	278.96	207.63	7.661E – 028
500.00	317.85	275.95	179.87	1.611E – 019
600.00	315.64	271.94	152.47	5.310E – 014
700.00	312.44	267.04	125.52	4.297E – 010
800.00	308.13	261.30	99.09	3.384E – 007
900.00	302.63	254.83	73.28	5.580E – 005
1000.00	295.84	247.69	48.15	3.054E – 003

TABLE 4.93
Thermodynamic Data of $NiSO_4$

T K	ΔHf° kJ mol⁻¹	S° J mol⁻¹ deg⁻¹	ΔGf° kJ mol⁻¹	Reference
298.15	–872.91	92.00	–900.34	Barin 1989
400.00	–858.64	133.15	–911.90	Barin 1989
500.00	–844.21	165.33	–926.87	Barin 1989
600.00	–829.36	192.38	–944.79	Barin 1989
700.00	–814.10	215.90	–965.23	Barin 1989
800.00	–798.43	236.83	–987.89	Barin 1989
900.00	–782.34	255.77	–1012.53	Barin 1989
1000.00	–765.83	273.16	–1038.99	Barin 1989
1100.00	–748.91	289.28	–1067.12	Barin 1989
1200.00	–731.57	304.36	–1096.81	Barin 1989

TABLE 4.94
$NiSO_4 = NiO + SO_3(g)$

T K	ΔH° kJ	ΔS° J deg⁻¹	ΔG° kJ	K
298.15	237.44	202.76	176.99	9.764E – 032
400.00	233.60	191.61	156.96	3.175E – 021
500.00	230.93	185.58	138.14	3.697E – 015
600.00	228.53	181.25	119.78	3.729E – 011
700.00	225.47	176.55	101.89	2.490E – 008
800.00	222.22	172.20	84.45	3.056E – 006
900.00	218.79	168.17	67.44	1.218E – 004
1000.00	215.19	164.38	50.81	2.216E – 003
1100.00	211.40	160.77	34.56	2.285E – 002
1200.00	207.41	157.30	18.66	1.514E – 001

TABLE 4.95
$NiSO_4 = NiO + SO_2(g) + 1/2\ O_2(g)$

T K	ΔH° kJ	ΔS° J deg⁻¹	ΔG° kJ	K
298.15	336.39	296.78	247.91	3.664E – 044
400.00	332.80	286.37	218.25	3.137E – 029
500.00	330.16	280.42	189.94	1.427E – 020
600.00	327.63	275.86	162.12	7.682E – 015
700.00	324.33	270.78	134.79	8.737E – 011
800.00	320.76	266.02	107.95	8.936E – 008
900.00	316.97	261.55	81.57	1.842E – 005
1000.00	312.96	257.34	55.63	1.242E – 003
1100.00	308.75	253.32	30.10	3.721E – 002
1200.00	304.31	249.46	4.96	6.082E – 001

TABLE 4.96
Density of K_2SO_4

Phase	T(K)	d	Reference
c,II (α)	291	2.663	68
c,II	298	2.662	69
c,I (β)	298	2.621	70
Liquid	1343	1.888	26
	1376	1.870	
	1418	1.848	
	1472	1.818	
	1520	1.792	
	1579	1.760	
	1670	1.737	
	1713	1.687	
	1763	1.660	
	1803	1.637	
	1859	1.607	
	1929	1.569	

TABLE 4.97
Transitions of K_2SO_4

Phase Change	T(K)	ΔH kJ/mol	ΔS J/(mol*K)	Reference
c,II → c,I	856	8.12	9.50	9,10,71
c,I → l	1342	36.6	27.3	

TABLE 4.98
Thermodynamic Data of K_2SO_4

T K	ΔHf° kJ mol^{-1}	S° J mol^{-1} deg^{-1}	ΔGf° kJ mol^{-1}	Reference
298.15	−1437.79	175.56	−1490.13	Barin 1993
400.00	−1423.25	217.37	−1510.20	Barin 1993
500.00	−1407.82	251.79	−1533.71	Barin 1993
600.00	−1391.53	281.45	−1560.40	Barin 1993
700.00	−1373.86	308.66	−1589.92	Barin 1993
800.00	−1353.94	335.21	−1622.11	Barin 1993
900.00	−1324.78	369.52	−1657.35	Barin 1993
1000.00	−1305.62	389.70	−1695.32	Barin 1993
1100.00	−1285.66	408.72	−1735.25	Barin 1993
1200.00	−1264.88	426.80	−1777.03	Barin 1993
1300.00	−1243.28	444.08	−1820.58	Barin 1993
1400.00	−1187.89	485.28	−1867.28	Barin 1993
1500.00	−1167.75	499.18	−1916.51	Barin 1993

TABLE 4.99
$K_2SO_4 = K_2O + SO_3(g)$

T K	ΔH° kJ	ΔS° J deg⁻¹	ΔG° kJ	K
298.15	680.53	183.22	625.90	2.166E – 110
400.00	679.35	179.86	607.40	4.731E – 080
500.00	678.02	176.91	589.57	2.529E – 062
600.00	676.62	174.36	572.01	1.579E – 050
700.00	674.42	171.00	554.73	4.004E – 042
800.00	670.45	165.73	537.87	7.548E – 036
900.00	657.63	150.66	522.04	5.004E – 031
1000.00	655.15	148.05	507.10	3.234E – 027
1100.00	680.54	173.18	490.05	5.341E – 024
1200.00	678.20	171.15	472.83	2.610E – 021
1300.00	675.13	168.69	455.83	4.819E – 019
1400.00	638.33	141.27	440.56	3.641E – 017
1500.00	636.84	140.24	426.48	1.404E – 015

TABLE 4.100
$K_2SO_4 = K_2O + SO_2(g) + 1/2\ O_2(g)$

T K	ΔH° kJ	ΔS° J deg⁻¹	ΔG° kJ	K
298.15	779.48	277.24	696.82	8.126E – 123
400.00	778.55	274.63	668.70	4.675E – 088
500.00	777.26	271.75	641.38	9.764E – 068
600.00	775.73	268.97	614.34	3.253E – 054
700.00	773.29	265.24	587.62	1.405E – 044
800.00	769.00	259.55	561.36	2.207E – 037
900.00	755.80	244.04	536.17	7.569E – 032
1000.00	752.92	241.01	511.91	1.812E – 027
1100.00	777.89	265.73	485.59	8.698E – 024
1200.00	775.11	263.31	459.13	1.030E – 020
1300.00	771.59	260.50	432.94	4.007E – 018
1400.00	734.33	232.74	408.50	5.720E – 016
1500.00	732.38	231.39	385.30	3.817E – 014

TABLE 4.101
$K_2SO_4 = 2\ K(g) + SO_2(g) + O_2$

T K	ΔH° kJ	ΔS° J deg⁻¹	ΔG° kJ	K
298.15	1318.98	598.49	1140.54	1.465E – 200
400.00	1315.97	589.90	1080.01	8.991E – 142
500.00	1312.33	581.80	1021.43	1.918E – 107
600.00	1308.23	574.34	963.63	1,265E – 084
700.00	1303.02	566.35	906.58	2.210E – 068
800.00	1295.80	556.73	850.41	2.946E – 056
900.00	1279.49	537.56	795.68	6.545E – 047
1000.00	1273.33	531.08	742.25	1.681E – 039
1100.00	1266.48	524.55	689.47	1.808E – 033
1200.00	1258.92	517.98	637.34	1.798E – 028
1300.00	1250.64	511.35	585.88	2.866E – 024
1400.00	1208.64	480.08	536.53	9.553E – 021
1500.00	1201.96	475.47	488.76	9.519E – 018

TABLE 4.102
Thermodynamic Data of $RaSO_4$

T K	ΔHf° kJ mol⁻¹	S° J mol⁻¹ deg⁻¹	ΔGf° kJ mol⁻¹	Reference
298.15	–1471.00	159.00	–1518.40	Phillips 88
400.00	–1460.51	189.27	–1536.22	Phillips 88
500.00	–1450.21	212.25	–1556.33	Phillips 88
600.00	–1439.91	231.03	–1578.53	Phillips 88
700.00	–1429.61	246.91	–1602.44	Phillips 88
800.00	–1419.31	260.67	–1627.84	Phillips 88
900.00	–1409.01	272.80	–1654.52	Phillips 88
1000.00	–1398.71	283.65	–1682.36	Phillips 88

TABLE 4.103
Thermodynamic Data of $Ce_2(SO_4)_3$

T K	ΔHf° kJ mol⁻¹	S° J mol⁻¹ deg⁻¹	ΔGf° kJ mol⁻¹	Reference
298.15	–3954.30	287.44	–4040.00	Barin et al. 1977
400.00	–3954.30	287.44	–4073.79	Barin et al. 1977
500.00	–3893.53	442.21	–4114.63	Barin et al. 1977
600.00	–3860.42	502.51	–4161.93	Barin et al. 1977
700.00	–3825.33	556.57	–4214.93	Barin et al. 1977
800.00	–3788.25	606.05	–4273.90	Barin et al. 1977
900.00	–3749.18	652.04	–4336.02	Barin et al. 1977
1000.00	–3708.13	695.28	–4403.41	Barin et al. 1977
1100.00	–3665.08	736.29	–4475.00	Barin et al. 1977
1200.00	–3620.05	775.46	–4550.60	Barin et al. 1977
1300.00	–3573.03	813.08	–4630.04	Barin et al. 1977
1400.00	–3524.03	849.39	–4713.18	Barin et al. 1977
1500.00	3473.04	884.56	–4799.88	Barin et al. 1977

TABLE 4.104
Thermodynamic Data of Ce_2O_3

T K	ΔHf° kJ mol^{-1}	S° J mol^{-1} deg^{-1}	ΔGf° kJ mol^{-1}	Reference
298.15	–1796.19	150.58	–1841.09	Barin 1993
400.00	–1783.61	186.79	–1858.33	Barin 1993
500.00	–1770.48	216.07	–1878.52	Barin 1993
600.00	–1756.87	240.89	–1901.40	Barin 1993
700.00	–1742.89	262.42	–1926.59	Barin 1993
800.00	–1728.62	281.48	–1953.80	Barin 1993
900.00	–1714.08	298.60	–1982.82	Barin 1993
1000.00	–1699.30	314.17	–2013.47	Barin 1993
1100.00	–1684.29	328.48	–2045.61	Barin 1993
1200.00	–1669.05	341.73	–2079.13	Barin 1993
1300.00	–1653.60	354.10	–2113.93	Barin 1993
1400.00	–1637.95	365.70	–2149.92	Barin 1993
1500.00	–1622.08	376.65	–2187.05	Barin 1993

TABLE 4.105
$Ce_2(SO_4)_3 = Ce_2O_3 + 3\ SO_3(g)$

T K	ΔH° kJ	ΔS° J deg^{-1}	ΔG° kJ	K
298.15	970.81	633.46	781.95	9.874E – 138
400.00	970.29	631.87	717.54	1.953E – 094
500.00	970.44	632.18	654.34	4.319E – 069
600.00	970.58	632.44	591.11	3.429E – 052
700.00	970.18	631.85	527.89	4.032E – 040
800.00	968.87	630.12	464.78	4.474E – 031
900.00	966.45	627.28	401.90	4.705E – 024
1000.00	962.74	623.39	339.36	1.872E – 018
1100.00	957.64	618.54	277.25	6.812E – 014
1200.00	951.07	612.83	215.68	4.084E – 010
1300.00	942.97	606.36	154.71	6.069E – 007
1400.00	933.30	599.19	94.43	2.996E – 004
1500.00	922.02	591.42	34.89	6.093E – 002

TABLE 4.106
$Ce_2(SO_4)_3 = Ce_2O_3 + 3\ SO_2(g) + 3/2\ O_2(g)$

T K	ΔH° kJ	ΔS° J deg⁻¹	ΔG° kJ	K
298.15	1267.67	915.53	994.71	5.216E – 175
400.00	1267.90	916.17	901.43	1.884E – 118
500.00	1268.14	916.72	809.78	2.486E – 085
600.00	1267.89	916.28	718.12	2.997E – 063
700.00	1266.77	914.57	626.57	1.741E – 047
800.00	1264.51	911.57	535.25	1.119E – 035
900.00	1260.98	907.43	444.29	1.628E – 028
1000.00	1256.06	902.56	353.80	3.294E – 019
1100.00	1249.68	896.19	263.87	2.942E – 013
1200.00	1241.78	889.32	174.59	2.510E – 008
1300.00	1232.33	881.77	86.03	3.491E – 004
1400.00	1221.30	873.60	–1.74	1.161E + 000
1500.00	1208.65	864.88	–88.67	1.225E + 003

TABLE 4.107
Thermodynamic Data of $La_2(SO_4)_3$

T K	ΔHf° kJ mol⁻¹	S° J mol⁻¹ deg⁻¹	ΔGf° kJ mol⁻¹	Reference
298.15	–3941.30	280.33	–4024.88	Phillips 88
400.00	–3912.79	362.58	–4057.82	Phillips 88
500.00	–3884.80	425.04	–4097.32	Phillips 88
600.00	–3856.81	476.08	–4142.46	Phillips 88
700.00	–3828.83	519.23	–4192.28	Phillips 88

TABLE 4.108
Thermodynamic Data of $Nd_2(SO_4)_3$

T K	ΔHf° kJ mol⁻¹	S° J mol⁻¹ deg⁻¹	ΔGf° kJ mol⁻¹	Reference
298.15	–3899.49	288.28	–3985.44	Barin et al. 1977
400.00	–3870.69	371.23	–4019.18	Barin et al. 1977
500.00	–3840.41	438.72	–4059.77	Barin et al. 1977
600.00	–3808.14	497.49	–4106.63	Barin et al. 1977
700.00	–3773.88	550.26	–4159.06	Barin et al. 1977
800.00	–3737.64	598.63	–4216.54	Barin et al. 1977
900.00	–3699.41	643.64	–4278.68	Barin et al. 1977
1000.00	–3659.19	685.99	–4345.18	Barin et al. 1977
1100.00	–3616.98	726.29	–4415.81	Barin et al. 1977
1200.00	–3572.79	764.64	–4490.36	Barin et al. 1977
1300.00	–3526.61	801.60	–4568.69	Barin et al. 1977
1400.00	–3478.44	837.29	–4650.64	Barin et al. 1977
1500.00	–3428.28	871.88	–4736.11	Barin et al. 1977

TABLE 4.109
Thermodynamic Data of Nd_2O_3

T K	ΔHf° kJ mol⁻¹	S° J mol⁻¹ deg⁻¹	ΔGf° kJ mol⁻¹	Reference
298.15	–1807.91	158.57	–1855.19	Knacke et al. 1991
400.00	–1796.07	192.66	–1873.13	Knacke et al. 1991
500.00	–1783.75	220.13	–1893.81	Knacke et al. 1991
600.00	–1770.94	243.48	–1917.02	Knacke et al. 1991
700.00	–1757.71	263.86	–1942.41	Knacke et al. 1991
800.00	–1744.12	282.00	–1962.72	Knacke et al. 1991
900.00	–1730.18	298.42	–1998.75	Knacke et al. 1991
1000.00	–1715.90	313.46	–2029.36	Knacke et al. 1991
1100.00	–1701.30	327.37	–2061.41	Knacke et al. 1991
1200.00	–1686.39	340.34	–2094.80	Knacke et al. 1991
1300.00	–1671.16	352.53	–2129.45	Knacke et al. 1991
1400.00	–1655.04	364.46	–2165.29	Knacke et al. 1991
1500.00	–1639.74	375.20	–2202.28	Knacke et al. 1991

TABLE 4.110
$Nd_2(SO_4)_3 = Nd_2O_3 + 3\ SO_3(g)$

T K	ΔH° kJ	ΔS° J deg⁻¹	ΔG° kJ	K
298.15	904.29	640.61	713.29	1.057E – 125
400.00	903.88	639.36	648.13	2.267E – 085
500.00	904.05	639.73	584.18	9.241E – 062
600.00	904.22	640.05	520.19	5.125E – 046
700.00	903.91	639.59	456.20	9.019E – 035
800.00	902.76	638.07	392.31	2.414E – 026
900.00	900.58	635.51	328.62	8.429E – 020
1000.00	897.20	631.96	265.24	1.394E – 014
1100.00	892.52	627.51	202.26	2.482E – 010
1200.00	886.47	622.26	139.76	8.237E – 007
1300.00	878.99	614.28	77.83	7.455E – 004
1400.00	870.62	610.06	16.53	2.417E – 001
1500.00	859.87	602.66	–44.11	3.483E + 001

TABLE 4.111
$Nd_2(SO_4)_3 = Nd_2O_3 + 3SO_2(g) + 3/2\ O_2(g)$

T K	ΔH° kJ	ΔS° J deg^{-1}	ΔG° kJ	K
298.15	1201.14	922.68	926.05	5.586E – 163
400.00	1201.49	923.65	832.02	2.187E – 109
500.00	1201.75	924.26	739.62	5.319E – 078
600.00	1201.54	923.89	647.20	4.479E – 057
700.00	1200.50	922.30	554.89	3.895E – 042
800.00	1198.40	919.52	462.78	6.036E – 031
900.00	1195.11	915.65	371.02	2.917E – 022
1000.00	1190.52	910.83	279.69	2.452E – 015
1100.00	1184.56	905.16	188.88	1.072E – 009
1200.00	1177.18	898.75	98.68	5.062E – 005
1300.00	1168.35	891.69	9.15	4.288E – 001
1400.00	1158.61	884.47	–79.64	9.370E + 002
1500.00	1146.50	876.12	–167.68	6.911E + 005

TABLE 4.112
Density of Rb_2SO_4

Phase	T(K)	d	Reference
c,II	293	3.614	25
c,II	298	3.613	8
c,II	333	3.594	25
Liquid	1359	2.538	26
	1385	2.521	
	1418	2.499	
	1468	2.466	
	1508	2.440	
	1562	2.403	
	1617	2.367	
	1670	2.331	
	1688	2.319	
	1755	2.275	
	1818	2.233	

TABLE 4.113
Transitions of Rb_2SO_4

Phase Change	T(K)	ΔH kJ/mol	ΔS J/(mol·K)	Reference
c,II → c,I	923			9
c,I → liquid	1347			9

TABLE 4.114
Thermodynamic Data of Rb_2SO_4

T K	ΔHf° kJ mol⁻¹	S° J mol⁻¹ deg⁻¹	ΔGf° kJ mol⁻¹	Reference
298.15	–1435.61	197.44	–1494.48	Barin 1989
400.00	–1423.22	233.08	–1516.45	Barin 1989
500.00	–1409.69	263.22	–1541.30	Barin 1989
600.00	–1395.05	289.88	–1568.98	Barin 1989
700.00	–1379.40	313.98	–1599.19	Barin 1989
800.00	–1362.79	336.16	–1631.71	Barin 1989
900.00	–1345.23	356.83	–1666.37	Barin 1989
1000.00	–1321.20	382.22	–1703.42	Barin 1989
1100.00	–1300.69	401.76	–1742.63	Barin 1989
1200.00	–1280.19	419.60	–1783.71	Barin 1989
1300.00	–1259.69	436.01	–1826.50	Barin 1989
1400.00	–1200.54	479.98	–1872.51	Barin 1989
1500.00	–1179.62	494.41	–1921.23	Barin 1989

TABLE 4.115
$Rb_2SO_4 = Rb_2O + SO_3(g)$

T K	ΔH° kJ	ΔS° J deg⁻¹	ΔG° kJ	K
298.15	700.85	184.85	645.73	7.262E – 114
400.00	701.69	187.32	626.76	1.403E – 082
500.00	702.12	188.29	607.97	3.023E – 064
600.00	703.11	190.12	589.04	5.196E – 052
700.00	707.13	196.71	560.44	3.196E – 043
800.00	727.56	222.94	549.21	1.373E – 036
900.00	726.57	221.78	526.96	2.590E – 031
1000.00	719.27	214.02	505.25	4.038E – 027
1100.00	715.61	210.53	484.03	1.032E – 023
1200.00	712.06	207.44	463.13	6.898E – 021
1300.00	708.59	204.66	442.53	1.650E – 018
1400.00	666.53	173.36	423.83	1.533E – 016
1500.00	662.76	170.76	406.62	6.902E – 015

TABLE 4.116
$Rb_2SO_4 = Rb_2O + SO_2(g) + 1/2\ O_2(g)$

T K	ΔH° kJ	ΔS° J deg^{-1}	ΔG° kJ	K
298.15	799.80	278.88	716.65	2.735E – 126
400.00	800.89	282.09	688.06	1.386E – 090
500.00	801.35	283.14	659.78	1.167E – 069
600.00	802.21	284.73	631.37	1.070E – 055
700.00	806.00	290.95	602.33	1.121E – 045
800.00	826.10	316.75	572.70	4.014E – 038
900.00	824.74	315.17	541.10	3.917E – 032
1000.00	817.04	306.97	510.07	2.263E – 027
1100.00	812.96	303.08	479.57	1.680E – 023
1200.00	808.96	299.60	449.43	2.722E – 020
1300.00	805.04	296.46	419.64	1.372E – 017
1400.00	762.53	264.83	391.77	2.408E – 015
1500.00	758.31	261.92	365.44	1.877E – 013

TABLE 4.117
Transitions of Ag_2SO_4

Phase Change	T(K)	ΔH kJ/mol	ΔS J/(mol˙K)	Reference
c.II → c,I	685	8.0	11.7	9, 10
c,I → liquid	933	17.		

TABLE 4.118
Thermodynamic Data of Ag_2SO_4

T K	ΔHf° kJ mol^{-1}	S° J mol^{-1} deg^{-1}	ΔGf° kJ mol^{-1}	Reference
298.15	–715.88	200.41	–775.64	Barin 1989
400.00	–701.89	340.71	–798.17	Barin 1989
500.00	–686.97	273.95	–823.94	Barin 1989
600.00	–670.89	303.24	–852.83	Barin 1989
700.00	–653.63	329.81	–884.50	Barin 1989
800.00	–616.55	381.05	–921.39	Barin 1989
900.00	–596.96	404.11	–960.66	Barin 1989
1000.00	–558.60	444.85	–1003.45	Barin 1989

TABLE 4.119
$Ag_2SO_4 = Ag_2O + SO_3(g)$

T K	ΔH° kJ	ΔS° J deg^{-1}	ΔG° kJ	K
298.15	289.07	177.66	236.13	4.048E – 042
400.00	287.63	173.51	218.23	3.160E – 029
500.00	286.23	170.37	201.04	9.901E – 022
600.00	284.63	167.47	184.15	9.270E – 017
700.00	282.68	164.47	167.55	3.134E – 013
800.00	261.59	134.57	153.93	8.883E – 011
900.00	258.62	131.08	140.65	6.862E – 009
1000.00	237.45	108.55	129.00	1.824E – 007

TABLE 4.120
$Ag_2SO_4 = Ag_2O + SO_2(g) + 1/2\ O_2(g)$

T K	ΔH° kJ	ΔS° J deg^{-1}	ΔG° kJ	K
298.15	388.02	271.68	307.02	1.611E – 054
400.00	417.59	334.08	283.96	8.238E – 038
500.00	415.54	329.53	250.77	6.305E – 027
600.00	412.81	324.58	218.07	1.033E – 019
700.00	409.32	319.21	185.87	1.346E – 014
800.00	386.34	286.80	156.90	5.682E – 011
900.00	381.17	280.72	128.52	3.468E – 008
1000.00	357.52	255.47	102.05	4.668E – 006

TABLE 4.121
Density of Na_2SO_4

Phase	T(K)	d	Reference
c,V	298	2.664	84
c,III	298	2.697	84
c,I	993	2.64 (X-ray)	85
Liquid	1173	2.061	26
	1218	2.039	
	1263	2.017	
	1305	1.997	
	1350	1.971	

TABLE 4.122
Transitions of Na_2SO_4

Phase Change	T(K)	ΔH kJ/mol	ΔS J/(mol*K)	Reference
c,V → c,III	450	3.10	6.9	10
c,III → c,I	515	7.49	14.56	10
c,I → liquid	1157	23.85	20.63	10

TABLE 4.123
Thermodynamic Data of Na_2SO_4

T K	ΔHf° kJ mol⁻¹	S° J mol⁻¹ deg⁻¹	ΔGf° kJ mol⁻¹	Reference
298.15	−1387.82	149.60	−1432.42	Knacke et al. 1991
400.00	−1373.86	189.73	−1449.76	Knacke et al. 1991
500.00	−1358.39	224.18	−1470.48	Knacke et al. 1991
600.00	−1330.38	276.56	−1496.31	Knacke et al. 1991
700.00	−1312.57	304.01	−1525.37	Knacke et al. 1991
800.00	−1294.14	328.59	−1557.02	Knacke et al. 1991
900.00	−1275.09	351.03	−1591.02	Knacke et al. 1991
1000.00	−1255.38	371.79	−1627.17	Knacke et al. 1991
1100.00	−1235.03	391.18	−1665.33	Knacke et al. 1991
1200.00	−1190.82	429.51	−1706.24	Knacke et al. 1991
1300.00	−1171.12	445.28	−1749.99	Knacke et al. 1991
1400.00	−1151.42	459.89	−1795.26	Knacke et al. 1991
1500.00	−1131.71	473.48	−1841.93	Knacke et al. 1991
1600.00	−1112.01	486.20	−1889.92	Knacke et al. 1991
1700.00	−1092.31	498.14	−1939.15	Knacke et al. 1991
1800.00	−1072.60	509.40	−1989.53	Knacke et al. 1991
1900.00	−1052.90	520.06	−2041.01	Knacke et al. 1991
2000.00	−1033.20	530.16	−2093.52	Knacke et al. 1991

TABLE 4.124
$Na_2SO_4 = Na_2O + SO_3(g)$

T K	ΔH° kJ	ΔS° J deg⁻¹	ΔG° kJ	K
298.15	574.07	182.22	519.74	8.627E − 092
400.00	573.07	179.34	501.33	3.370E − 066
500.00	571.54	175.97	483.56	3.011E − 051
600.00	558.41	150.70	467.99	1.797E − 041
700.00	556.18	147.27	453.09	1.538E − 034
800.00	553.90	144.23	438.52	2.318E − 029
900.00	551.43	141.32	424.25	2.374E − 025
1000.00	548.68	138.42	410.26	3.703E − 022
1100.00	547.33	137.18	396.43	1.491E − 019
1200.00	520.60	114.05	383.74	1.972E − 017
1300.00	530.49	122.02	371.87	1.140E − 015
1400.00	528.61	120.62	359.74	3.773E − 014
1500.00	574.99	153.65	344.51	1.004E − 012
1600.00	573.74	152.84	329.19	1.787E − 011
1700.00	572.53	152.11	313.95	2.253E − 010
1800.00	571.35	151.43	298.77	2.134E − 009
1900.00	570.20	150.81	283.66	1.589E − 008
2000.00	569.06	150.23	268.61	9.642E − 008

TABLE 4.125
$Na_2SO_4 = Na_2O + SO_2(g) + 1/2\ O_2(g)$

T K	ΔH° kJ	ΔS° J deg^{-1}	ΔG° kJ	K
298.15	673.02	276.24	590.66	3.237E – 104
400.00	672.27	274.11	562.63	3.329E – 074
500.00	670.78	270.81	535.37	1.163E – 056
600.00	657.51	245.31	510.33	3.702E – 045
700.00	655.05	241.51	485.99	5.395E – 037
800.00	652.45	238.04	462.01	6.778E – 031
900.00	649.61	234.70	438.38	3.590E – 026
1000.00	646.45	231.38	415.07	2.075E – 022
1100.00	644.67	229.73	391.97	2.428E – 019
1200.00	617.50	206.22	370.04	7.783E – 017
1300.00	626.95	213.82	348.98	9.477E – 015
1400.00	624.61	212.09	327.69	5.928E – 013
1500.00	670.53	244.80	303.33	2.731E – 011
1600.00	668.83	243.71	278.90	7.834E – 010
1700.00	667.18	242.70	254.58	1.503E – 008
1800.00	665.66	241.78	230.36	2.063E – 007
1900.00	663.98	240.92	206.23	2.138E – 006
2000.00	662.43	240.13	182.17	1.745E – 005

TABLE 4.126
Transitions of $SrSO_4$

Phase Change	T(K)	ΔH kJ/mol	ΔS J/(mol*K)	Reference
c,II → c,I	1425			9
c,I → liquid	1878			9

TABLE 4.127
Thermodynamic Data of $SrSO_4$(celestite)

T K	ΔHf° kJ mol^{-1}	S° J mol^{-1} deg^{-1}	ΔGf° kJ mol^{-1}	Reference
298.15	–1458.96	121.75	–1495.26	Barin 1993
400.00	–1448.31	152.42	–1509.28	Barin 1993
500.00	–1437.30	176.98	–1529.79	Barin 1993
600.00	–118.47	198.05	–1544.56	Barin 1993
700.00	–1413.61	216.73	–1565.32	Barin 1993
800.00	–1400.93	233.66	–1587.85	Barin 1993
900.00	–1387.69	249.24	–1612.01	Barin 1993
1000.00	–1373.89	263.77	–1637.67	Barin 1993
1100.00	–1359.54	277.45	–1664.73	Barin 1993
1200.00	–1344.63	290.41	–1693.13	Barin 1993
1300.00	–1329.16	302.79	–1722.80	Barin 1993
1400.00	–1313.15	314.66	–1753.67	Barin 1993
1500.00	–1286.28	333.29	–1786.22	Barin 1993

TABLE 4.128
$SrSO_4 = SrO + SO_3(g)$

T K	ΔH° kJ	ΔS° J deg^{-1}	ΔG° kJ	K
298.15	471.16	190.54	414.35	2.524E – 073
400.00	470.83	189.56	395.00	2.591E – 052
500.00	470.82	189.53	376.05	5.137E – 040
600.00	470.92	189.72	357.09	8.126E – 032
700.00	470.97	189.79	338.11	5.855E – 026
800.00	470.83	189.62	319.14	1.447E – 021
900.00	470.44	189.16	300.20	3.762E – 018
1000.00	469.74	188.42	281.33	2.015E – 015
1100.00	468.68	187.41	262.53	3.409E – 013
1200.00	467.24	186.16	243.85	2.426E – 011
1300.00	465.40	184.69	225.30	8.842E – 010
1400.00	463.14	183.02	206.91	1.902E – 008
1500.00	450.17	173.97	189.21	2.574E – 007

TABLE 4.129
$SrSO_4 = SrO + SO_2(g) + O_2(g)$

T K	ΔH° kJ	ΔS° J deg^{-1}	ΔG° kJ	K
298.15	570.11	284.56	485.31	8.436 – 086
400.00	570.03	284.33	456.30	2.560E – 060
500.00	570.05	284.37	427.87	1.984E – 045
600.00	570.03	284.33	399.43	1.674E – 035
700.00	569.83	284.03	371.01	2.054E – 028
800.00	569.38	283.43	342.63	4.231E – 023
900.00	568.62	282.54	314.33	5.690E – 019
1000.00	567.51	281.37	286.14	1.129E – 015
1100.00	566.02	279.96	258.07	5.552E – 013
1200.00	564.14	278.33	230.15	9.572E – 011
1300.00	561.85	276.50	202.41	7.354E – 009
1400.00	559.14	274.49	174.86	2.989E – 007
1500.00	545.71	265.13	148.02	6.999E – 006

TABLE 4.130
Density of Tl_2SO_4

Phase	T(K)	d	Reference
$Tl_2SO_4(c)$	298	6.77	93
$Tl_2SO_4(c)$	303	6.768	94
$Tl_2(SO_4)_3(c)$	288	5.22	14

TABLE 4.131
Transitions of Tl_2SO_4

Phase Change	T(K)	ΔH kJ/mol	ΔS J/(mol*K)	Reference
II → I	765			95
I → liquid	905	23	25.5	9

TABLE 4.132
Thermodynamic Data of Tl_2SO_4

T K	ΔHf° kJ mol^{-1}	S° J mol^{-1} deg^{-1}	ΔGf° kJ mol^{-1}	Reference
298.15	–931.78	230.54	–1000.51	Barin 1989
400.00	–917.09	272.83	–1026.22	Barin 1989
500.00	–901.40	307.79	–1055.29	Barin 1989
600.00	–175.73	338.65	–1087.64	Barin 1989
700.00	–866.25	366.68	–1122.93	Barin 1989
800.00	–846.80	392.64	–1160.91	Barin 1989
900.00	–826.09	417.02	–1201.41	Barin 1989
1000.00	–781.69	465.02	–1246.71	Barin 1989
1100.00	–761.19	484.56	–1294.21	Barin 1989
1200.00	–740.69	502.40	–1343.57	Barin 1989
1300.00	–720.19	518.81	–1394.64	Barin 1989
1400.00	–699.69	534.01	–1447.29	Barin 1989
1500.00	–679.18	548.15	–1501.41	Barin 1989

TABLE 4.133
Thermodynamic Data of Tl_2O

T K	ΔHf° kJ mol^{-1}	S° J mol^{-1} deg^{-1}	ΔGf° kJ mol^{-1}	Reference
298.15	–169.03	145.19	–212.32	Knacke et al. 1991
400.00	–160.83	168.84	–228.36	Knacke et al. 1991
500.00	–152.49	187.44	–246.21	Knacke et al. 1991
600.00	–143.91	203.08	–265.75	Knacke et al. 1991
700.00	–135.10	216.65	–286.75	Knacke et al. 1991
800.00	–126.07	228.70	–309.03	Knacke et al. 1991
900.00	–85.48	276.34	–334.19	Knacke et al. 1991
1000.00	–73.71	288.74	–362.45	Knacke et al. 1991
1100.00	–61.94	299.96	–391.90	Knacke et al. 1991
1200.00	–50.17	310.21	–422.41	Knacke et al. 1991
1300.00	–38.39	319.63	–453.91	Knacke et al. 1991
1400.00	–26.62	328.35	–486.32	Knacke et al. 1991
1500.00	14.85	336.47	–519.56	Knacke et al. 1991

TABLE 4.134
$Tl_2SO_4 = Tl_2O + SO_3(g)$

T K	ΔH° kJ	ΔS° J deg^{-1}	ΔG° kJ	K
298.15	366.98	171.42	315.87	4.532E – 056
400.00	366.01	168.65	298.55	1.023E – 039
500.00	364.71	165.75	281.83	3.590E – 030
600.00	362.89	162.45	265.42	7.792E – 024
700.00	360.40	158.63	249.36	2.462E – 019
800.00	357.14	154.29	233.71	5.486E – 016
900.00	384.39	186.23	216.78	2.613E – 013
1000.00	359.29	158.55	200.73	3.265E – 011
1100.00	358.20	157.52	184.93	1.650E – 009
1200.00	357.21	156.66	169.22	4.298E – 008
1300.00	356.31	155.93	153.60	6.729E – 007
1400.00	355.47	155.31	138.03	7.071E – 006
1500.00	354.68	154.77	122.53	5.404E – 005

TABLE 4.135
$Tl_2SO_4 = Tl_2O + SO_2(g) + 1/2\ O_2(g)$

T K	ΔH° kJ	ΔS° J deg^{-1}	ΔG° kJ	K
298.15	366.98	171.42	315.87	4.532E – 056
400.00	366.01	168.65	298.55	1.023E – 039
500.00	364.71	165.75	281.83	3.590E – 030
600.00	362.89	162.45	265.42	7.792E – 024
700.00	360.40	158.63	249.36	2.462E – 019
800.00	357.14	154.29	233.71	5.486E – 016
900.00	384.39	186.23	216.78	2.613E – 013
1000.00	359.29	158.55	200.73	3.265E – 011
1100.00	358.20	157.52	184.93	1.650E – 009
1200.00	357.21	156.66	169.22	4.298E – 008
1300.00	356.31	155.93	153.60	6.729E – 007
1400.00	355.47	155.31	138.03	7.071E – 006
1500.00	354.68	154.77	122.53	5.494E – 005

TABLE 4.136
Thermodynamic Data of $Th(SO_4)_2$

T K	ΔHf° kJ mol⁻¹	S° J mol⁻¹ deg⁻¹	ΔGf° kJ mol⁻¹	Reference
298.15	–2542.46	159.00	–2590.01	Barin 1993
400.00	–2523.74	213.26	–2609.04	Barin 1993
500.00	–2502.88	259.70	–2632.73	Barin 1993
600.00	–2479.72	301.87	–2660.84	Barin 1993
700.00	–2454.25	341.08	–2693.01	Barin 1993
800.00	–2426.47	378.15	–2728.98	Barin 1993
900.00	–2396.37	413.56	–2768.58	Barin 1993
1000.00	–2363.97	447.68	–2811.65	Barin 1993
1100.00	–2329.26	480.75	–2858.08	Barin 1993
1200.00	–2292.24	512.94	–2907.77	Barin 1993
1300.00	–2252.91	544.41	–2960.65	Barin 1993
1400.00	–2211.27	575.26	–3016.63	Barin 1993
1500.00	–2167.33	605.57	–3075.68	Barin 1993

TABLE 4.137
Thermodynamic Data of ThO_2 (thorianite)

T K	ΔHf° kJ mol⁻¹	S° J mol⁻¹ deg⁻¹	ΔGf° kJ mol⁻¹	Reference
298.15	–1226.41	65.23	–1245.86	Knacke et al. 1991
400.00	–1219.77	84.35	–1253.51	Knacke et al. 1991
500.00	–1212.82	99.84	–1262.74	Knacke et al. 1991
600.00	–1205.62	112.97	–1273.40	Knacke et al. 1991
700.00	–1198.24	124.34	–1285.28	Knacke et al. 1991
800.00	–1190.72	134.38	–1298.23	Knacke et al. 1991
900.00	–1183.09	143.37	–1312.12	Knacke et al. 1991
1000.00	–1175.35	151.52	–1326.87	Knacke et al. 1991
1100.00	–1167.52	159.00	–1342.40	Knacke et al. 1991
1200.00	–1159.59	165.89	–1358.65	Knacke et al. 1991
1300.00	–1151.57	172.30	–1375.56	Knacke et al. 1991
1400.00	–1143.47	178.30	–1393.10	Knacke et al. 1991
1500.00	–1135.29	183.95	–1411.21	Knacke et al. 1991

TABLE 4.138
$Th(SO_4)_2 = ThO_2 + 2\ SO_3(g)$

T K	ΔH° kJ	ΔS° J deg^{-1}	ΔG° kJ	K
298.15	524.66	419.77	399.50	1.007E – 070
400.00	523.47	416.37	356.92	2.439E – 047
500.00	521.65	412.36	315.48	1.096E – 033
600.00	518.78	407.15	274.49	1.264E – 024
700.00	514.50	400.59	234.09	3.392E – 018
800.00	508.57	392.69	194.42	2.018E – 013
900.00	500.85	383.62	155.59	9.310E – 010
1000.00	491.23	373.51	117.73	7.081E – 007
1100.00	479.65	362.48	80.92	1.436E – 004
1200.00	466.04	350.65	45.26	1.071E – 002
1300.00	450.37	338.12	10.81	3.676E – 001
1400.00	432.62	324.97	–22.35	6.820E + 000
1500.00	412.74	311.27	–54.16	7.695E + 001

TABLE 4.139
$Th(SO_4)_2 = ThO_2 + 2\ SO_2(g) + O_2(g)$

T K	ΔH° kJ	ΔS° J deg^{-1}	ΔG° kJ	K
298.15	722.56	607.82	541.34	1.418E – 095
400.00	721.87	605.90	479.51	2.381E – 063
500.00	720.13	602.05	419.10	1.634E – 044
600.00	716.99	596.38	359.17	5.362E – 032
700.00	712.23	589.06	299.88	4.176E – 023
800.00	705.66	580.32	241.40	1.725E – 016
900.00	687.20	570.38	183.86	2.130E – 011
1000.00	688.78	559.42	227.36	2.223E – 007
1100.00	674.33	547.58	77.00	3.808E – 004
1200.00	659.84	534.98	17.87	1.668E – 0001
1300.00	643.27	521.73	–34.97	2.543E + 001
1400.00	624.61	507.91	–86.46	1.683E + 003
1500.00	603.83	493.58	–136.54	5.689E + 004

TABLE 4.140
Thermodynamic Data of $SnSO_4$

T K	ΔHf° kJ mol⁻¹	S° J mol⁻¹ deg⁻¹	ΔGf° kJ mol⁻¹	Reference
298.15	–1014.60	138.57	–1055.92	Barin 1993
400.00	–1003.36	170.97	–1071.75	Barin 1993
500.00	–991.61	197.15	–1090.19	Barin 1993
600.00	–978.86	220.38	–1111.08	Barin 1993
700.00	–964.97	241.76	–1134.20	Barin 1993
800.00	–949.90	261.87	–1159.39	Barin 1993
900.00	–933.59	281.06	–1186.54	Barin 1993
1000.00	–916.04	299.54	–1215.58	Barin 1993
1100.00	–897.23	317.46	–1246.43	Barin 1993
1200.00	–877.14	334.93	–1279.06	Barin 1993
1300.00	–855.78	352.02	–1313.41	Barin 1993
1400.00	–833.14	368.79	–1349.45	Barin 1993
1500.00	–809.22	385.29	–1387.16	Barin 1993

TABLE 4.141
Thermodynamic Data of $Sn(SO_4)_2$

T K	ΔHf° kJ mol⁻¹	S° J mol⁻¹ deg⁻¹	ΔGf° kJ mol⁻¹	Reference
298.15	–1648.50	149.79	–1693.16	Barin 1993
400.00	–1629.52	204.45	–1711.30	Barin 1993
500.00	–1609.63	248.79	–1734.02	Barin 1993
600.00	–1588.50	287.28	–1760.86	Barin 1993
700.00	–1566.12	321.75	–1791.34	Barin 1993
800.00	–1542.50	353.28	–1825.12	Barin 1993
900.00	–1517.63	382.55	–1861.92	Barin 1993
1000.00	–1491.52	410.05	–1901.57	Barin 1993
1100.00	–1464.16	436.11	–1943.89	Barin 1993
1200.00	–1435.56	460.99	–1988.75	Barin 1993
1300.00	–1405.72	484.87	–2036.05	Barin 1993
1400.00	–1374.63	507.90	–2085.70	Barin 1993
1500.00	–1342.30	530.20	–2137.61	Barin 1993

TABLE 4.142
Thermodynamic Data of SnO

T K	ΔHf° kJ mol^{-1}	S° J mol^{-1} deg^{-1}	ΔGf° kJ mol^{-1}	Reference
298.15	–285.77	56.48	–302.61	Barin 1993
400.00	–280.72	71.04	–309.13	Barin 1993
500.00	–275.50	82.67	–316.84	Barin 1993
600.00	–270.11	92.50	–325.61	Barin 1993
700.00	–264.58	101.02	–335.29	Barin 1993
800.00	–258.93	108.57	–345.78	Barin 1993
900.00	–253.16	115.35	–356.98	Barin 1993
1000.00	–247.30	121.53	–368.83	Barin 1993
1100.00	–241.33	127.22	–381.27	Barin 1993
1200.00	–235.26	132.50	–394.26	Barin 1993
1300.00	–229.10	137.43	–407.76	Barin 1993
1400.00	–222.84	142.07	–421.74	Barin 1993
1500.00	–216.48	146.46	–436.17	Barin 1993

TABLE 4.143
Thermodynamic Data of SnO_2

T K	ΔHf° kJ mol^{-1}	S° J mol^{-1} deg^{-1}	ΔGf° kJ mol^{-1}	Reference
298.15	–580.28	52.34	–595.89	Barin 1989
400.00	–574.38	69.29	–602.10	Barin 1989
500.00	–567.79	83.98	–609.78	Barin 1989
600.00	–560.68	96.92	–618.84	Barin 1989
700.00	–553.20	108.45	–629.12	Barin 1989
800.00	–545.42	118.83	–640.49	Barin 1989
900.00	–537.41	128.27	–652.85	Barin 1989
1000.00	–529.19	136.93	–666.12	Barin 1989
1100.00	–520.81	144.92	–680.21	Barin 1989
1200.00	–512.29	152.32	–695.08	Barin 1989
1300.00	–503.68	159.22	–710.66	Barin 1989
1400.00	–494.99	165.65	–726.91	Barin 1989
1500.00	–486.26	171.68	–743.78	Barin 1989

TABLE 4.144
$SnSO_4 = SnO + SO_3(g)$

T K	ΔH° kJ	ΔS° J deg^{-1}	ΔG° kJ	K
298.15	333.07	174.68	280.98	5.873E – 050
400.00	332.39	172.71	263.31	4.094E – 035
500.00	331.91	171.63	246.09	1.943E – 026
600.00	331.09	170.15	229.00	1.154E – 020
700.00	329.64	167.93	212.09	1.487E – 016
800.00	327.38	164.93	195.44	1.730E – 013
900.00	324.21	161.20	179.13	4.007E – 011
1000.00	320.05	156.83	163.22	2.975E – 009
1100.00	314.85	151.88	147.78	9.590E – 008
1200.00	308.57	146.42	132.86	1.645E – 006
1300.00	301.20	140.53	118.51	1.729E – 005
1400.00	292.71	134.24	104.77	1.232E – 004
1500.00	283.09	127.61	91.68	6.416E – 004

TABLE 4.145
$SnSO_4 = SnO + SO_2(g) + 1/2\ O_2(g)$

T K	ΔH° kJ	ΔS° J deg^{-1}	ΔG° kJ	K
298.15	432.02	268.70	351.90	2.204E – 062
400.00	431.60	267.47	324.61	4.045E – 043
500.00	431.13	266.47	297.91	7.503E – 032
600.00	430.19	264.76	271.34	2.376E – 024
700.00	428.50	262.17	244.99	5.218E – 019
800.00	425.93	258.24	218.93	5.058E – 015
900.00	422.39	254.58	193.26	6.061E – 012
1000.00	417.82	249.78	168.04	1.667E – 009
1100.00	412.19	244.42	143.32	1.562E – 007
1200.00	405.47	238.59	119.17	6.491E – 006
1300.00	397.65	232.33	95.62	1.438E – 004
1400.00	388.71	225.71	72.71	1.935E – 003
1500.00	378.63	218.76	50.49	1.745E – 002

TABLE 4.146
$SnSO_4 = SnO_2 + SO_2(g)$

T K	ΔH° kJ	ΔS° J deg^{-1}	ΔG° kJ	K
298.15	137.50	161.99	89.21	2.345E – 016
400.00	136.40	158.74	72.91	3.012E – 010
500.00	135.75	157.29	57.11	1.080E – 006
600.00	134.90	155.75	41.45	2.461E – 004
700.00	133.51	153.62	25.98	1.152E – 002

TABLE 4.147
$Sn(SO_4)_2 = SnO_2 + 2\ SO_2(g) + O_2(g)$

T K	ΔH° kJ	ΔS° J deg^{-1}	ΔG° kJ	K
298.15	474.59	604.14	294.46	2.554E – 052
400.00	473.05	599.65	233.19	3.519E – 031
500.00	471.91	597.10	173.36	7.728E – 019
600.00	470.70	594.91	113.76	1.247E – 010
700.00	469.13	592.50	54.38	8.742E – 005
800.00	466.99	589.65	–4.73	2.036E + 000

TABLE 4.148
Thermodynamic Data of $U(SO_4)_2$

T K	ΔHf° kJ mol^{-1}	S° J mol^{-1} deg^{-1}	ΔGf° kJ mol^{-1}	Reference
298.15	–2317.94	161.08	–2365.96	Barin 1989
400.00	–2299.07	215.35	–2385.21	Barin 1989
500.00	–2278.22	261.78	–2409.11	Barin 1989
600.00	–2255.06	303.95	–2437.43	Barin 1989
700.00	–2229.58	343.17	–2469.80	Barin 1989
800.00	–2201.80	380.23	–2505.99	Barin 1989
900.00	–2171.71	415.65	–2545.79	Barin 1989
1000.00	–2131.31	449.76	–2589.07	Barin 1989
1100.00	–2104.60	482.83	–2635.71	Barin 1989
1200.00	–2067.58	515.03	–2685.61	Barin 1989
1300.00	–2028.25	546.49	–2738.69	Barin 1989
1400.00	–1986.61	577.34	–2794.89	Barin 1989
1500.00	–1942.66	607.65	–2854.14	Barin 1989

TABLE 4.149
Thermodynamic Data of UO_2

T K	ΔHf° kJ mol^{-1}	S° J mol^{-1} deg^{-1}	ΔGf° kJ mol^{-1}	Reference
298.15	–1064.91	77.82	–1108.11	Barin et al. 1977
400.00	–1077.90	97.98	–1117.09	Barin et al. 1977
500.00	–1070.39	114.72	–1127.75	Barin et al. 1977
600.00	–1062.54	129.03	–1139.96	Barin et al. 1977
700.00	–1054.46	141.48	–1153.50	Barin et al. 1977
800.00	–1046.21	152.49	–1168.21	Barin et al. 1977
900.00	–1037.84	162.36	–1183.96	Barin et al. 1977
1000.00	–1029.34	171.30	–1200.65	Barin et al. 1977
1100.00	–1020.75	179.49	–1218.19	Barin et al. 1977
1200.00	–1012.06	187.05	–1236.53	Barin et al. 1977
1300.00	–1003.29	194.08	–1255.58	Barin et al. 1977
1400.00	–994.43	200.64	–1275.32	Barin et al. 1977
1500.00	–985.49	206.81	–1295.70	Barin et al. 1977

TABLE 4.150
$U(SO_4)_2 = UO_2 + 2\ SO_3(g)$

T K	ΔH° kJ	ΔS° J deg⁻¹	ΔG° kJ	K
298.15	441.50	430.29	313.21	1.327E – 055
400.00	440.67	425.15	269.51	6.354E – 036
500.00	439.42	425.15	226.85	1.994E – 024
600.00	437.20	421.13	184.52	8.606E – 017
700.00	433.62	415.64	142.67	2.254E – 011
800.00	428.42	408.72	101.44	2.377E – 007
900.00	421.44	400.53	60.97	2.892E – 004
1000.00	412.58	391.20	21.37	7.648E – 002
1100.00	401.75	380.90	–17.24	6.588E + 000
1200.00	388.90	369.73	–54.78	2.425E + 002

TABLE 4.151
$U(SO_4)_2 = UO_2 + 2\ SO_2(g) + O_2(g)$

T K	ΔH° kJ	ΔS° J deg⁻¹	ΔG° kJ	K
298.15	639.40	618.33	455.05	1.868E – 080
400.00	639.08	617.45	392.10	6.202E – 052
500.00	637.89	614.84	330.47	2.972E – 035
600.00	635.41	610.35	269.20	3.651E – 024
700.00	631.34	604.12	208.46	2.775E – 016
800.00	625.51	596.35	148.42	2.033E – 010
900.00	617.79	587.29	89.23	6.617E – 006
1000.00	608.12	577.12	31.00	2.401E – 002
1100.00	596.44	566.00	–26.16	1.747E + 001
1200.00	582.71	554.06	–82.17	3.776E + 003

TABLE 4.152
Thermodynamic Data of $ZnSO_4$ (zinkosite)

T K	ΔHf° kJ mol⁻¹	S° J mol⁻¹ deg⁻¹	ΔGf° kJ mol⁻¹	Reference
298.15	–982.80	110.50	–1015.75	Barin 1993
400.00	–972.32	140.69	–1028.59	Barin 1993
500.00	–961.25	165.35	–1043.93	Barin 1993
600.00	–949.43	186.88	–1061.56	Barin 1993
700.00	–936.85	206.27	–1081.23	Barin 1993
800.00	–923.50	224.08	–1102.76	Barin 1993
900.00	–909.39	240.69	–1126.01	Barin 1993
1000.00	–894.52	256.35	–1150.87	Barin 1993
1100.00	–859.51	290.38	–1178.92	Barin 1993
1200.00	–844.99	303.01	–1208.60	Barin 1993
1300.00	–830.47	314.63	–1239.49	Barin 1993
1400.00	–815.95	325.39	–1271.50	Barin 1993
1500.00	–801.43	335.41	–1304.55	Barin 1993

TABLE 4.153
$ZnSO_4 = ZnO + SO_3(g)$

T K	ΔH° kJ	ΔS° J deg⁻¹	ΔG° kJ	K
298.15	236.58	189.91	179.95	2.952E – 032
400.00	236.00	188.23	160.71	1.028E – 021
500.00	235.56	187.24	141.94	1.481E – 015
600.00	235.02	186.27	123.26	1.855E – 011
700.00	234.21	185.03	104.69	1.538E – 008
800.00	233.01	104.69	86.27	2.327E – 006
900.00	231.35	181.47	68.02	1.127E – 004
1000.00	229.16	179.18	49.98	2.449E – 003
1100.00	207.04	157.43	33.86	2.465E – 002
1200.00	205.59	156.17	18.19	1.616E – 001
1300.00	204.30	155.14	2.62	7.846E – 001
1400.00	203.16	154.29	–12.85	3.016E + 000
1500.00	202.16	153.60	–28.24	9.629E + 000

TABLE 4.154
$ZnSO_4 = ZnO + SO_2(g) + 1/2\ O_2(g)$

T K	ΔH° kJ	ΔS° J deg⁻¹	ΔG° kJ	K
298.15	335.53	283.94	250.87	1.108E – 044
400.00	335.20	282.99	222.01	1.015E – 029
500.00	334.79	282.08	193.75	5.720E – 021
600.00	334.13	280.88	165.60	3.821E – 015
700.00	333.08	279.27	137.59	5.398E – 011
800.00	331.56	277.25	109.76	6.806E – 008
900.00	329.52	274.86	82.15	1.705E – 005
1000.00	326.93	272.13	54.80	1.372E – 003
1100.00	304.38	249.98	29.40	4.014E – 002
1200.00	302.49	248.34	4.49	6.376E – 001
1300.00	300.76	246.95	–20.27	6.525E + 000
1400.00	299.16	245.76	–44.91	4.738E + 001
1500.00	297.70	244.75	–69.43	2.618E + 002

TABLE 4.155
Thermodynamic Data of $Zr(SO_4)_2$

T K	ΔHf° kJ mol⁻¹	S° J mol⁻¹ deg⁻¹	ΔGf° kJ mol⁻¹	Reference
298.15	–2243.46	133.05	–2283.13	DeKock 1984
400.00	–2224.34	187.87	–2299.49	DeKock 1984
500.00	–2202.47	236.60	–2320.77	DeKock 1984
600.00	–2179.57	278.34	–2346.57	DeKock 1984
700.00	–2156.50	313.90	–2376.23	DeKock 1984
800.00	–2133.71	344.34	–2409.18	DeKock 1984
900.00	–2111.43	370.59	–2444.96	DeKock 1984
1000.00	–2089.83	393.36	–2483.18	DeKock 1984

TABLE 4.156
Thermodynamic Data of ZrO_2 (Baddelite)

T K	ΔHf° kJ mol⁻¹	S° J mol⁻¹ deg⁻¹	ΔGf° kJ mol⁻¹	Reference
298.15	–1097.46	50.36	–1112.48	Barin 1993
400.00	–1091.31	68.07	–1118.53	Barin 1993
500.00	–1084.71	82.78	–1126.10	Barin 1993
600.00	–1077.80	95.37	–1135.02	Barin 1993
700.00	–1070.68	106.33	–1145.12	Barin 1993
800.00	–1063.41	116.05	–1156.24	Barin 1993
900.00	–1056.00	124.77	–1168.29	Barin 1993
1000.00	–1048.48	132.69	–1181.17	Barin 1993

TABLE 4.157
$Zr(SO_4)_2 = ZrO_2 + 2\ SO_3(g)$

T K	ΔH° kJ	ΔS° J deg⁻¹	ΔG° kJ	K
298.15	354.47	430.85	226.01	2.516E – 040
400.00	352.54	425.49	182.34	1.536E – 024
500.00	349.36	418.39	140.16	2.272E – 015
600.00	346.45	413.07	98.60	2.600E – 009
700.00	344.31	409.77	57.48	5.136E – 005
800.00	343.13	408.17	16.59	8.250E – 002
900.00	343.00	408.00	–24.20	2.540E + 001
1000.00	343.96	409.00	–65.05	2.500E + 003

TABLE 4.158
$Zr(SO_4)_2 = ZrO_2 + 2\ SO_2(g) + O_2(g)$

T K	ΔH° kJ	ΔS° J deg⁻¹	ΔG° kJ	K
298.15	552.37	618.90	367.85	3.542E – 065
400.00	550.94	615.01	304.94	1.499E – 040
500.00	547.83	608.08	243.79	3.387E – 026
600.00	544.66	602.30	183.28	1.103E – 016
700.00	542.04	598.24	123.27	6.323E – 010
800.00	540.22	595.80	63.58	7.054E – 005
900.00	539.35	594.76	4.06	5.811E – 001
1000.00	539.50	594.92	–55.41	7.848E + 002

REFERENCES

1. Louisnathan, S. J., Hills, R. J., and Gibbs, G. V., *Phys. Chem. Minerals,* 1, 53, 1977.
2. See 1.2 for description of HSC program.
3. Pechkovski, V. V. and Gaisinovich, M. S., *Russ. J. Inorg. Chem*., 9, 1245, 1964.
4. Knutsen, G. F. and Searcy, A. W., *J. Electrochem. Soc.,* 125, 327, 1978.
5. Ingraham, T. R., *Trans. Metall. Soc. AIME*, 242, 1299, 1968.
6. Chou, K. and Soong, C., *Thermochim. Acta,* 78, 285, 1984.
7. Nilson, L. F. and Petterson, O., *Berichte*, 13B, 1459, 1880.
8. *International Critical Tables*, McGraw Hill, New York, 1926 to 1930.
9. Rossini, F. D., Wagman, D. D., Evans, W. H., Levine, S., and Jaffe, I., *Selected Values of Chemical Thermodynamic Properties*, U.S. Government Printing Office, Washington, D.C., 1952.
10. Wagman, D. D., Evans, W., Halow, I., and Dudley, R., *American Institute of Physics Handbook*, McGraw Hill, New York, 1963, 170.
11. Johnson, D. W. and Gallagher, P. K., *J. Phys. Chem.,* 76, 1474, 1972.
12. Grund, A., *Tschermak's Mineralog. u. Petrog. Mitt*., 5(3), 227, 1955.
13. Margulis, E. V., Grishankina, N. S., and Kopylov, N. I., *Russ. J. Inorg. Chem.,* 10, 1253, 1965.
14. Beck, G., *Z. Anorg. Chem*., 174, 31, 1928.
15. Hoschek, G., *Monatsh. Chem*., 93, 826, 1912.
16. Panetier, G., Bregeault, J., and Guenot, J., *Mem. Pres. Soc. Chim. France, Bull*., 1850, 1963.
17. Marchal, G., *J. Chim. Phys*., 22, 495, 1925.
18. Spiess, M. and Gruehn, R., *Z. Anorg. Allgem. Chem*., 445, 16, 1979.
19. Rakshit, J. N., *Z. Elektrochem*., 32, 276, 1936.
20. Neumann, E., *J. Res. Nat. Bur. Stds*., 27, 191, 1941.
21. Mamantov, V. A., Menshikov, V. V., and Aziev, R. Z., *Khimia,* 39, 497, 1984.
22. Hanik, F., Galikova, L., Havlika, J., Kapralik, I., and Ambrus, V., *Br. Ceram. Soc. Trans.,* 84, 22, 1985.
23. Lau, K. H., Brittain, R. D., Lamoreau, R. H., and Hildenbrand, D. L., *J. Electrochem. Soc.,* 132, 3041, 1985.
24. Tutton, A. E. H., *J. Chem. Soc.,* 87, 1188, 1905.
25. Tutton, A. E. H., *J. Chem. Soc.,* 65, 654, 1894.
26. Jaeger, F. M., *Z. Anorg. Chem.,* 101, 178, 1917.
27. Moles, E. and Crespi, M., *Z. Phys. Chem.,* 130, 337, 1927.
28. Pechkovsky, V. V., *J. Appl. Chem. USSR,* 31, 1130, 1958.
29. Pechkovsky, V. V., Zvedin, A. G., and Beresneva, T. L., *Kinetika i Kataliz,* 4, 208, 1963.
30. Tagawa, H. and Saijo, H., *Thermochim. Acta,* 91, 67, 1985.
31. Coing-Boyat, J., *Compt. Rend.,* 248, 2109, 1959.
32. Pistorius, C.W.F.T., *Z. Phys. Chem.,* N.F. 27, 326, 1961.
33. Ketov, A. N., Pechkovsky, V. V., Varskoi, B. N., and Starkoe, N. P., *J. Appl. Chem. USSR,* 34, 498, 1961.
34. Terem, H. V., and Tugtepe, M., *Rev. Fac. Sci. Univ. Istanbul*, 22C, 271, 1957.
35. Ingraham, T. R. and Marier, P., *Trans. Metall. Soc. AIME,* 223, 363, 1964.
36. Neumann, B., Kroeger, C., and Iwanovski, R., *Z. Elektrochem*., 41, 821, 1935.
37. Reinders, W. and Goudrian, F., *Z. Anorg. Chem.,* 126, 85, 1923.
38. Pistorius, C. W. F. T., Z. *Krist*., 116, 220, 1961.
39. Beck, G., *Z. Praktt. Chem*., 156, 227, 1940.
40. Von Hevesy, G. and Cremer, E., *Z. Anorg. Allgem. Chem*., 195, 339, 1931.
41. Seward, R. P., *J. Am. Chem. Soc.,* 55, 2740, 1933.
42. Khvorostukhima, N. A., Rumyantsev, Ya. V., and Skobeev, I. K., *Trudy Vost - Sibirsk. Filiala Akad. Nauk SSSR Sibirsk, Otdel*., 41, 83, 1962.
43. Neumann, B. and Heintke, G., *Z. Elektrochem*., 43, 246, 1937.
44. Warner, N. A. and Ingraham, T. R., *Can. J. Chem.,* 38, 2196, 1960.
45. Woehler, L., Plueddemann, W., and Woehler, P., *Berichte*, 41, 703, 1908.
46. Neumann, B., and Heintke, G., *Z. Elektrochem.,* 43, 246, 1937.
47. Pechkovsky, V. V., *J. Appl. Chem. USSR*, 32, 2691, 1959.
48. Pozin, M.E. and Ginstling, A. M., *J. Appl. Chem. USSR*, 23, 1227, 1323, 1950; 24, 151, 1951.
49. Pechkovsky, V. V., Zvzdin, A. G., and Beresneva, T. L., *Kinetika i Kataliz,* 4, 208, 1963.

50. Warner, N. A. and Ingraham, T. R., *Canad. J. Chem. Eng.*, 40, 262, 1962.
51. Terem, H. N. and Akalan, S., *Rev. Fac. Sci. Univ. Istanbul,* 16A, 150, 1951.
52. Kelley, K. K., Contributions to the data on theoretical metallurgy VII. The thermodynamic properties of sulfur and its inorganic compounds. *U.S. Bur. Mines Bull.*, 406, 1937.
53. Hoschek, G., *Monatsh. Chem.*, 93, 826, 1926.
54. Miura, M., *J. Sci. Hiroshima Univ. Ser.*, 14, 70, 1949.
55. Foerland, T. and Krogh-Moe, I., *Acta Chem. Scand.*, 11, 565, 1956.
56. Ducarroir, M., Romero-Parredes, H., Steinmetz, D., Sibiende, F., and Tmar, M., *J. Hydrogen Energy,* 9, 579, 1984.
57. Pechkovsky, V. V., *J. Appl. Chem. USSR*, 29, 1229, 1956.
57a. Thorpe, T. E. and Watts, J. I., *J. Chem. Soc.,* 37, 102, 1880.
58. Cueilleron, J. and Hartmannshenn, J. I., *Bull. Soc. Chim. France*, 168, 1959.
59. Ostroff, A. G. and Sanderson, R. T., *J. Inorg. Nucl. Chem.*, 9, 45, 1959.
60. Pechkovsky, V. V., *J. Appl. Chem.*, 28, 217, 1955.
61. Beck, G., *Z. Anorg. Chem.,* 174, 31, 1928.
62. Pascal, P., Ed., *Nouveau Traite de Chimique Mineral*, 2nd ed., Masson et Cie., Paris, 1956.
63. Aurivillius, K. and Malmros, B., *Acta Chem. Scand.,* 15, 1932, 1961.
64. Caillere, S. and Pobeguin, T., *Bull. Soc. Chim. France,* 85, 48, 1962.
65. Cueilleron, J., and Hartsmannshenn, O., *Bull. Soc. Chim. France,* 168, 1959.
66. Malard, C., *Bull. Soc. Chim. France*, 2296, 1961.
67. Lau, K. H., Cubicciotti, D., and Hildenbrand, D. L., *J. Electrochem. Soc.,* 126, 490, 1979.
68. Heydweiller, A., *Z. Phys.*, 3, 308, 1920.
69. Andrae, J. L., *Z. Phys. Chem.,* 82, 112, 1913.
70. Robinson, M. T., *J. Phys. Chem.,* 62, 925, 1958.
71. Kelley, K. K., Contributions to the data on theoretical metallurgy XIII. High temperature heat content, heat capacity, and entropy data for the elements and inorganic compounds. *U.S. Bur. Mines Bull.*, 584, 1960.
72. Woehler, L. and Gruenzweig, M., *Berichte,* 46, 1587, 1913.
73. ibid., p. 1726.
74. Nathans, M.W. and Wendlandt, W. W., *J. Inorg. Nucl. Chem.,* 24, 869, 1962.
75. Ivakin, A. A., *Izv. Sibirsk. Otdel. Akad. Nauk SSSR*, 49, 1962.
76. Woehler, L. and Gruenzweig, M., *Berichte*, 46, 1726, 1913.
77. Duval, C., *Inorganic Thermogravimetric Analysis*, 2nd ed., Elsevier, Amsterdam, 1963.
78. Pokrovskii, A. N. and Kovba, L. M., Russ. *J. Inorg. Chem.*, 21, 305, 1976.
79. Sklyarenko, Yu. S., Stroganova, N. S., and Ivanov, V. I., *Russ. J. Inorg. Chem.*, 10, 1203, 1965.
80. *Gmelin Handbook of Inorganic Chemistry, Lanthanum volume.*
81. Beck, G., *Z. Prakt. Chem.*, 156, 227, 1940.
82. Komissarova, L. N., Shatskii, V. M., and Moiseichenko, G., *Russ. J. Inorg. Chem.*, 10, 407, 1965.
83. Wulff, P. and Schaller, D., *Z. Krist.*, 87, 43, 1934.
84. Kracek, F. C. and Gibson, R. E., *J. Phys. Chem.,* 33, 1304, 1929.
85. Simanov, Yu. P. and Kirkina, D. F., *J. Inorg. Chem. USSR,* 2, 364, 1957.
86. Thorpe, E., and Francis, A. G., *Z. Anorg. Chem.,* 66, 411, 1910.
87. Stern K. H., *High Temperature Properties of Inorganic Salts, Part 1, Sulfates. NSRDS - NBS 7*, 1966; section on chromium..
88. *Gmelin Handbook of Inorganic Chemistry. Tantalum volume.*
89. ibid. *Thallium volume.*
90. Peltier, S. and Duval, C., *Anal. Chim. Acta*, 2, 210, 1948.
91. Hegedues, A.J. and Fukker, K., *Z. Anorg. Allgem. Chem.*, 284, 20, 1956.
92. Favorskaya, L. V. and Ponomaryeva, E. I., *Akad. Nauk Kaz. SSR, Ser. Met. Obogashch. i Ogneup.*, 2, 30, 1959.
93. Gattow, *Naturwissenschaften*, 47, 442, 1960.
94. Cohen, E. and Kooy, J., *Z. Phys. Chem.,* 139A, 320. 1928.
95. Donaldson, J.D. and Moser, W., *J. Chem. Soc.,* 4000, 1960.
96. Hayek, E. and Engelbrecht, A., *Monatshefte*, 80, 640, 1949.
97. *Gmelin Handbook of Inorganic Chemistry, Uranium volume.*

98. Rivenq, F., *Bull. Soc. Chim. France,* 4, 1697, 1937.
99. Udupa, M. R., *Thermochim. Acta,* 51, 169, 1981.
100. Neumann, B. and Sonntag, A., *Z. Elektrochem.,* 39, 799, 1933.
101. Hayek, E. and Engelbrecht, A., *Monatshefte*, 80, 640, 1949.
102. Brittain, R. D., Lau, K. H., Knittel, D. F., and Hildenbrand, D. L., *J. Phys. Chem.,* 90, 2259, 1986.
103. Weigl, F. and Trinkl, A., *Radiochim. Acta,* 9, 140, 1968.
104. Chuklantsev, V. G., *J. Inorg. Chem. USSR,* 2, 2014, 1957.
105. Woehler, L., Plueddemann, W., and Woehler, P., *Berichte*, 46, 1726, 1908.

THE FOLLOWING REFERENCES APPEARS IN SOME TABLES:

Phillips 88: Phillips, S. L., Haley, F. V., and Silvester, L. F., *Thermodynamic Tables for Nuclear Waste Isolation*, Sandia National Laboratory, Albuquerque, NM, 1988.

DeKock, C.W., *Thermodynamic Properties of Selected Metal Sulfates and Their Hydrates*, Bur. Mines Inf. Circular, 9081, 1984.

5 Nitrites and Nitrates

5.1 GASEOUS EQUILIBRIA IN THE N–O SYSTEM

When nitrites and nitrates are thermally decomposed, nitrogen, oxygen, and their compounds are produced. A literature survey produces the following list of substances which appear in one or another decomposition: O_2, N_2, NO, NO_2, N_2O_3, N_2O_4, N_2O_5. Which of these compounds are produced in any specific case depends on the salt, temperature, and experimental conditions. Since reactions between these compounds occur quite readily, thermodynamic functions for them have been listed (Tables 5.1 to 5.8). The matter of reactions between all these compounds is somewhat complicated, because several of them are thermodynamically quite unfavorable and thus do not need to be considered. On that basis the following conclusions can be drawn:

N_2O_5 is unstable with respect to dissociation to either N_2O_4 or N_2O_3 from 298 K up.

The equilibrium $N_2O_4 = 2\ NO_2$ shifts markedly to the right with increasing temperature. Above 400 K the concentration of N_2O_4 is negligible.

Although N_2O_3 has been reported as a product of nitrate decomposition, it is unstable with respect to dissociation into NO_2 and NO above 400 K; it could therefore exist in equilibrium mixtures only near ambient temperatures.

The equilibrium $NO + 1/2\ O_2 = NO_2$ is very temperature dependent; up to 500 K, NO_2 predominates markedly; above that temperature both species are significant and must always be considered in the salt decomposition.

N_2 and O_2 are very stable with respect to both NO and NO_2; thus if N_2 is formed during salt decomposition, it will not react further; O_2 may react with NO and with nitrites.

If all equilibria were established quickly during decomposition of the salts, N_2 and O_2 would be the only gaseous components; however, the reaction $NO_2 = 1/2\ N_2 + O_2$ is very slow below 1000 K; thus, if NO_2 is the primary decomposition product, it will not decompose appreciably by this reaction and only the equilibrium $NO + 1/2\ O_2 = NO_2$ needs to be considered.

For similar reasons, the decomposition of NO to the elements can be neglected. The same conclusions were reached by Kelley[1] in examining the decomposition of a few nitrates. (The reader may investigate the thermodynamics of all the conceivable reactions in the N–O system by using the data in Tables 5.1 to 5.8).

5.2 THE NITRITES

Relatively little is known about anhydrous nitrites except the alkali metal and alkaline earth salts. (See Figure 5.1.) The nitrite ion is less stable than the nitrate, so the nitrites of magnesium and beryllium, small and highly charged cations, are quite unstable.

The nitrite ion is bent, with bond angles of 116° and 132°, and N–O distances of 1.13 and 1.23Å having been reported.[2,3] Cotton and Wilkinson[4] describe the bent structure in terms of resonance or, in MO theory, by assuming that N forms three sp^2 hybrid orbitals, one housing an unshared pair, and the other two being used in σ bonding to the oxygens. The p_z orbitals of N and O are used for π bond formation.

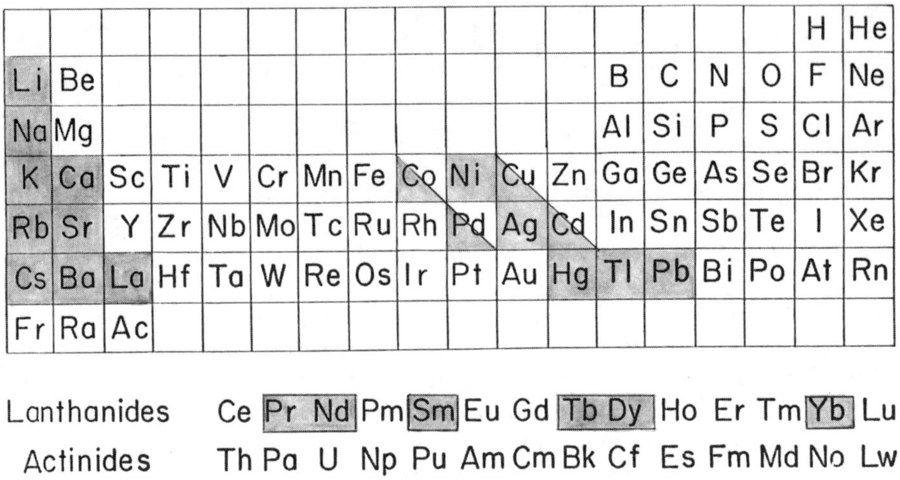

FIGURE 5.1 The known nitrites.

The coordination of the nitrite ion appears to differ with the salt. For example, in $AgNO_3$, Ag^+ is closer to N than to O, whereas in $NaNO_2$, Na^+ is closer to O than to N.[2] The periodic table existence chart (Figure 5.1) which lists the known nitrites omits bonding information.

Nitrites decompose at lower temperatures than the corresponding nitrates. The course of decomposition depends on the nature of the salt and the gas phase composition. Nitrites of the more electropositive elements decompose to form the oxide directly by reactions of the type $2\ MNO_2 = M_2O + NO + NO_2$, but if the gases are not immediately removed, the nitrite is easily oxidized to the nitrate by NO_2, if the latter salt is stable at the decomposition temperature. In some cases, particularly the rare-earth nitrites, the first decomposition product is an oxynitrite. According to Addison and Sutton,[5] simple nitrites will be stable only if a higher valent-state of the metal is not readily available. Otherwise, decomposition to oxynitrites will occur to achieve the more stable valence of the metal. This view is supported somewhat by the nonexistence of many nitrites of metals in the first two long periods of the periodic table.

5.3 THE NITRATES

5.3.1 General

Although anhydrous nitrates of the alkali and alkaline earth metals have been known and widely used for a long time, the anhydrous nitrates of many other metals have been prepared only relatively recently. (See Figure 5.2.) The existence of some is still in doubt. Methods for preparing these nitrates have been described by Addison and Logan.[6] These authors, as well as Hardy and Field,[8] have also written extensive reviews of nitrate chemistry. Since not much has changed since they were written, only a brief summary of factors relevant to thermal decomposition is provided here.

In the hydrated salts and in aqueous solution the nitrate ion is a distinct entity. This is also true for the anhydrous nitrates of the electropositive metals, but in many other salts the nitrate ion is bonded covalently to the metal. The distinction between ionic and covalent nitrates can most easily be made on the basis of their infrared spectra. Chasan and Norwitz[9] prepared a listing of ionic and covalent salts based on an examination of the literature, and covalent nitrate complexes were reviewed by Addison and Sutton.[5] Spectra of liquid nitrates were discussed by Wait and Janz.[10] Figure 5.2 represents the author's evaluation of the literature with respect to the existence and bonding of anhydrous nitrates, but no attempt was made to classify the covalent nitrates with respect to the number of oxygens bonded to the metal atom. This topic, which is of great structural interest,

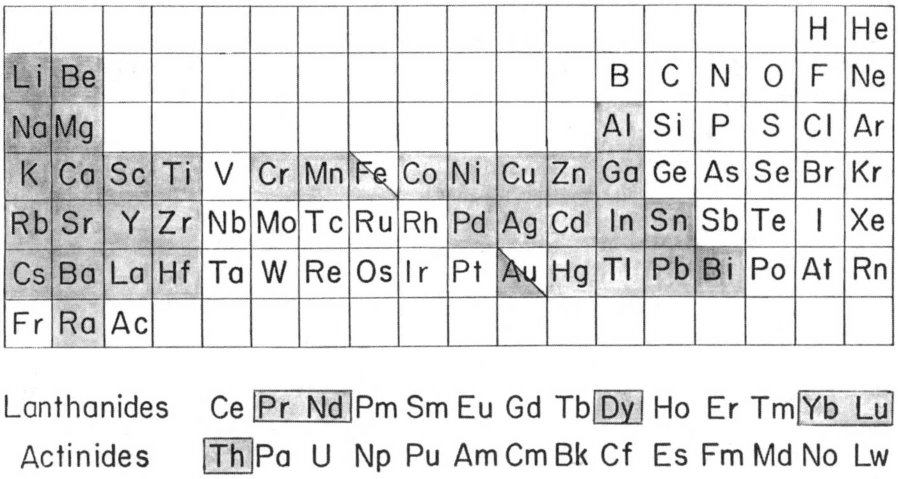

FIGURE 5.2 The known nitrates.

is discussed in the review by Addison and Logan.[7] In some cases the type of bonding has not been clearly established. In others, which are discussed in this chapter, evidence for the existence of the compound is shaky.

The alkali and alkaline earth nitrates, except $Be(NO_3)_2$, are ionic, but otherwise there are no obvious criteria. For example, $AgNO_3$ and $Cd(NO_3)_2$ are ionic, but $Cu(NO_3)_2$ and $Zn(NO_3)_2$ are covalent. $Co(NO_3)_2$ exhibits characteristics of both ionic and covalent bonding.

The ionic nitrates generally melt to liquids which are stable to various degrees above their melting points. These liquids can be distilled under reduced pressure.[11] The covalent nitrates are generally not stable as liquids. When heated, they first sublime, frequently giving molecular vapors, and then decompose. The cation influences the stability of the anion through its ability to distort its structure in the same manner as for carbonates and sulfates, as discussed in Chapters 2 and 4.

Since the nitrite of a given metal is generally much less stable than the nitrate, the former can appear only as an unstable intermediate in the decomposition of the nitrate. This is particularly true for covalent nitrates. In the case of ionic nitrates, however, both salts may be more or less equally unstable over some temperature range, so that the decomposition reactions can become quite complex. This is particularly so since the salts may be oxidized or reduced by the gaseous decomposition products. For example, NO_2 produced by the decomposition of the nitrate may oxidize the nitrite ion also formed back to the nitrate. Since the experimental arrangement usually determines the gas-phase composition, reports by different authors frequently conflict. In such cases the common features have been emphasized in this chapter. A consequence of the complexity of most nitrate decompositions is that kinetic studies have usually been restricted to identifying the reactions. Even when rate constants and activation energies are reported it is frequently not clear with which particular reactions they are identified.

5.3.2 Decomposition Equilibria

In contrast to the decomposition of carbonates, which yield only a single gaseous product, and of sulfates, for which useful generalizations about gas composition can be made (Chapter 4), the situation for nitrates is more complex. Partial pressures of the gaseous constituents must be calculated by a rather complicated method (see below) and, even so, some compromises are required. For example, although it is known that over some temperature ranges alkali metal and alkaline earth nitrates decompose to nitrites as a first step, no high- temperature thermodynamic data for the latter salts are available. Therefore the following procedure has been adopted; when good quality experimental decomposition studies are available, they are discussed. When thermodynamic data are available, the decomposition is treated as if the oxide is the final product, analogous to the treatment of carbonates (Chapter 2) and sulfates (Chapter 4); e.g., for a divalent salt

$$M(NO_3)_2 = MO + 2\ NO_2 + 1/2\ O_2 \tag{5.1}$$

and the gas equilibrium

$$2\ NO_2 = 2\ NO + O_2 \tag{5.2}$$

The reasons for neglecting other gaseous species have already been discussed in Chapter 5.1. The approach is the one used previously by Kelley,[1] and in all cases where nitrates are not stable represents the best approximation to the physical situation. The equilibrium constants for Equations 5.1 and 5.2 are

$$K_1 = \frac{P_{NO_2}^2 P_{O_2}^{1/2} a_{MO}}{a_{M(NO_3)_2}} \tag{5.3}$$

$$K_2 = \frac{P_{NO_2}^2 P_{O_2}}{P_{NO_2}^2} \tag{5.4}$$

The partial pressures of the three gases can be calculated if the oxide and nitrate do not form solid solutions (unit activity of each) and if the only source of the gases is the salt. In addition to Equations 5.3 and 5.4, a third relation is derived from the fact that N and O must be liberated in a fixed atomic ratio. Thus

$$P_{O_2} = \frac{1}{4}(P_{NO_2} + P_{NO}) + \frac{1}{2}P_{NO} \tag{5.5}$$

or

$$4P_{O_2} = P_{NO_2} + 3P_{NO}$$

If the condensed phases are pure, a condition which can hold only if both the salt and the oxide are solids, the term $a(MO)/a(MNO_3)_2 = 1$ and appropriate substitution yields an equation in $P(O_2)$ alone:

$$\frac{K_1^{1/2}}{P_{O_2}^{1/4}} + \frac{3(K_1K_2)^{1/2}}{P_{O_2}^{3/4}} = 4P_{O_2} \tag{5.6}$$

Making the substitution $X = P^{1/4}(O_2)$, this equation can be written in the form

$$X^7 - a\,X^2 - b = 0 \tag{5.7}$$

where $a = K_1^{1/2}$, $b = 3/4\ (K_1.K_2)^{1/2}$ .This equation has only one real positive root for positive values of a and b, but cannot be solved in closed form. Newton's method was used to solve it by iteration. After the solution for $P(O_2)$ had been obtained, $P(NO_2)$ and $P(NO)$ were calculated from the equations

$$P_{NO_2} = K_1^{1/2} / P_{O_2}^{1/4}$$

$$P_{NO} = (K_1 K_2)^{1/2} / P_{O_2}^{3/4}$$

The pressures thus obtained are those to be expected if the pure nitrate is initially placed into a closed container containing no other reactants, and the system is allowed to reach equilibrium.

Although the pressures in the real system may differ from those calculated by the above procedure, particularly if nitrites are formed, the values obtained are a useful guide to the stability of the salts.

All the partial pressures calculated from the equilibrium constants assume unit activity for the condensed-phase components. This assumption is good when they are solid. Above the melting points of the salts, however, continued decomposition of the salt will result in a solution containing dissolved oxide and the partial pressures will depend on the melt composition, and will therefore change as the decomposition proceeds. Because of the form of K_1, the partial pressure calculation will be worst for small oxide concentrations. An examination of the various tables shows that O_2 and NO are the major products of nitrate decomposition, the concentration of NO_2 being rather minor. This results from the fact that the equilibrium $2\ NO_2 = 2\ NO + O_2$ lies to the right for low pressures.

5.4 INDIVIDUAL NITRITES AND NITRATES

5.4.1 Aluminum

$Al(NO_3)_3$ has been prepared in nearly pure form by vacuum sublimation of its adduct with N_2O_5. It condensed as a white powder at –78°.[12] None of its properties have been measured. Its classification as a covalent was made on the basis of the spectrum of the N_2O_5 adduct.

5.4.2 Barium

5.4.2.1 $Ba(NO_2)_2$

The melting point of the salt is very much in doubt. Centnerszwer and Piekelny[13] give 262°C. Protsenko and Bordyushkova[14] give 284°C. Neither paper describes any experimental detail. Similarly, solid transitions were reported[15] at 203 and 230°C, but no details are available.

Decomposition of the salt has been studied by three groups of researchers[13,16,17] at somewhat different temperatures and under various atmospheres. Although they differ in detail, they generally agree on the main features.

The primary decomposition reaction, which is reversible, is

$$Ba(NO_2)_2 = BaO + NO_2 + NO \qquad (5.1)$$

The reaction begins as low as 250°C. In the temperature range in which $Ba(NO_3)_2$ is stable, the NO_2 formed oxidizes the nitrite[18]

$$Ba(NO_2)_2 + 2\ NO_2 = Ba(NO_3)_2 + 2\ NO \qquad (5.2)$$

Above 400°C , BaO reacts with NO to form N_2.

$$BaO + 3\ NO = Ba(NO_2)_2 + 1/2\ N_2 \tag{5.3}$$

"Overall" reactions given by the various authors are obtained from various combinations of Reactions 5.1 to 5.3. The speed of the overall reaction and the distribution of products are evidently affected by the composition of the gas phase and the manner of its removal. For example, Protsenko and Bordyushkova[17] observed that in the range 410 to 440°C the rate varied in the order air > vacuum > argon. The rate in air may be the greatest since O_2 can oxidize NO which may increase the rate of 5.2. In argon the Arrhenius activation energy for a first-order reaction is 65.3 kJ.

The effect of gas composition and pressure on the temperature range over which the decomposition reaction occurs can be seen by comparing the above results with a TGA study under vacuum.[19] The first stage of the decomposition, corresponding to

$$2\ Ba(NO_2)_2 = Ba(NO_3)_2 + BaO + NO + 1/2\ N_2$$

occurs between 90 and 150°C. The second stage, beginning at 450°C and complete at 600°C, is the decomposition of the nitrate

$$Ba(NO_3)_2 = BaO + 2\ NO_2 + 1/2\ O_2$$

The melting point has been reported as 546 ± 11 K.[13,17] The density as a function of temperature is given in Table 5.9 and thermodynamic data in Table 5.10.

5.4.2.2 $Ba(NO_3)_2$

Decomposition has been observed as low as 525°C, but it is still slow at 550°.[20] In fact, even the liquid salt decomposes only slowly just above its melting point.[21,22] The first stage of the reaction is the formation of the nitrite

$$Ba(NO_3)_2 = Ba(NO_2)_2 + O_2$$

but this salt is very unstable so that the overall decomposition is

$$Ba(NO_3)_2 = BaO + 2\ NO_2 + 1/2\ O_2$$

As can be observed from the calculated dissociation pressures in Table 5.12, the dissociation might be expected to be inhibited in air. This may account for the stability observed by Addison and Coldrey.[21] The salt appears to have no solid-state transitions. Ambient temperature density is 3.23.[57]

Thermodynamic data are given in Tables 5.11 and 5.12.

5.4.3 Beryllium

$Be(NO_3)_2$ is a white, hygroscopic solid. The bonding between the metal and the nitrate group is covalent. On being heated above 125°C it rapidly loses NO_2 with the formation of the oxide nitrate $Be_4O(NO_3)_6$.[24] Obviously, for that reason, there are no high-temperature thermodynamic data. Data at 298.15 K are shown in Table 5.13.

5.4.4 Bismuth

The preparation of $Bi(NO_3)_2$ has been reported. Bonding appears to be covalent.[27]

5.4.5 Cadmium

5.4.5.1 $Cd(NO_2)_2$

The pure anhydrous salt has apparently never been prepared, although it has been reported in reaction mixtures.[28]

5.4.5.2 $Cd(NO_3)_2$

The infrared spectrum[29] indicates the solid to be typically ionic. It decomposes rapidly above 573 K,[30] consistent with the thermodynamic data. The reported melting point is uncertain. The transition is given in Table 5.14, and thermodynamic data in Tables 5.15 to 5.17. Decomposition is reported to be reversible and to yield no intermediates:[31]

$$Cd(NO_3)_2 = CdO + NO_2 + 1/2\ O_2$$

5.4.6 Calcium

5.4.6.1 $Ca(NO_2)_2$

The solid-state transitions were obtained from heating curves (Table 5.18). The nature of the phases was not characterized.[32]

The decomposition reactions of $Ca(NO_2)_2$ are similar to those for the Ba and Sr salts. Centnerszwer and Piekelny[13] measured the decomposition pressure over the temperature range 540 to 588 K. In an inert atmosphere log P is linear in 1/T; p varied from 0 at 540 K to 693 torr at 588 K. This suggests that a single reaction is responsible for the measured pressure, but it is not at all clear what this reaction is, since the enthalpy change calculated by the authors from the pressure data (147.7 kJ) differs considerably from $\Delta H°$ at 298 K for the reaction cited by the authors.

$$Ca(NO_2)_2 + 2\ NO_2 = CaO + NO_2 + NO$$

for which ΔH = 234.4 kJ. The lack of high-temperature data for $Ca(NO_2)_2$ prevents a comparison of $\Delta H°$ at the experiment temperature. At somewhat higher temperatures only NO appears as the product, probably because NO_2 oxidizes the nitrite

$$Ca(NO_2)_2 + 2\ NO_2 = Ca(NO_3)_2 + 2\ NO$$

The decomposition was also studied by Protsenko and Bordyushkova[14] at somewhat higher temperatures, 693 and 723 K, which are still sufficiently low for the nitrate to be stable, or at least for its decomposition to be negligible compared to that of the nitrite. They noted that the rate of decomposition varied with the gas phase in the order: argon < air < vacuum. From an analysis of the gas phase and the weight loss of the sample, they conclude that in argon the main reaction is the formation of the nitrate with partial reduction of the NO to N_2:

$$4\ Ca(NO_2)_2 = 2\ Ca(NO_3)_2 + 2\ CaO + 2\ NO + N_2$$

which is fairly consistent with the above reactions. These have also been found by Oza and Oza,[33] who find that they are reversible. Thus, since the nitrate is unstable in the presence of NO, the production of N_2 is inhibited and the above reaction will be minor unless the product gases are continually removed from the system.

Interpretation of the pressure measurements of Centnerszwer and Pieckielny would therefore need to take into account all the above reactions. In reaction schemes as complicated as the present one, equilibrium pressure measurements alone are likely to be of limited utility, particularly since the partial pressures of the component gases depend not only on the equilibrium constants of the reactions, but also on the composition of the condensed phases.

5.4.6.2 $Ca(NO_3)_2$

The salt exhibits no crystalline transitions.[34] Thermodynamic data are shown in Table 5.19. Its decomposition has not been studied in detail. Some indication of its stability may be gained from the thermodynamic data for the decomposition reaction (Table 5.20 and 5.21). Although the calculated dissociation pressures are substantial above 700 K, Addison and Coldrey[35] found the liquid to be clear at 828 K and small bubbles to form at 833 K, but both of these temperatures lie below the generally accepted melting point of 834 K. The behavior of the salt near the melting point evidently needs more careful study.

A kinetic study[36] carried out in the range 858 to 1000 K indicates the evolution of NO_2, with the initial formation of a supersaturated solution of CaO in $Ca(NO_3)_2$.

5.4.7 Cesium

5.4.7.1 $CsNO_2$

The salt is stable to 450°C and decomposes to the oxide above that temperature if the product gases, NO and NO_2, are removed.[37] In their presence the oxide is partially oxidized to the nitrate.

Below 700 K, $CsNO_2$ evaporates without decomposition and the vapor contains both monomers and dimers. ΔH of sublimation is 161.5 kJ, and 208.8 kJ for the dimer, recalculated to 298 K. The enthalpy of dimerization is 477 kJ.[38]

5.4.7.2 $CsNO_3$

The salt exhibits one solid–solid transition. The hysteresis, which is frequently exhibited by this transformation leads to some uncertainty in the enthalpy change. The problem is discussed in detail by Rao and Rao.[39]

A semi-quantitative study[40] indicates that the salt decomposes only slowly up to 828 K, 0.05% nitrite appearing in the melt after heating for 10 min. A mass spectrometric study[41] indicates that, although some decomposition occurs as low as 775 K, prolonged heating at 835 K does not permit detection of decomposition in the powder pattern of the residue. These results were extended by Bagaratian et al.[38] who also found decomposition to nitrites to be slight and who give sublimation enthalpies, calculated to 298 K, as 195 kJ for the monomer and 234 kJ for the dimer. The enthalpy of dimerization is –158 kJ.

Densities of $CsNO_2$ and $CsNO_3$ are given in Tables 5.22 and 5.23, transitions in Table 5.24, and thermodynamic data in Tables 5.25 to 5.29.

5.4.8 Chromium

Anhydrous $Cr(NO_3)_2$ has been prepared. The compound is covalent. When the pale green powder is heated in a vacuum or an atmosphere of nitrogen, it begins to decompose rapidly at 60°C, and the decomposition rate is a maximum at 100°C. Decomposition may proceed through a series of oxide nitrates, but none of these is stable enough to give a plateau on TGA. The first plateau is attained at 200°C and corresponds to Cr_3O_8.[42]

5.4.9 Cobalt

5.4.9.1 $Co(NO_2)_2$

The existence of this salt has been reported,[43] but it must still be regarded as doubtful. It supposedly begins to evolve NO near 100°C.

5.4.9.2 $Co(NO_3)_2$

The salt decomposes to Co_3O_4 in a single step without formation of an intermediate.[44] The decomposition begins at 100 to 105°C,[44,45] but there is some question as to the temperature at which decomposition is maximum. Both Addison and Sutton[44] and Addison and Logan[46] state that the rate is maximum at 270°C, but the experimental curve published by Addison and Sutton[44] shows this maximum approximately 100°C lower.

Thermodynamic data are given in Tables 5.30 and 5.31.

5.4.10 Copper

5.4.10.1 $Cu(NO_2)_2$

Reports about the existence of this salt are conflicting.[47] Its preparation is doubtful.

5.4.10.2 $Cu(NO_3)_2$

Below 500 K heating of the solid salt results primarily in the vaporization of the covalent molecule as a monomer. In this temperature range the only thermal decomposition, into NO_2 and O_2, is exhibited by the solid. The vapor is more stable. The vapor pressure of $Cu(NO_3)_2$ was determined by Addison and Hathaway[48] by extrapolating pressure-time curves to zero time in order to subtract the pressures of NO_2 and O_2. These vapor pressures increased from 0.32 torr at 430 K to 3.59 torr at 405 K. A plot of log P vs. 1/T is linear and yields a sublimation enthalpy of 67.0 kJ. Above 500 K both the solid and the vapor phase decompose to $NO_2 + O_2$.

No high-temperature thermodynamic data are available. Transitions are listed in Table 5.32 and thermodynamic data at 298 K in Table 5.33.

5.4.11 Gallium

$Ga(NO_3)_2$ was probably prepared by Dupre,[50] using vacuum dehydration of the hydrate at 40°C. A large fraction of the salt had decomposed to Ga_2O_3 when the temperature reached 200°C.

Thermodynamic data at 298 K are given in Table 5.34.

5.4.12 Gold

Evidence for the existence of $Au(NO_3)_3$ seems to be based on a report by Field and Hardy[52] who treated hydrated gold nitrates successively with HNO_3 and N_2O_5. Although the gold content corresponded to the trinitrate, no quantitative nitrate analysis was performed. Further information is lacking.

5.4.13 Hafnium

Although Addison and Logan[46] indicated the existence of an anhydrous nitrate in their periodic table, there is some question as to whether such a compound has actually been prepared. Field and Hardy[52] succeeded in preparing the addition compound $Hf(NO_3)_4.N_2O_5$, but this compound sublimed as such, rather than losing N_2O_5. More recently, Tuseev, Izmailovich, and Komissarova[53] succeeded in preparing $Hf(NO_3)_4$ and studying its decomposition. Melting and decomposition occurred simultaneously at 383 to 403 K, followed by a two-stage decomposition leading first to the formation of an oxynitrate in the successive temperature ranges 443 to 448 and 483 to 488 K.

5.4.14 Indium

The infrared spectrum shows that $In(NO_3)_3$ is covalent and stable to at least 90°C, since this was the temperature of its preparation.[54] The thermodynamic data available at 298 K are listed in Table 5.35.

5.4.15 Iron

$Fe(NO_3)_3$ has only been prepared as the volatile adduct $Fe(NO_3)_3.N_2O_5$, although it is possible that $Fe(NO_3)_3$ exists in the vapor state.[55]

5.4.16 Lead

5.4.16.1 $Pb(NO_2)_2$

Very little is known about this salt. It has been prepared by treating the chloride with $AgNO_2$ and measuring decomposition by TGA.[56] It occurs in two stages, the first one beginning at about 110%C being consistent with the stoichiometry

$$3\ Pb(NO_2)_2 = Pb(NO_3)_2 + 2\ PbO + 4\ NO$$

The second stage is the decomposition of the nitrate. However, this occurs in the range 320 to 450°C and seems to proceed directly to PbO, a result which is inconsistent with other studies of nitrate decomposition (see below).

5.4.16.2 $Pb(NO_3)_2$

The thermal decomposition of this salt has been studied by a number of researchers.[57–59] There is general agreement that decomposition occurs without oxidation of a Pb(II) ion to a higher valent state, but it is not clear whether a nitrite may be an intermediate in the decomposition. In the course of the decomposition several oxynitrates are formed, but there is no general agreement as to the composition of these salts. The most careful study seems to have been carried out by Neumann and Sonntag[58,61] who measured the total pressure involved in three reversible equilibria and analyzed the solid phases. The equilibria (renotated from the original papers) and the temperature ranges over which they were measured are

$$\text{I. 539–632 K: } 3Pb(NO_3)_2 = [Pb(NO_3)_2.2\ PbO] + 4\ NO_2 + O_2$$

$$\text{II. 625–719 K: } 2\ [Pb(NO_3)_2.2PbO] = Pb(NO_3)_2.5\ PbO] + 2\ NO_2 + 1/2\ O_2$$

$$\text{III. 688–809 K: } Pb(NO_3)_2.5\ PbO = 6\ PbO + 2\ NO_2 + 1/2\ O_2$$

The overlapping of the temperature ranges is accounted for by the fact that the continuous removal of product gas will result in the disappearance of a lower oxide and the formation of a higher one, with the resulting establishment of a new equilibrium.

Nothing seems to be known of the structures of the oxynitrates, so the writing of the formulas is somewhat arbitrary, e.g., $Pb(NO_3)_2.2\ PbO = 3PbO.N_2O_5$.

Neumann and Sonntag calculated equilibrium constants for the above three reactions by calculating the partial pressures of NO_2 and O_2, using the equilibrium constants of Bodenstein[60] for the equilibrium reaction $NO_2 = NO + 1/2\ O_2$. Since these differ slightly from current data, they were used to reanalyze the three reactions. It was noted that a plot of ln *K* of reaction I vs. 1/T could only very approximately be described as linear, but rather consisted of three linear portions. A plot of ln *K* vs. 1/T for reaction III could also more accurately be described as consisting of two linear portions. For reaction II the plot was linear over the entire temperature range. **ΔH°** values calculated for the various portions are

Reaction	T(K)	ΔH (kJ)
I	539–583	694
	583–603	629
	613–632	758
II	625–719	337
III	688–763	390
	763–809	324

There seems to be no simple way to choose among the enthalpies for the various slopes, although the high-temperature plot of I does not join smoothly to the lower two and may represent experimental error. The low-temperature plot of I shows more scatter than the plot at the higher temperature.

Thermodynamic data for Pb $(NO_3)_2$ are given in Table 5.36, and for its decomposition in Tables 5.37 and 5.38, but there are no data for the oxynitrates.

In addition to the decompositions discussed above, an autocatalytic decomposition resulting in the formation of Pb(IV) has also been reported,[62] but no details are available.

5.4.17 Lithium

5.4.17.1 $LiNO_2$

The thermal decomposition of $LiNO_2$ results in the formation of $LiNO_3$, N_2, O_2, and oxides of nitrogen.[63] In order to disentangle the various reactions giving rise to these products, Lee and Johnson[64] studied the decomposition by sweeping out product gases from a thin, quiescent layer of molten nitrite, and using only the results from the first few percent of decomposition. The exit gases were analyzed by starting with fresh batches each time. From the results, the following first-order reaction steps are proposed:

$$3\ LiNO_2 = LiNO_3 + Li_2O + 2\ NO$$

$$2\ LiNO_2 = Li_2O + N_2O + O_2$$

$$2\ LiNO_2 + 2\ NO = 2\ LiNO_3 + N_2$$

$$2\ LiNO_3 = 2\ LiNO_2 + O_2$$

Decomposition begins at 433 K. Up to 623 K no NO_2 is detected, but it was produced at 723 K.

5.4.17.2 $LiNO_3$

The isothermal decomposition of $LiNO_3$ proceeds according to the following three reactions:

$$2\ LiNO_3 = 2LiNO_2 + O_2$$

$$2\ LiNO_3 = Li_2O + N_2 + 5/2\ O_2$$

$$2\ LiNO_2 = Li_2O + (NO + NO_2)$$

The first reaction begins at 853 to 893 K, with the others beginning only at 893 K.[65,66]

Densities of the two salts are given in Tables 5.39 and 5.40 transitions in Table 5.41, and thermodynamic data in Tables 5.42 and 5.44.

5.4.18 Magnesium

$Mg(NO_3)_2$ is stable to nearly 600 K. Its decomposition was studied from 575 to 725 K.[72] Initially, NO_2 is liberated and the decomposition reactions are inferred from the NO_2 liberated. The curves follow first-order topochemical kinetics. Thermodynamic data are given in Tables 5.45 to 5.47.

5.4.19 Manganese

Reports about the decomposition of $Mn(NO_3)_2$ differ somewhat. According to the older work of Guntz and Martin,[73] decomposition begins near 160°C and produces the dioxide

$$Mn(NO_3)_2 = MnO_2 + 2\ NO_2$$

More careful analysis of the remaining solid phase[74] shows that it is $MnO_{1.655}$ and that both NO_2 and O_2 are evolved. Under dry N_2 the decomposition rate reaches a maximum near 230°C. A very elaborate study by a Dutch group[75] concludes that decomposition proceeds in a single step to MnO_2, but since their starting material was an aqueous solution, and water vapor frequently affects the course of these reactions, their kinetic analysis is probably not relevant.

The salt is characterized as covalent by Addison and Gatehouse,[76] but Dehnicke and Strähle[77] find that it, like $Co(NO_3)_2$, possesses considerable ionic character.

Thermodynamic data are given in Tables 5.48 and 5.49.

5.4.20 Mercury

5.4.20.1 $Hg(NO_2)_2$

The salt appears to have been prepared for the first time by Ray[78] and in the pure state by Oza, Jha, and Ezekiel.[79] It is a yellow powder which begins to decompose at 50°C and readily decomposes at 90°C. HgO and $Hg(NO_3)_2$ appear as products and so the reactions probably are

$$2\ Hg(NO_2)_2 = 2\ HgNO_3 + 2\ NO$$

$$Hg(NO_3)_2 = HgO + N_2O_3$$

5.4.20.2 $HgNO_3$

The salt has been prepared only relatively recently in the anhydrous state.[80] It decomposes to a yellow solid with the evolution of brown fumes when heated above 100°C. It appears to be covalent.

5.4.20.3 $Hg(NO_3)_2$

The anhydrous salt has been prepared,[81] but aside from infrared studies which indicate that the compound is covalent,[81,82] little has been done to determine its properties.

Decomposition of the salt is appreciable at 160°C and appears to lead to a basic nitrate, $HgO.Hg(NO_3)_2$. At 300°C this compound is stable and the final product is HgO. In the presence of NO some nitrite and NO_2 are formed.[79]

Thermodynamic data at 298 K are listed in Table 5.50.

5.4.21 Nickel

5.4.21.1 $Ni(NO_2)_2$

The salt is stable up to 260°C in an argon atmosphere and decomposes at 220°C in a vacuum. It is slightly volatile.[83]

5.4.21.2 $Ni(NO_3)_2$

The salt decomposes on heating, the initial decomposition, near 260°C, resulting in formation of the nitrite.[84] Thermodynamic data are given in Tables 5.51 and 5.52.

5.4.22 Palladium

5.4.22.1 $Pd(NO_2)_2$

A compound of approximately this composition was reported.[85]

5.4.22.2 $Pd(NO_3)_2$

The anhydrous salt has been prepared.[86] The infrared spectrum indicates it is covalent and sufficiently stable in the vapor to permit sublimation in a vacuum.

No entropy data are available; enthalpies at 298 K are listed in Table 5.53.

5.4.23 Potassium

5.4.23.1 KNO_2

No solid-phase transitions have been reported. The salt begins to decompose noticeably at 410°C. The first step in the decomposition sequence is probably[87,88]

$$2\ KNO_2 = K_2O + NO_2 + NO$$

Between 410 and 460°C the NO produced reacts with KNO_2:

$$KNO_2 + 2\ NO = 2\ KNO_3 + N_2$$

The NO_2 produced in the first reaction reacts further:

$$K_2O + 2\ NO_2 = KNO_2 + KNO_3$$

$$KNO_2 + NO_2 = KNO_3 + NO$$

Which reactions are favored depend largely on the atmospheric composition. For example, in an inert atmosphere which removes the gaseous products the first reaction will predominate. Otherwise, since KNO_3 is stable in this temperature range and N_2 is unreactive, the net result of the above reactions is the production of KNO_3 and N_2. In an oxygen atmosphere, particularly in the temperature range 550 to 600°C, the reaction $KNO_2 + 1/2\ O_2 = KNO_3$ goes to completion. In the range

650 to 750°C, KNO_3 becomes increasingly unstable and the reaction attains equilibrium. Above 800°C the reaction $2\ KNO_2 = K_2O + N_2 + 3/2\ O_2$ goes to completion.[87–89] These results are consistent with endothermic peaks observed by differential scanning calorimetry.[90]

5.4.23.2 KNO_3

The phase transitions are fairly complicated and have been discussed by several authors.[91–93] In addition to the stable low-temperature orthorhombic form II which transforms at 401 K to the rhombohedral I, there exists a metastable form III which can be obtained by cooling I to 396 K; III then transforms to II on further cooling to 386 K.

KNO_3 melts without decomposition to a liquid which is stable in air to at least 803 K.[94] When heated in air, it begins to decompose near 923 K.[95] Between 823 and 1023 K the quasi-equilibrium $KNO_3 = KNO_2 + 1/2\ O_2$ is set up.[89,95,96] The activation energy for the forward reaction is 274 kJ. If the O_2 is removed, KNO_2 decomposes, various oxides of nitrogen are produced, and the reaction is more complicated (see discussion under KNO_2). As pointed out above, the two salts tend to interconvert over a limited temperature range. Above the melting points this leads to a solution containing both salts. For some specified oxygen partial pressure, e.g., air, the solution composition is then a unique function of temperature if equilibrium is established. In the range 823 to 1023 K the equilibrium has been studied by several authors.[89,95,96] All these results have been summarized by Kramer, Munir, and Volponi[90] and are consistent with their DSC results.

Densities are listed in Tables 5.54 and 5.55, transitions in Table 5.56 and thermodynamic data in Tables 5.57 to 5.61.

5.4.24 Radium

$Ra(NO_3)_2$ has a cubic lattice, isomorphous with $Ba(NO_3)_2$. Its density is calculated to be 4.11 gcm^{-3}.[97] Thermodynamic properties are listed in Table 5.62.

5.4.25 Rare Earths

5.4.25.1 Nitrites

Several anhydrous nitrites have been prepared and their decomposition studied by TGA.[98] They all decompose at lower temperatures than the corresponding nitrates. With the exception of $Pr(NO_2)_3$ whose decomposition product is Pr_6O_{11}, the nitrites decompose according to

$$4\ R(NO_2)_3 = 2\ R_2O_3 + 12\ NO + 3\ O_2$$

Below are listed the initial decomposition and minimum oxide formation temperatures as given by Freeman.[89]

Salt	Initial Decomposition °C	Minimum Oxide Formation °C
$La(NO_2)_3$	240	580
$Pr(NO_2)_3$	220	450
$Nd(NO_2)_3$	200	510
$Sm(NO_2)_3$	200	500
$Dy(NO_2)_3$	110	500
$Yb(NO_2)_3$	90	350

These authors also measured the decomposition kinetics of Nd, Dy, and Tb nitrites. The reactions were first order with very low (4 to 15 kJ) activation energies which they attributed to metal-oxygen interaction.

5.4.25.2 Nitrates

There has been considerable discussion as to (a) whether the anhydrous nitrates are stable intermediates in the thermal decomposition of the corresponding hydrates and (b) whether some of the anhydrous nitrates are ionic or covalent.

A study of the infrared spectra by Walker and Ferraro[99] seems to have established rather firmly that the anhydrous nitrates are covalent. Unfortunately, there appear to have been no TGA or other thermal decomposition studies of rare-earth nitrates originally prepared in the anhydrous state. Both Wendlandt[100,101] and Patil, Gosavi, and Rao[98] studied the thermal decomposition of the hydrated salts. They agree that oxynitrates of composition $RONO_3$ are formed as intermediates in nearly all cases, but differ as to whether the anhydrous nitrates are also intermediates. Wendlandt[101] found that the lighter rare earths, La, Pr, Nd, gave the anhydrous salt as a stable intermediate:

$$R(NO_3)_3.6H_2O \Rightarrow R(NO_3)_3 \Rightarrow RONO_3 \Rightarrow R_2O_3$$

whereas the heavier rare earths decomposed directly to the oxynitrate. However, Patil, Gosavi, and Rao[98] claimed that most rare earths on careful decomposition do give the anhydrous nitrate as an intermediate, but only in the limited temperatures given below:

Salt	Stability Range of Anhydrous Nitrate (°C)	Stability Range of $RONO_3$ (°C)
$La(NO_3)_3$	240–420	525–575
$Nd(NO_3)_3$	280–350	460–500
$Dy(NO_3)_3$	260–280	420–440
$Yb(NO_3)_3$	260–270	340–400
$Lu(NO_3)_3$	220–230	370–390

$Pr(NO_3)_3$ is exceptional in that the final oxide is Pr_6O_{11}. The hydrated cerium salt shows no stable intermediates, but goes to CeO_2. The above uncertainties could probably be cleared up by TGA studies of anhydrous nitrates. The above data indicate that these should be stable to about 300°C.

The kinetics of the nitrate decompositions of Nd, Dy, and Yb were found to be first order with activation energies ranging from 23 to 46 kJ higher than those of the corresponding nitrites.[98]

5.4.26 Rubidium

5.4.26.1 $RbNO_2$

The salt appears to be quite stable up to 450°C and decomposes with increasing rapidity above that temperature.[102] Bagaratian et al.[38] give the sublimation enthalpy of the monomer as 162 kJ and 209 kJ for the dimer.

Decomposition reactions appear to be similar to those of the other alkali metal nitrites, i.e., the course of the decomposition depends on the gas phase. If the gas is continually removed, the reaction proceeds directly to the oxide:

$$2RbNO_2 = Rb_2O + NO + NO_2$$

In the presence of the nitrogen oxides the salt may be oxidized to the nitrate, particularly in the temperature range where the nitrate is stable.

5.4.26.2 $RbNO_3$

The salt is stable up to 600°C and begins to lose weight above that temperature.[103] The phase transitions and high temperature heat capacity have been studied by Mustajoki,[104] but the lack of a reliable entropy value for the salt and high-temperature data for the oxide prevent the calculation of decomposition equilibria. Although no experimental studies of the decomposition seem to have been reported, it is expected to resemble that of the other alkali metal nitrates. Like them, the salt can be vacuum distilled above 350°C.[105]

A mass-spectrometric study[38] gives 183 kJ for the sublimation enthalpy of the monomer and 198 kJ for the dimer. The dimerization enthalpy is –167 kJ.

Densities of the salts are given in Tables 5.63 and 5.64, transitions in Table 5.65, and thermodynamic data in Tables 5.66 to 5.68.

5.4.27 Scandium

Mellor[108] summarizes earlier work which claims preparation of $Sc(NO_3)_3$ by heating the 4-hydrate at 100°C in a vacuum. The melting point is said to lie in the range 125 to 150°C.

5.4.28 Silver

5.4.28.1 $AgNO_2$

The thermal decomposition of this salt has not been studied in detail, but experiments[109] indicate that it decomposes at 120°C to give metallic silver and NO_2 with first-order kinetics and an activation energy of about 25 kJ.

5.4.28.2 $AgNO_3$

The decomposition equilibria are complicated by the thermal instability of Ag_2O. This means that instead of the usual two equilibria the system can be considered in terms of four reactions only three of which are independent:

$$(1)\ 2\ AgNO_3 = Ag_2O + 2\ NO_2 + 1/2\ O_2$$

$$(2)\ 2\ NO_2 = 2\ NO + O_2$$

$$(3)\ 2\ AgNO_3 = 2\ Ag + 2\ NO_2 + O_2$$

$$(4)\ Ag_2O = 2\ Ag + 1/2\ O_2$$

Equilibrium constants for (1), (3), and (4) have been listed. Ag_2O is quite unstable above 400 K and hence reactions (1) and (4) are not in equilibrium for reasonable pressures of O_2. Below 500 K, $P(O_2)$ is fixed by reaction (4). $P(NO_2)$ and $P(NO)$ can then be calculated from K_1, K_3, and K_4. Qualitatively, decomposition is negligible in the solid state, but becomes appreciable at 30 to 40°C above the melting point.[110]

Densities are listed in Table 5.69, transitions in Table 5.70, and thermodynamic data in Tables 5.71 to 5.73.

5.4.29 Sodium

5.4.29.1 $NaNO_2$

$NaNO_2$ is unstable above 330°C.[111] The first step in the decomposition is probably 2 $NaNO_2$ = $Na_2O + NO_2 + NO$ (N_2O_5 is too unstable at and above this temperature to be significant). Between 330 and 380°C the reaction sequence seems to be as follows:

$$2\ NaNO_2 = Na_2O + NO_2 + NO \qquad (1)$$

The NO produced in (1) reacts with $NaNO_2$

$$2\ NaNO_2 + 2\ NO = 2\ NaNO_3 + N_2 \quad (2)$$

The NO_2 produced in (1) reacts further[112]

$$Na_2O + 2\ NO_2 = NaNO_2 + NaNO_3 \quad (3)$$

$$NaNO_2 + NO_2 = NaNO_3 + NO \quad (4)$$

Of these, at least reaction (4) is reversible.[113] Since $NaNO_3$ is stable in this temperature range and N_2 is unreactive, the net result of reactions (1) to (4) is the production of $NaNO_3$ and N_2. If the gas phase is continually removed, only (1) will occur.

At higher temperatures, 600 to 750°C, $NaNO_2$ reacts with O_2 to establish the equilibrium with $NaNO_3$:

$$NaNO_3 = NaNO_2 + 1/2\ O_2 \quad (5)$$

If the decomposition is carried out in an inert atmosphere the first decomposition product is N_2, accompanied by increasing concentrations of O_2 in the later stages of the reaction.[114] The stoichiometry of this reaction can be written as[115]

$$2\ NaNO_2 = Na_2O + N_2 + 1/2\ O_2 \quad (6)$$

Freeman[114] attributes the evolution of N_2 to the formation of a superoxide which later decomposes to Na_2O and O_2. This view is supported by Bond and Jacobs[115] who report that this reaction also proceeds in air. For reaction (5) the Arrhenius activation energy is given by Freeman[114] as 187 kJ and by Bond and Jacobs[115] as 169 kJ. For reaction (6) it is 179 kJ.[115] Both reactions are fitted by a contracting sphere model which was originally derived for solids.

Equations of this type generally describe sigmoid α vs. t curves. In view of the complexity of these reactions, calculation of these equilibria, except for reaction (5), would probably not be useful.

5.4.29.2 $NaNO_3$

$NaNO_3$ melts without decomposition to a liquid which is stable in air to at least 500°C[116] and begins to decompose slowly at 600°C.[114] Between 600 and 750°C a pseudoequilibrium is established between air and a liquid containing $NaNO_3$ and $NaNO_2$ whose composition is temperature dependent. As an example of the stability of $NaNO_3$ in the solid state some calculations of its decomposition to Na_2O and $NaNO_2$ have been made. In the liquid state nitrate and nitrite are completely miscible; the solution is virtually ideal; and the liquid composition depends on $p(O_2)$.

Densities are given in Tables 5.74 and 5.75, transitions in Table 5.76, and thermodynamic data in Tables 5.77 to 5.82.

5.4.30 Strontium

5.4.30.1 $Sr(NO_2)_2$

The melting point of this salt is in doubt since the reported values[117,118] vary by 37°C. The solid phases have not been characterized.

The thermal decomposition has been studied by several authors.[117–119] It is similar to those of the barium and calcium salts. A measurable decomposition pressure has been reported at 537 K, but most studies have been carried out at higher temperatures. At 675 to 775 K, where $Sr(NO_3)_2$ is stable, the overall reaction is given as[118]

$$4\ Sr(NO_2)_2 = 2\ Sr(NO_3)_2 + 2\ SrO + 2\ NO + N_2$$

From a careful analysis of the condensed- and vapor-phase components in the range 813 to 973 K, Oza and Patel[119] deduce the following reaction sequence:

$$Sr(NO_2)_2 = SrO + NO + NO_2 \quad (1)$$

A part of the NO_2 reacts with the remaining nitrite

$$Sr(NO_2)_2 + 2\ NO_2 = Sr(NO_3)_2 + 2\ NO \quad (2)$$

In a closed system this reaction may be reversible. The rest of the NO_2 reacts with the oxide, if oxygen is present

$$SrO + 2\ NO_2 + 1/2\ O_2 = Sr(NO_3)_2 \quad (3)$$

The oxygen may be produced by the dissociation of NO_2

$$NO_2 = NO + 1/2O_2 \quad (4)$$

Up to 873 K, reaction (3) produces nitrate while above 915 K, $Sr(NO_3)_2$ decomposes. N_2 is produced by

$$Sr(NO_2)_2 + 2\ NO = Sr(NO_3)_2 + N_2 \quad (5)$$

Reaction (5) probably occurs only at the higher temperatures since it was not reported below 642 K.[117]

There are no high-temperature thermodynamic data which would permit the calculation of equilibria for any of the above reactions.

5.4.30.2 $Sr(NO_3)_2$

The melting point of this salt is uncertain. The value listed is that selected by Lumsden,[120] but Gordon and Campbell[121] report 891 K, and Addison and Coldrey[122] give 878 K. Similarly, reports on the stability of the salt above its melting point vary from slight "bubbling at 888 K" to 945 K.[123] Crystalline transitions appear to be absent.[123]

Transitions are listed in Table 5.83, and thermodynamic data in Tables 5.84 to 5.86.

5.4.31 Thallium

5.4.31.1 $TlNO_2$

Very little is known about this salt. Phase transitions have been reported, but no details are given by the authors.[124] The salt has been reported to decompose above its melting point (182°C) according to Protsenko and Brykova[15]

$$2TlNO_2 = Tl_2O + NO_2 + NO$$

5.4.31.2 $TlNO_3$

The heat of fusion of $TlNO_3$ has also been determined by Kleppa and McCarthy[125] who obtained 9.47 kJ, but their melting point was 5°C lower than the commonly accepted one.

The vaporization of $TlNO_3$ was studied by Cubiciotti[126] in the range 539 to 612 K. He found that, in addition to NO as the decomposition product, $TlNO_3$ also vaporized as the salt. In the range 490 to 612 K the vapor pressure is given by log P(atm) = –7486/T – log T + 21.77 and the enthalpy of vaporization by ΔH ($Jmol^{-1}$) 143,344 – 41.84 T, and the entropy of vaporization by ΔS ($Jdeg^{-1}mol^{-1}$) = 370.9 – 96 log T.

Densities are given in Table 5.87, transitions in Table 5.88, and thermodynamic data in Tables 5.89 to 5.91.

5.4.32 Thorium

$Th(NO_3)_4$ decomposes without the formation of basic salts according to Misciatelli[144]

$$Th(NO_3)_4 = ThO_2 + NO_2 + O_2$$

Thermodynamic functions are given in Tables 5.92 and 5.93.

5.4.33 Tin

$Sn(NO_3)_4$ melts at 364 K and begins to decompose at 371 K. No intermediate oxide nitrate appears to form, and the final product is SnO_2.[127] There appears to be no information on the decomposition of $Sn(NO_3)_2$, but thermodynamic calculations indicate that it is unstable above 500 K.

Thermodynamic data of $Sn(NO_3)_2$ are given in Tables 5.94 and 5.95.

5.4.34 Titanium

$Ti(NO_3)_4$ is a covalent salt, stable at least to its melting point at 331 K.[128] It decomposes to TiO_2 with the evolution of O_2 and NO_2.[129]

5.4.35 Zinc

$Zn(NO_3)_2$ is stable up to 373 K, and decomposes only slightly even at 513 K. Above this temperature, decomposition becomes increasingly rapid, consistent with the equation[130]

$$Zn(NO_3)_2 = ZnO + 2\ NO_2 + 1/2\ O_2$$

Thermodynamic data are only available at 298 K. These are listed in Table 5.96.

5.4.36 Zirconium

Anhydrous $Zr(NO_3)_4$ has been prepared. It can be purified by sublimation. The compound is covalent and very hygroscopic.[131]

TABLE 5.1
Thermodynamic Data of O_2

T K	ΔHf° kJ mol^{-1}	S° J mol^{-1} deg^{-1}	ΔGf° kJ mol^{-1}	Reference
298.15	0.000	205.15	–61.17	Barin 1993
400.00	3.06	213.96	–82.53	Barin 1993
500.00	6.21	220.98	–104.28	Barin 1993
600.00	9.44	226.87	–126.68	Barin 1993
700.00	12.74	231.95	–149.63	Barin 1993
800.00	16.09	236.42	–173.05	Barin 1993
900.00	19.48	240.42	–196.90	Barin 1993
1000.00	22.92	244.04	–221.12	Barin 1993
1100.00	26.39	247.35	–245.70	Barin 1993
1200.00	29.91	250.41	–270.59	Barin 1993

TABLE 5.2
Thermodynamic Data of N_2

T K	ΔHf° kJ mol^{-1}	S° J mol^{-1} deg^{-1}	ΔGf° kJ mol^{-1}	Reference
298.15	0.00	191.61	–57.13	JANAF 1985
400.00	2.97	200.19	–77.10	JANAF 1985
500.00	5.91	206.74	–97.46	JANAF 1985
600.00	8.89	212.13	–118.41	JANAF 1985
700.00	11.94	216.87	–139.82	JANAF 1985
800.00	15.05	221.02	–161.77	JANAF 1985
900.00	18.22	224.76	–184.06	JANAF 1985
1000.00	21.46	228.17	–206.71	JANAF 1985
1100.00	24.76	231.31	–229.69	JANAF 1985
1200.00	28.11	234.23	–252.97	JANAF 1985

TABLE 5.3
Thermodynamic Data of NO

T K	ΔHf° kJ mol^{-1}	S° J mol^{-1} deg^{-1}	ΔGf° kJ mol^{-1}	Reference
298.15	90.29	210.76	27.45	Knacke et al. 1991
400.00	93.34	219.57	5.52	Knacke et al. 1991
500.00	96.41	226.40	–16.79	Knacke et al. 1991
600.00	99.53	232.11	–39.72	Knacke et al. 1991
700.00	102.73	237.02	–63.19	Knacke et al. 1991
800.00	105.97	241.35	–87.11	Knacke et al. 1991
900.00	109.26	245.23	–111.44	Knacke et al. 1991
1000.00	112.61	248.75	–136.14	Knacke et al. 1991
1100.00	116.00	251.98	–161.18	Knacke et al. 1991
1200.00	119.43	254.97	–186.53	Knacke et al. 1991

TABLE 5.4
Thermodynamic Data of NO_2

T K	ΔHf° kJ mol⁻¹	S° J mol⁻¹ deg⁻¹	ΔGf° kJ mol⁻¹	Reference
298.15	33.10	240.02	–38.47	Knacke et al. 1991
400.00	37.06	251.44	–63.51	Knacke et al. 1991
500.00	41.30	260.88	–89.14	Knacke et al. 1991
600.00	45.78	269.04	–115.65	Knacke et al. 1991
700.00	50.46	276.25	–142.92	Knacke et al. 1991
800.00	55.30	282.71	–170.87	Knacke et al. 1991
900.00	60.29	288.59	–199.44	Knacke et al. 1991
1000.00	65.41	293.99	–228.57	Knacke et al. 1991
1100.00	70.65	298.97	–258.22	Knacke et al. 1991
1200.00	75.97	303.81	–288.36	Knacke et al. 1991

TABLE 5.5
Thermodynamic Data of N_2O

T K	ΔHf° kJ mol⁻¹	S° J mol⁻¹ deg⁻¹	ΔGf° kJ mol⁻¹	Reference
298.15	82.05	219.98	16.46	Knacke et al. 1991
400.00	86.15	231.79	–6.57	Knacke et al. 1991
500.00	90.54	241.57	–30.25	Knacke et al. 1991
600.00	95.28	250.20	–54.84	Knacke et al. 1991
700.00	100.30	257.94	–80.36	Knacke et al. 1991
800.00	105.52	264.90	–100.40	Knacke et al. 1991
900.00	110.87	273.21	–135.22	Knacke et al. 1991
1000.00	116.34	276.97	–160.63	Knacke et al. 1991
1100.00	121.90	282.27	–188.59	Knacke et al. 1991
1200.00	127.54	287.17	–217.07	Knacke et al. 1991

TABLE 5.6
Thermodynamic Data of N_2O_3

T K	ΔHf° kJ mol⁻¹	S° J mol⁻¹ deg⁻¹	ΔGf° kJ mol⁻¹	Reference
298.15	82.84	309.34	–9.39	Knacke et al. 1991
400.00	89.84	329.51	–41.96	Knacke et al. 1991
500.00	97.36	346.25	–75.77	Knacke et al. 1991
600.00	105.47	361.03	–111.14	Knacke et al. 1991
700.00	114.06	374.25	–147.92	Knacke et al. 1991
800.00	122.97	386.14	–185.95	Knacke et al. 1991
900.00	133.00	396.70	–225.11	Knacke et al. 1991
1000.00	141.42	406.72	–265.30	Knacke et al. 1991
1100.00	150.88	415.73	–306.43	Knacke et al. 1991
1200.00	160.45	424.06	–348.42	Knacke et al. 1991

TABLE 5.7
Thermodynamic Data of N_2O_4

T K	ΔHf° kJ mol⁻¹	S° J mol⁻¹ deg⁻¹	ΔGf° kJ mol⁻¹	Reference
298.15	9.08	304.37	–81.67	Knacke et al. 1991
400.00	17.48	328.53	–113.93	Knacke et al. 1991
500.00	26.72	349.12	–147.84	Knacke et al. 1991
600.00	36.84	367.55	–183.69	Knacke et al. 1991
700.00	47.65	384.20	–221.29	Knacke et al. 1991
800.00	58.92	399.24	–260.47	Knacke et al. 1991
900.00	70.50	412.88	–301.09	Knacke et al. 1991
1000.00	82.36	425.37	–345.01	Knacke et al. 1991
1100.00	94.41	436.86	–386.13	Knacke et al. 1991
1200.00	106.61	447.47	–430.35	Knacke et al. 1991

TABLE 5.8
Thermodynamic Data of N_2O_5

T K	ΔHf° kJ mol⁻¹	S° J mol⁻¹ deg⁻¹	ΔGf° kJ mol⁻¹	Reference
298.15	11.30	346.55	–92.03	Barin 1989
400.00	22.09	377.94	–128.93	Barin 1989
500.00	34.03	404.16	–168.05	Barin 1989
600.00	46.70	427.24	–209.64	Barin 1989
700.00	59.84	447.47	–253.40	Barin 1989
800.00	73.30	465.45	–299.06	Barin 1989
900.00	87.01	481.60	–346.43	Barin 1989
1000.00	100.92	496.25	–395.33	Barin 1989
1100.00	115.00	509.67	–445.64	Barin 1989
1200.00	129.22	522.04	–497.23	Barin 1989

TABLE 5.9
Density of $Ba(NO_3)_2$

Phase	T(K)	d	Reference
c	298	3.232	132
l	550	3.254	133
	560	3.247	133
	570	3.240	133
	580	3.233	133
	590	3.226	133
	600	3.219	133
	610	3.212	133
	620	3.205	133

TABLE 5.10
Thermodynamic Data of $Ba(NO_3)_2$

T K	ΔHf° kJ mol⁻¹	S° J mol⁻¹ deg⁻¹	ΔGf° kJ mol⁻¹	Reference
298.15	–992.07	213.80	–1055.81	Barin et al. 1977
400.00	–975.39	261.77	–1080.09	Barin et al. 1977
500.00	–956.93	302.87	–1108.37	Barin et al. 1977
600.00	–936.70	339.71	–1140.53	Barin et al. 1977
700.00	–914.82	373.41	–1176.20	Barin et al. 1977
800.00	–891.34	404.73	–1215.13	Barin et al. 1977
900.00	–866.31	434.20	–1257.09	Barin et al. 1977
1000.00	–839.73	462.19	–1301.92	Barin et al. 1977

TABLE 5.11
$Ba(NO_3)_2 = BaO + 2\ NO_2(g) + 1/2\ O_2(g)$

T K	ΔH° kJ	ΔS° J deg⁻¹	ΔG° kJ	K
298.15	504.71	439.10	373.797	3.213E – 066
400.00	502.47	432.70	329.388	9.610E – 044
500.00	499.16	425.37	286.478	1.173E – 030
600.00	494.77	417.38	244.336	5.332E – 022
700.00	489.26	408.93	203.017	7.070E – 016
800.00	482.64	400.092	162.563	2.426E – 011
900.00	474.09	390.95	123.009	7.247E – 008
1000.00	465.91	381.53	84.383	3.908E – 005

TABLE 5.12
Decomposition Pressures of $Ba(NO_3)_2$

T K	P (O_2) bar	P (NO_2) bar	P (NO) bar	P (Tot) bar
298.15	$<10^{-15}$	$<10^{-15}$	$<10^{-15}$	$<10^{-15}$
400.00	3.83E – 015	$<10^{-15}$	5.03E – 015	8.94E – 015
500.00	1.46E – 010	3.05E – 013	1.94E – 010	3.40E – 010
600.00	1.64E – 007	1.11E – 009	2.18E – 007	3.84E – 007
700.00	2.40E – 005	3.67E – 007	3.18E – 005	5.61E – 005
800.00	9.75E – 004	2.69E – 005	1.29E – 003	2.29E – 003
900.00	2.2E – 002	1.1E – 003	2.8E – 002	5.1E – 002
1000.00	0.15	9.1E – 003	0.20	0.36

TABLE 5.13
Thermodynamic Data of $Be(NO_3)_2$ at 298.15K

	ΔHf° kJmol⁻¹	S° Jmol⁻¹	Reference
$Be(NO_3)_2$	–602 ± 8	184	133
BeO	–606.09	14.14	133

TABLE 5.14
Transitions of $Cd(NO_3)_2$

Phase Change	T(K)	ΔH kJ/mol	ΔS J/(mol*K)	Reference
c → l	633	18.3	28.8	125

TABLE 5.15
Thermodynamic Data of $Cd(NO_3)_2$

T K	ΔHf° kJ mol⁻¹	S° J mol⁻¹ deg⁻¹	ΔGf° kJ mol⁻¹	Reference
298.15	–456.96	207.95	–518.95	Knacke et al. 1991
400.00	–438.63	260.82	–542.96	Knacke et al. 1991
500.00	–415.40	312.73	–571.77	Knacke et al. 1991
600.00	–394.06	351.63	–605.04	Knacke et al. 1991
700.00	–372.73	384.53	–641.89	Knacke et al. 1991
800.00	–351.39	413.02	–681.80	Knacke et al. 1991
900.00	–330.05	438.15	–724.39	Knacke et al. 1991
1000.00	–308.71	460.63	–769.34	Knacke et al. 1991

TABLE 5.16
$Cd(NO_3)_2 = CdO + 2\ NO_2(g) + 1/2\ O_2$

T K	ΔH° kJ	ΔS° J deg⁻¹	ΔG° kJ	K
298.15	264.16	429.48	136.106	1.422E – 024
400.00	259.92	417.16	93.053	7.040E – 013
500.00	251.50	398.27	52.366	3.380E – 006
600.00	245.65	387.59	13.098	7.238E – 002
700.00	240.34	379.39	–25.233	7.649E + 001
800.00	235.38	372.90	–62.836	1.268E + 004

TABLE 5.17
Decomposition Pressures of $Cd(NO_3)_2$

T K	P (O_2) bar	P (NO_2) bar	P (NO) bar	P (Tot) bar
298.15	1.09E – 010	3.69E – 010	2.27E – 10	7.05E – 010
400.00	5.60E – 006	1.72E – 005	1.72E – 006	2.45E – 005
500.00	2.85E – 003	7.96E – 003	1.15E – 003	1.20E – 002
600.00	0.167	0.421	8.21E – 002	0.670
700.00	2.92	6.69	1.66	11.27

TABLE 5.18
Transitions of $Ca(NO_3)_2$

Phase Change	T(K)	ΔH kJ/mol	ΔS J/(mol*K)	Reference
$Ca(NO_2)_2$ s → s	539			15
s → s	633			15
s → l	665			39
$Ca(NO_3)_2$ s → l	834	23.4	28.1	120

TABLE 5.19
Thermodynamic Data of $Ca(NO_3)_2$

T K	ΔHf° kJ mol^{-1}	S° J mol^{-1} deg^{-1}	ΔGf° kJ mol^{-1}	Reference
298.15	–938.39	193.30	–966.03	Knacke et al. 1991
400.00	–921.88	240.78	–1018.19	Knacke et al. 1991
500.00	–903.52	281.66	–1044.35	Knacke et al. 1991
600.00	–883.34	318.41	–1074.38	Knacke et al. 1991
700.00	–861.45	352.11	–1107.93	Knacke et al. 1991
800.00	–837.92	383.51	–1144.73	Knacke et al. 1991
900.00	–812.78	413.10	–1184.57	Knacke et al. 1991
1000.00	–786.05	441.25	–1227.30	Knacke et al. 1991

TABLE 5.20
$Ca(NO_3)_2 = CaO + 2\ NO_2(g) + 1/2\ O_2(g)$

T K	ΔH° kJ	ΔS° J deg^{-1}	ΔG° kJ	K
298.15	369.49	427.39	242.068	3.866E – 043
400.00	367.07	420.26	198.881	1.063E – 026
500.00	363.57	412.71	157.216	.753E – 017
600.00	358.94	404.30	116.361	7.397 – 011
700.00	353.15	395.40	76.372	1.998E – 006
800.00	346.16	386.08	37.295	3.670E – 003
900.00	337.95	376.43	–0.833	1.118E + 000
1000.00	328.50	366.48	–37.981	9.640E + 001

TABLE 5.21
Decomposition Pressures of $Ca(NO_3)_2$

T K	P (O_2) bar	P (NO_2) bar	P (NO) bar	P (Tot) bar
298.15	$<10^{-15}$	$<10^{-15}$	$<10^{-15}$	$<10^{-15}$
400.00	3.04E – 010	2.68E – 011	3.96E – 010	7.27E – 010
500.00	1.14E – 006	2.02E – 007	1.45E – 006	2.80E – 006
600.00	2.73E – 004	7.08E – 005	3.41E – 004	6.85E – 004
700.00	1.33E – 002	4.40E – 003	1.62E – 002	3.39E – 002
800.00	0.235	9.16E – 002	0.283	0.609

TABLE 5.22
Density of $CsNO_2$

Phase	T(K)	d (gcm^{-3})	Reference
Liquid	681	2.844	106
	691	2.835	106
	701	2.826	106
	711	2.815	106
	721	2.805	106
	732	2.793	106
	742	2.783	106
	752	2.772	106

TABLE 5.23
Density of $CsNO_3$

Phase	T(K)	d (gcm^{-3})	Reference
c	293	3.643	107
l	690	2.8160	69
	700	2.8044	69
	710	2.7927	69
	720	2.7810	69
	730	2.7694	69
	740	2.7577	69
	750	2.7461	69
	760	2.734	69

TABLE 5.24
Transitions of $CsNO_2$ and $CsNO_3$

Phase Change	T(K)	ΔH kJ/mol	ΔS J/(mol*K)	Reference
$CsNO_2$s → l	679			134
$CsNO_3$ hexagonal → cubic	425	3.7	8.7	134
cubic → l	679	14.1	20.8	134
s → g	298	28.2		38

TABLE 5.25
Thermodynamic Data of $CsNO_2$

T K	ΔHf° kJ mol^{-1}	S° J mol^{-1} deg^{-1}	ΔGf° kJ mol^{-1}	Reference
298.15	–369.45	181.17	–423.46	Knacke et al. 1991
400.00	–360.10	208.13	–443.35	Knacke et al. 1991
500.00	–350.93	228.61	–465.23	Knacke et al. 1991
600.00	–341.75	245.33	–488.95	Knacke et al. 1991
700.00	–332.58	259.48	–514.21	Knacke et al. 1991
800.00	–323.40	271.73	–540.78	Knacke et al. 1991
900.00	–314.22	282.54	–568.51	Knacke et al. 1991
1000.00	–305.05	292.21	–597.25	Knacke et al. 1991

TABLE 5.26
Thermodynamic Data of $CsNO_3$

T K	ΔHf° kJ mol^{-1}	S° J mol^{-1} deg^{-1}	ΔGf° kJ mol^{-1}	Reference
298.15	–505.85	153.72	–551.68	Knacke et al. 1991
400.00	–495.53	183.41	–568.89	Knacke et al. 1991
500.00	–479.59	219.32	–589.25	Knacke et al. 1991
600.00	–466.96	242.35	–612.37	Knacke et al. 1991
700.00	–440.11	282.72	–638.02	Knacke et al. 1991
800.00	–426.52	300.88	–667.22	Knacke et al. 1991
900.00	–412.92	316.90	–698.12	Knacke et al. 1991
1000.00	–399.32	331.22	–730.54	Knacke et al. 1991

TABLE 5.27
$2\ CsNO_3 = Cs_2O + 2\ NO_2(g) + 1/2\ O_2$

T K	ΔH° kJ	ΔS° J deg^{-1}	ΔG° kJ	K
298.15	731.91	422.03	606.076	6.443E – 107
400.00	728.65	412.70	563.567	2.510E – 075
500.00	714.94	381.37	524.250	1.688E – 055
600.00	708.68	369.95	486.714	4.209E – 043
700.00	674.77	319.68	450.994	2.206E – 034
800.00	668.05	310.69	419.493	4.052E – 028
900.00	661.97	303.53	388.80	2.710E – 023
1000.00	656.50	297.76	358.74	1.818E – 019

TABLE 5.28
$CsNO_3 = CsNO_2 + 1/2\ O_2(g)$

T K	ΔH° kJ	ΔS° J deg^{-1}	ΔG° kJ	K
298.15	136.40	130.02	97.632	7.830E – 018
400.00	136.95	131.70	84.275	9.862E – 012
500.00	131.77	119.78	71.880	3.091E – 008
600.00	129.93	116.42	60.076	5.882E – 006
700.00	113.91	92.73	48.993	2.207E – 004
800.00	111.16	89.06	39.908	2.478E – 003
900.00	108.43	85.85	31.165	1.553E – 002
1000.00	105.73	83.00	22.725	6.499E – 002

TABLE 5.29
Decomposition Pressures of $CsNO_3$

T K	P (O_2) bar	P (NO_2) bar	P (NO) bar	P (Tot) bar
298.15	$<10^{-15}$	$<10^{-15}$	$<10^{-15}$	$<10^{-15}$
400.00	$<10^{-15}$	$<10^{-15}$	$<10^{-15}$	$<10^{-15}$
500.00	$<10^{-15}$	$<10^{-15}$	$<10^{-15}$	$<10^{-15}$
600.00	1.56E – 013	$<10^{-15}$	2.08E – 013	3.62E – 013
700.00	1.26E – 010	4.45E – 015	1.68E – 010	2.94E – 010
800.00	1.59E – 008	1.79E – 010	2.12E – 008	3.73E – 008
900.00	6.65E – 007	1.82E – 010	8.86E – 007	1.55E – 006
1000.00	1.29E – 005	7.13E – 009	1.71E – 005	3.01E – 005

TABLE 5.30
Thermodynamic Data of $Co(NO_3)_2$

T K	ΔHf° kJ mol^{-1}	S° J mol^{-1} deg^{-1}	ΔGf° kJ mol^{-1}	Reference
298.15	–421.55	176.98	–474.31	Knacke et al. 1991
400.00	–405.15	224.24	–494.84	Knacke et al. 1991
500.00	–388.20	262.01	–519.21	Knacke et al. 1991
600.00	–370.42	294.41	–547.07	Knacke et al. 1991
700.00	–351.80	323.10	–577.97	Knacke et al. 1991

TABLE 5.31
$Co(NO_3)_2 = Co + 2\ NO_2(g) + 1/2\ O_2(g)$

T K	ΔH° kJ	ΔS° J deg^{-1}	ΔG° kJ	K
298.15	249.79	458.60	113.06	1.550E – 20
400.00	248.39	454.55	66.57	2.024E – 009
500.00	246.86	451.15	21.29	5.971E – 003
600.00	245.03	447.81	–23.66	1.148E + 002
700.00	242.81	444.41	–68.27	1.245E + 010

TABLE 5.32
Transitions of $CuNO_3$

Phase Change	T(K)	ΔH kJ/mol	ΔS J/(mol*K)	Reference
c → l	528			48
c → g	453–493	65.3	87.0	48

TABLE 5.33
Thermodynamic Data of $Cu(NO_3)_2$ (298.15 K)

	ΔHf° kJ mol⁻¹	S° J mol⁻¹ deg⁻¹	Reference
$Cu(NO_3)_2$	–302.9	193.3	71, 133
Cu_2O	–168.6	93.14	133
CuO	–157.3	42.64	133

TABLE 5.34
Thermodynamic Data of $Ga(NO_3)_2$ (298.15 K)

	ΔHf° kJ mol⁻¹	S° J mol⁻¹ deg⁻¹	Reference
$Ga(NO_3)_2$	–619 ± 17		51
GaO(g)	279	231	133
Ga_2O_3	–1089	84.98	133

TABLE 5.35
Thermodynamic Data of $In(NO_3)_3$

	ΔHf° kJ mol⁻¹	S° J mol⁻¹ deg⁻¹	Reference
$In(NO_3)_3$	–632		51
In_2O_3	925.79	104	133

TABLE 5.36
Thermodynamic Data of $Pb(NO_3)_2$

T K	ΔHf° kJ mol⁻¹	S° J mol⁻¹ deg⁻¹	ΔGf° kJ mol⁻¹	Reference
298.15	–456.85	224.68	–523.84	Knacke et al. 1991
400.00	–440.14	272.72	–549.23	Knacke et al. 1991
500.00	–421.66	313.88	–578.60	Knacke et al. 1991
600.00	–401.41	350.75	–611.86	Knacke et al. 1991
700.00	–379.51	384.48	–648.64	Knacke et al. 1991
800.00	–356.01	415.84	–688.68	Knacke et al. 1991

TABLE 5.37
$Pb(NO_3)_2 = PbO + 2\ NO_2(g) + 1/2\ O_2(g)$

T K	ΔH° kJ	ΔS° J deg⁻¹	ΔG° kJ	K
298.15	304.97	426.63	177.774	7.115E – 032
400.00	302.55	419.71	134.663	2.590E – 018
500.00	299.06	411.98	93.073	1.888E – 010
600.00	294.52	403.72	52.285	2.804E – 005
700.00	288.90	395.09	12.342	1.199E – 001
800.00	282.28	386.16	–26.723	5.559E + 001

TABLE 5.38
Decomposition Pressures of $Pb(NO_3)_2$

T K	P (O_2) bar	P (NO_2) bar	P (NO) bar	P (Tot) bar
298.15	3.63E – 013	3.44E – 013	3.68E – 013	1.08E – 012
400.00	8.24E – 008	9.50E – 008	7.82E – 008	2.56E – 007
500.00	1.08E – 004	1.35E – 004	9.95E – 005	3.42E – 004
600.00	1.24E – 002	1.58E – 003	1.13E – 002	2.53E – 003
700.00	0.120	0.353	0.449	0.922
800.00	55.6	4.17	5.22	64.99
900.00	27.3	33.4	25.3	86.0
1000.00	118.	140.8	111.4	370.2

TABLE 5.39
Density of $LiNO_2$

Phase	T(K)	d (gcm^{-3})	Reference
l	493	1.638	67
	498	1.636	67
	503	1.632	67
	512	1.629	67
	518	1.626	67
	532	1.620	67
	543	1.615	67

TABLE 5.40
Density of $LiNO_3$

Phase	T(K)	d (gcm^{-3})	Reference
c	293	2.366	68
l	550	1.768	68
	560	1.762	68
	570	1.757	68
	580	1.751	68
	590	1.746	68
	600	1.740	68
	610	1.735	68
	620	1.729	68
	630	1.724	68
	640	1.719	68
	650	1.713	68
	660	1.708	68
	670	1.702	68
	680	1.697	68
	690	1.691	68

TABLE 5.41
Transitions of $LiNO_3$

Phase Change	T(K)	ΔH kJ/mol	ΔS J/(mol*K)	Reference
$LiNO_2$ c → l	493			69
$LiNO_3$ c → l	526	26.7	50.8	70

TABLE 5.42
Thermodynamic Data of $LiNO_3$

T K	ΔHf° kJ mol^{-1}	S° J mol^{-1} deg^{-1}	ΔGf° kJ mol^{-1}	Reference
298.15	–483.13	90.00	–509.96	Knacke et al. 1991
400.00	–473.59	117.46	–520.58	Knacke et al. 1991
500.00	–463.33	140.31	–533.49	Knacke et al. 1991
600.00	–426.35	209.76	–552.20	Knacke et al. 1991
700.00	–415.22	226.92	–574.06	Knacke et al. 1991
800.00	–404.09	241.78	–597.51	Knacke et al. 1991
900.00	–392.96	254.89	–622.36	Knacke et al. 1991
1000.00	–381.83	266.61	–548.44	Knacke et al. 1991

TABLE 5.43
$2\ LiNO_3 = Li_2O + 2\ NO_2(g) + 1/2\ O_2(g)$

T K	ΔH° kJ	ΔΔS° J deg^{-1}	ΔG° kJ	K
298.15	433.72	440.50	302.385	1.045E – 053
400.00	430.21	430.35	258.069	1.918E – 034
500.00	426.54	422.19	215.447	3.094E – 023
600.00	370.45	315.87	180.926	1.769E – 016
700.00	366.86	310.32	149.633	6.812E – 012
800.00	363.92	306.39	118.809	1.745E – 008
900.00	361.56	303.60	88.318	7.477E – 006
1000.00	359.70	301.64	58.062	9.266E – 004

TABLE 5.44
Decomposition Pressures of $LiNO_3$

T K	P (O_2) bar	P (NO_2) bar	P (NO) bar	P (Tot) bar
298.15	$<10^{-15}$	$<10^{-15}$	$<10^{-15}$	$<10^{-15}$
400.00	1.69E – 012	1.23E – 014	2.25E – 012	3.95E – 012
500.00	1.95E – 008	4.70E – 010	2.59E – 008	4.59E – 008
600.00	6.35E – 006	2.65E – 007	8.38E – 006	1.54E – 005
700.00	3.38E – 004	1.92E – 005	4.44E – 004	8.01E – 004
800.00	6.56E – 003	4.65E – 004	8.59E – 003	1.56E – 002
900.00	6.47E – 002	5.42E – 003	8.45E – 002	0.155
1000.00	0.401	3.83E – 002	0.521	0.960

TABLE 5.45
Thermodynamic Data of $Mg(NO_3)_2$

T K	ΔHf° kJ mol⁻¹	S° J mol⁻¹ deg⁻¹	ΔGf° kJ mol⁻¹	Reference
298.15	–790.65	164.00	–839.55	Barin 1989
400.00	–774.87	209.35	–858.61	Barin 1989
500.00	–756.62	243.95	–881.59	Barin 1989
600.00	–735.52	288.34	–908.52	Barin 1989
700.00	–711.51	325.39	–939.22	Barin 1989
800.00	–684.56	361.23	–973.55	Barin 1989

TABLE 5.46
$Mg(NO_3)_2 = MgO + 2\ NO_2(g) + 1/2\ O_2(g)$

T K	ΔH° kJ	ΔS° J deg⁻¹	ΔG° kJ	K
298.15	255.60	445.54	122.762	3.096E – 022
400.00	253.39	439.25	77.692	7.140E – 011
500.00	249.63	430.93	34.162	2.696E – 004
600.00	243.77	420.31	–8.418	5.406E + 000
700.00	235.58	407.73	–49.835	5.237E + 003
800.00	224.92	393.54	–89.911	7.431E + 005

TABLE 5.47
Decomposition Pressures of $Mg(NO_3)_2$

T K	P (O_2) bar	P (NO_2) bar	P (NO) bar	P (Tot) bar
298.15	8.65E – 010	3.25E – 009	7.10E – 011	3.25E – 005
400.00	3.17E – 005	1.13E – 004	4.73E – 006	1.49E – 004
500.00	2.70E – 004	1.42E – 002	4.76E – 002	6.21E – 002
600.00	0.784	2.47	0.222	3.48
700.00	12.95	38.27	4.51	55.73
800.00	99.1	273.2	41.1	413.4
900.00	452.6	1160	216.3	1830
1000.00	1440	3410	776.3	5626

TABLE 5.48
Thermodynamic Data of $Mn(NO_3)_2$

T K	ΔHf° kJ mol⁻¹	S° J mol⁻¹ deg⁻¹	ΔGf° kJ mol⁻¹	Reference
298.15	–576.30	168.62	–626.57	CRC 1994
400.00	–576.30	168.62	–643.75	CRC 1994
500.00	–576.30	168.62	–660.61	CRC 1994
600.00	–576.30	168.62	–677.47	CRC 1994
700.00	–576.30	168.62	–694.33	CRC 1994

TABLE 5.49
$Mn(NO_3)_2 = MnO + 2\ NO_2(g) + 1/2\ O_2(g)$

T K	ΔH° kJ	ΔS° J deg^{-1}	ΔG° kJ	K
298.15	257.27	473.71	116.03	4.673E – 021
400.00	271.45	514.52	65.64	2.679E – 009
500.00	286.32	547.67	12.48	4.966E – 002
600.00	301.87	576.00	–43.737	6.427E + 003
700.00	317.96	600.80	–102.60	4.539E + 007

TABLE 5.50
Thermodynamic Data of $Hg(NO_3)_2$ (298.15 K)

	ΔHf° kJ mol^{-1}	S° J mol^{-1} deg^{-1}	Reference
$Hg(NO_3)$	–285 + 17	230 + 8	25
HgO(red)	–90.83	70.29	133

TABLE 5.51
Thermodynamic Data of $Ni(NO_3)_2$

T K	ΔHf° kJ mol^{-1}	S° J mol^{-1} deg^{-1}	ΔGf° kJ mol^{-1}	Reference
298.15	–415.00	192.05	–472.26	Fabricius et al. 1994
400.00	–415.00	192.05	–491.82	Fabricius et al. 1994
500.00	–415.00	192.05	–511.02	Fabricius et al. 1994
600.00	–415.00	192.05	–530.23	Fabricius et al. 1994
700.00	–415.00	192.00	–549.43	Fabricius et al. 1994

TABLE 5.52
$Ni(NO_3)_2 = NiO + 2\ NO_2(g) + 1/2\ O_2$

T K	ΔH° kJ	ΔS° J deg^{-1}	ΔG° kJ	K
298.15	241.49	428.56	113.71	1.192E – 020
400.00	255.87	469.92	67.90	1.358E – 009
500.00	271.62	505.00	19.12	1.005E – 002
600.00	288.10	535.08	–32.95	7.388E + 002
700.00	304.40	560.20	–87.74	3.530E + 006

TABLE 5.53
Thermodynamic Data of $Pd(NO_3)_2$

	ΔHf° kJ mol^{-1}	S° J mol^{-1} deg^{-1}	Reference
$Pd(NO_3)_2$	–238 ± 17	—	25
PdO	–85.4	—	133

TABLE 5.54
Density of KNO_2

Phase	T(K)	d (gcm^{-3})	Reference
l	710	1.797	69
	720	1.692	69
	730	1.685	69
	740	1.679	69
	750	1.672	69

TABLE 5.55
Density of KNO_3

Phase	T(K)	d (gcm^{-3})	Reference
l	700	1.805	69
	710	1.797	69
	720	1.790	69
	730	1.783	69
	740	1.776	69
	750	1.768	69
	760	1.761	69
	770	1.754	69
	780	1.746	69
	790	1.739	69
	800	1.732	69
	810	1.725	69
	820	1.717	69
	830	1.710	69
	840	1.703	69
	850	1.695	69
	860	1.688	69
	870	1.681	69

TABLE 5.56
Transitions of KNO_3

Phase Change	T(K)	ΔH kJ/mol	ΔS J/(mol*K)	Reference
KNO_2, c → l	713			135
KNO_3III, rhombohedral → II, orthorhombic	386	2.34	6.1	91
II → I	401	5.42	13.3	91
I → liquid	607	9.62	15.8	70

TABLE 5.57
Thermodynamic Data of KNO_2

T K	ΔHf° kJ mol⁻¹	S° J mol⁻¹ deg⁻¹	ΔGf° kJ mol⁻¹	Reference
298.15	–359.18	152.09	–415.16	Knacke et al. 1991
400.00	–359.18	182.82	–432.30	Knacke et al. 1991
500.00	–348.62	206.35	–451.80	Knacke et al. 1991
600.00	–337.46	226.68	–473.47	Knacke et al. 1991
700.00	–325.49	245.12	–497.07	Knacke et al. 1991
800.00	–312.59	262.33	–522.45	Knacke et al. 1991
900.00	–298.71	278,67	–549.51	Knacke et al. 1991
1000.00	–283.81	294.36	–578.17	Knacke et al. 1991

TABLE 5.58
Thermodynamic Data of KNO_3

T K	ΔHf° kJ mol⁻¹	S° J mol⁻¹ deg⁻¹	ΔGf° kJ mol⁻¹	Reference
298.15	–496.63	133.05	–534.30	Knacke et al. 1991
400.00	–484.19	163.07	–549.42	Knacke et al. 1991
500.00	–467.05	202.64	–568.38	Knacke et al. 1991
600.00	–455.00	224.62	–589.77	Knacke et al. 1991
700.00	–433.06	259.44	–614.67	Knacke et al. 1991
800.00	–420.73	275.91	–641.46	Knacke et al. 1991
900.00	–408.39	290.45	–669.79	Knacke et al. 1991
1000.00	–396.05	303.45	–699.49	Knacke et al. 1991

TABLE 5.59
$2\ KNO_3 = K_2O + 2\ NO_2(g) + 1/2\ O_2(g)$

T K	ΔH° kJ	ΔS° J deg⁻¹	ΔG° kJ	K
298.15	693.95	418.52	569.167	1.888E – 100
400.00	690.38	408.31	527.061	1.469E – 069
500.00	674.22	369.54	489.449	7.301E – 052
600.00	669.04	360.07	452.994	3.632E – 040
700.00	644.73	320.58	420.319	4.294E – 032
800.00	640.19	314.52	388.576	4.231E – 026
900.00	636.17	309.77	357.371	1.807E – 021
1000.00	632.61	306.02	326.589	8.696E – 018

TABLE 5.60
$KNO_3 = KNO_2 + 1/2\ O_2(g)$

T K	ΔH° kJ	ΔS° J deg^{-1}	ΔG° kJ	K
298.15	124.81	121.61	88.551	3.055E – 016
400.00	126.55	126.73	75.855	1.240E – 010
500.00	121.53	114.19	64.437	1.853E – 007
600.00	122.26	115.59	52.962	2.448E – 005
700.00	113.94	101.66	42.784	6.415E – 004
800.00	116.18	104.63	32.477	7.573E – 003
900.00	119.42	108.44	21.830	5.407E – 002
1000.00	123.70	112.94	10.766	2.739E – 001

TABLE 5.61
Decomposition Pressures of KNO_3

T K	P (O_2) bar	P (NO_2) bar	P (NO) bar	P (Tot) bar
298.15	$<10^{-15}$	$<10^{-15}$	$<10^{-15}$	$<10^{-15}$
400.00	$<10^{-15}$	$<10^{-15}$	$<10^{-15}$	$<10^{-15}$
500.00	$<10^{-15}$	$<10^{-15}$	$<10^{-15}$	$<10^{-15}$
600.00	1.08E – 012	$<10^{-15}$	1.43E – 012	2.51E – 012
700.00	5.67E – 010	4.24E – 014	7.56E – 010	1.33E – 009
800.00	6.01E – 008	1.31E – 011	8.02E – 008	1.40E – 007
900.00	2.21E – 006	1.10E – 009	2.94E – 006	5.16E – 006
1000.00	3.88E – 005	3.74E – 008	5.17E – 005	9.05E – 005

TABLE 5.62
Thermodynamic Data of $Ra(NO_3)_2$

	ΔHf° kJ mol^{-1}	S° J mol^{-1} deg^{-1}	Reference
$Ra(NO_3)_2$	–992	222	133
RaO	–523	—	133

TABLE 5.63
Density of $RbNO_2$

Phase	T(K)	d (gcm^{-3})	Reference
l	696	2.347	106
	706	2.339	106
	715	2.333	106
	726	2.322	106
	734	2.316	106
	745	2.306	106
	755	2.299	106
	759	2.295	106

TABLE 5.64
Density of $RbNO_3$

Phase	T(K)	d (gcm^{-3})	Reference
c	293	3.112	23
l	590	2.476	69
	600	2.466	69
	610	2.456	69
	620	2.446	69
	630	2.437	69
	640	2.427	69
	650	2.417	69
	660	2.407	69
	670	2.398	69
	680	2.388	69

TABLE 5.65
Transitions of $RbNO_3$

Phase Change	T(K)	ΔH kJ/mol	ΔS J/(mol*K)	Reference
rhombic(IV) → cubic(III)	433	3.90	9.00	104
III → trigonal(II)	488	3.21	6.58	104
II → I	554	0.96	1.74	104
I → l	583	4.64	7.96	104

TABLE 5.66
Thermodynamic Data of $RbNO_2$

T K	ΔHf° kJ mol^{-1}	S° J mol^{-1} deg^{-1}	ΔGf° kJ mol^{-1}	Reference
298.15	–367.00	162.00	–415.30	Gurvich 1990
400.00	–358.30	187.08	–433.13	Gurvich 1990
500.00	–349.48	206.75	–452.86	Gurvich 1990
600.00	–340.38	223.33	–474.38	Gurvich 1990
700.00	–318.86	255.25	–497.54	Gurvich 1990
800.00	–308.46	269.14	–523.77	Gurvich 1990
900.00	–298.06	281.39	–551.31	Gurvich 1990
1000.00	–287.66	292.35	–580.01	Gurvich 1990

TABLE 5.67
Thermodynamic Data of $RbNO_3$

T K	ΔHf° kJ mol^{-1}	S° J mol^{-1} deg^{-1}	ΔGf° kJ mol^{-1}	Reference
298.15	–494.70	144.90	–537.90	Gurvich 1990
400.00	–484.03	175.60	–554.26	Gurvich 1990
500.00	–467.79	211.96	–573.77	Gurvich 1990
600.00	–447.78	244.40	–594.42	Gurvich 1990
700.00	–433.18	266.91	–620.02	Gurvich 1990
800.00	–418.58	286.40	–647.71	Gurvich 1990
900.00	–403.98	303.60	–677.22	Gurvich 1990
1000.00	–389.38	318.98	–708.37	Gurvich 1990

TABLE 5.68
$2\ RbNO_3 = Rb_2O + 2\ NO_2(g) + 1/2\ O_2(g)$

T K	ΔH° kJ	ΔS° J deg^{-1}	ΔG° kJ	K
298.15	716.59	418.33	591.86	1.992E – 104
400.00	712.43	406.43	549.86	1.550E – 071
500.00	697.91	373.74	511.04	4.049E – 054
600.00	677.56	344.68	470.75	1.034E – 041
700.00	672.13	336.67	436.46	2.683E – 033
800.00	684.17	351.69	402.81	4.976E – 027
900.00	675.85	341.90	368.14	4.282E – 022
1000.00	667.82	333.43	334.39	3.403E – 018

TABLE 5.69
Density of $AgNO_3$

Phase	T(K)	d (gcm^{-3})	Reference
l	490	3.954	69
	500	3.954	69
	510	3.934	69
	520	3.924	69
	530	3.913	69
	540	3.903	69
	550	3.893	69
	560	3.888	69
	570	3.873	69
	580	3.862	69
	590	3.852	69
	600	3.842	69

TABLE 5.70
Transitions of $AgNO_3$

Phase Change	T(K)	ΔH kJ/mol	ΔS J/(mol*K)	Reference
c, α (orthorhombic) → C, β (rhombohedral)	433	2.4	5.5	136
c, β → l	483	12.49	25.9	137

TABLE 5.71
Thermodynamic Data of $AgNO_3$

T K	ΔHf° kJ mol⁻¹	S° J mol⁻¹ deg⁻¹	ΔGf° kJ mol⁻¹	Reference
298.15	–124.39	140.63	–166.32	Barin 1993
400.00	–113.93	170.66	–182.20	Barin 1993
500.00	–87.57	227.73	–201.43	Barin 1993
600.00	–74.77	251.07	–225.41	Barin 1993
700.00	–61.96	270.80	–251.53	Barin 1993
800.00	–49.16	287.90	–287.90	Barin 1993

TABLE 5.72
$2\ AgNO_3 = Ag_2O + 2\ NO_2(g) + 1/2\ O_2(g)$

T K	ΔH° kJ	ΔS° J deg⁻¹	ΔG° kJ	K
298.15	283.92	422.66	157.906	2.154E – 028
400.00	279.51	410.10	115.472	8.310E – 016
500.00	244.30	335.01	76.798	9.470E – 009
600.00	237.21	322.06	43.978	1.483E – 004
700.00	231.00	312.47	12.275	1.213E – 001
800.00	225.59	305.23	–18.594	1.637E + 001

TABLE 5.73
$2\ AgNO_3 = 2\ Ag + 2\ NO_2(g) + O_2(g)$

T K	ΔH° kJ	ΔS° J deg⁻¹	ΔG° kJ	K
298.15	314.97	489.29	169.090	2.364E – 030
400.00	310.27	475.91	119.902	2.193E – 016
500.00	274.38	399.32	74.718	1.562E – 008
600.00	266.29	384.56	35.556	8.023E – 004
700.00	258.78	372.97	–2.299	1.484E + 000
800.00	251.80	363.64	–39.114	3.582E + 002

TABLE 5.74
Density of $NaNO_2$

Phase	T(K)	d (gcm^{-3})	Reference
c	293	2.266	115
l	570	1.801	69
	580	1.793	69
	590	1.786	69
	600	1.778	69
	610	1.771	69
	620	1.763	69
	630	1.756	69
	640	1.749	69
	650	1.741	69
	660	1.734	69
	670	1.726	69
	680	1.719	69
	690	1.711	69
	700	1.704	69
	710	1.696	69
	720	1.686	69

TABLE 5.75
Density of $NaNO_3$

Phase	T(K)	d (gcm^{-3})	Reference
l	570	1.898	69
	580	1.891	69
	590	1.884	69
	600	1.877	69
	610	1.870	69
	620	1.862	69
	630	1.855	69
	640	1.848	69
	650	1.841	69
	660	1.834	69
	670	1.827	69
	680	1.820	69

TABLE 5.76
Transitions of $NaNO_3$

Phase Change	T(K)	ΔH kJ/mol	ΔS J/(mol*K)	Reference
$NaNO_2$, c $\rightarrow$ l	554	10.4	18.8	138
$NaNO_3$, c,II $\rightarrow$ c,I	548	3.95	7.21	139
c,I $\rightarrow$ l	579	15.5	26.8	70, 120

TABLE 5.77
Thermodynamic Data of $NaNO_2$

T K	ΔHf° kJ mol⁻¹	S° J mol⁻¹ deg⁻¹	ΔGf° kJ mol⁻¹	Reference
298.15	–358.99	121.34	–395.16	Barin 1993
400.00	–350.94	144.42	–408.70	Barin 1993
500.00	–341.05	166.39	–424.25	Barin 1993
600.00	–314.39	214.55	–443.12	Barin 1993
700.00	–302.26	233.26	–465.53	Barin 1993
800.00	–290.12	249.46	–489.69	Barin 1993
900.00	–277.99	263.75	–515.36	Barin 1993
1000.00	–265.85	276.53	–542.39	Barin 1993

TABLE 5.78
Thermodynamic Data of $NaNO_3$

T K	ΔHf° kJ mol⁻¹	S° J mol⁻¹ deg⁻¹	ΔGf° kJ mol⁻¹	Reference
298.15	–467.98	116.32	–502.66	Barin 1989
400.00	–457.33	146.87	–516.08	Barin 1989
500.00	–444.60	175.19	–532.20	Barin 1989
600.00	–414.96	227.92	–551.71	Barin 1989
700.00	–399.40	251.90	–575.74	Barin 1989
800.00	–383.84	272.68	–601.99	Barin 1989
900.00	–368.28	291.01	–630.19	Barin 1989
1000.00	–352.72	307.40	–660.13	Barin 1989

TABLE 5.79
$2\ NaNO_2 = Na_2O + N_2(g) + 3/2\ O_2(g)$

T K	ΔH° kJ	ΔS° J deg⁻¹	ΔG° kJ	K
298.15	299.99	331.70	201.096	5.833E – 036
400.00	298.89	328.72	167.398	1.375E – 022
500.00	294.68	319.47	134.945	7.965E – 015
600.00	257.51	252.60	105.952	5.961E – 010
700.00	249.92	240.89	81.298	8.570E – 007
800.00	242.76	231.32	57.703	1.706E – 004
900.00	235.98	223.34	34.981	9.324E – 003
1000.00	229.54	216.54	12.996	2.095E – 001

TABLE 5.80
$NaNO_3 = NaNO_2 + 1/2\ O_2(g)$

T K	ΔH° kJ	ΔS° J deg^{-1}	ΔG° kJ	K
298.15	108.99	107.60	76.914	3.341E – 014
400.00	107.92	104.53	66.114	2.321E – 009
500.00	106.65	101.68	55.804	1.478E – 006
600.00	105.29	100.07	45.253	1.148E – 004
700.00	103.52	97.33	35.387	2.286E – 003
800.00	101.76	94.99	25.774	2.075E – 002
900.00	100.04	92.95	16.380	1.120E – 001
1000.00	98.33	91.15	7.176	4.218E – 001

TABLE 5.81
$2\ NaNO_3 = Na_2O + 2\ NO_2(g) + 1/2\ O_2(g)$

T K	ΔH° kJ	ΔS° J deg^{-1}	ΔG° kJ	K
298.15	584.17	425.02	457.447	7.087E – 081
400.00	579.77	412.55	414.752	6.829E – 055
500.00	572.25	395.90	374.298	7.837E – 040
600.00	531.89	324.81	336.994	4.613E – 030
700.00	520.46	307.27	305.367	1.627E – 023
800.00	509.68	292.87	275.383	1.042E – 018
900.00	499.45	280.82	246.716	4.784E – 015
1000.00	489.72	270.56	219.16	3.559E – 012

TABLE 5.82
Decomposition Pressures of $NaNO_3$

T K	P (O_2) bar	P (NO_2) bar	P (NO) bar	P (Tot) bar
298.15	$<10^{-15}$	$<10^{-15}$	$<10^{-15}$	$<10^{-15}$
400.00	$<10^{-15}$	$<10^{-15}$	$<10^{-15}$	$<10^{-15}$
500.00	3.53E – 013	$<10^{-15}$	4.70E – 013	8.23E – 013
600.00	8.30E – 010	4.00E – 013	1.11E – 009	1.94E – 009
700.00	1.60E – 007	2.02E – 010	2.14E – 007	3.74E – 007
800.00	7.78E – 006	1.93E – 008	1.04E – 005	1.82E – 005
900.00	1.51E – 004	6.24E – 007	2.01E – 004	3.52E – 004
1000.00	1.56E – 003	9.50E – 006	2.07E – 003	3.63E – 003

TABLE 5.83
Transitions of $Sr(NO_3)_2$

Phase Change	T(K)	ΔH kJ/mol	ΔS J/(mol*K)	Reference
$Sr(NO_2)_2\ \gamma \rightarrow \beta$	547			15
$\beta \rightarrow \alpha$	558			15
$\alpha \rightarrow l$	676			13, 14
$Sr(NO_3)_2c \rightarrow l$	918	44.6	48.6	140

TABLE 5.84
Thermodynamic Data of $Sr(NO_3)_2$

T K	ΔHf° kJ mol⁻¹	S° J mol⁻¹ deg⁻¹	ΔGf° kJ mol⁻¹	Reference
298.15	–988.72	194.56	–1046.56	Ruzinov 1975
400.00	–972.40	241.65	–1069.06	Ruzinov 1975
500.00	–956.38	277.41	–1095.08	Ruzinov 1975
600.00	–940.35	306.62	–1124.32	Ruzinov 1975
700.00	–924.33	331.32	–1156.25	Ruzinov 1975
800.00	–908.30	352.72	–1190.48	Ruzinov 1975
900.00	–892.28	371.60	–1226.71	Ruzinov 1975
1000.00	–876.25	388.48	–1264.73	Ruzinov 1975

TABLE 5.85
$Sr(NO_3)_2 = SrO + 2\ NO_2(g) + 1/2\ O_2(g)$

T K	ΔH° kJ	ΔS° J deg⁻¹	ΔG° kJ	K
298.15	462.88	443.58	330.622	1.179E – 058
400.00	460.82	437.55	285.796	4.740E – 038
500.00	459.79	435.24	242.176	4.988E – 026
600.00	459.47	434.63	198.693	5.021E – 018
700.00	459.72	435.00	155.217	2.610E – 012
800.00	460.44	435.96	111.673	5.104E – 008
900.00	461.58	437.29	68.013	1.128E – 004
1000.00	463.08	438.87	24.207	5.438E – 002

TABLE 5.86
Decomposition Pressures of $Sr(NO_3)_2$

T K	P (O_2) bar	P (NO_2) bar	P (NO) bar	P (Tot) bar
298.15	$<10^{-15}$	$<10^{-15}$	$<10^{-15}$	$<10^{-15}$
400.00	1.56E – 013	$<10^{-15}$	2.08E – 011	3.64E – 013
500.00	3.10E – 009	2.99E – 011	4.13E – 009	7.24E – 009
600.00	2.29E – 006	5.76E – 008	3.03E – 006	5.33E – 006
700.00	2.57E – 004	1.28E – 005	3.38E – 004	6.08E – 004
800.00	8.92E – 003	7.35E – 004	1.16E – 002	2.12E – 002
900.00	0.141	1.73E – 002	0.183	0.344
1000.00	1.30	0.219	1.66	3.18

TABLE 5.87
Density of $TlNO_2$

Phase	T(K)	d (gcm^{-3})	Reference
l	480	4.905	69
	490	4.885	69
	500	4.867	69
	510	4.849	69
	520	4.830	69
	530	4.811	69
	540	4.792	69
	550	4.773	69
	560	4.755	69

TABLE 5.88
Transitions of $TlNO_2$

0.00	T(K)	ΔH kJ/mol	ΔS J/(mol*K)	Reference
$TlNO_2$s → s	442			15
s → l	459			15
$TlNO_3$ orthorhombic →	334.2	1.0	3.0	135
hexagonal → cubic	416	3.8	9.1	141
cubic → liquid	483	8.83	18.3	143

TABLE 5.89
Thermodynamic Data of $TlNO_3$

T K	ΔHf° kJ mol^{-1}	S° J mol^{-1} deg^{-1}	ΔGf° kJ mol^{-1}	Reference
298.15	–243.93	164.43	–292.95	Ruzinov 1975
400.00	–232.07	198.54	–311.49	Ruzinov 1975
500.00	–205.23	257.53	–334.00	Ruzinov 1975
600.00	–192.21	281.28	–360.97	Ruzinov 1975
700.00	–179.18	301.75	–390.13	Ruzinov 1975
800.00	–166.15	318.75	–421.15	Ruzinov 1975
900.00	–153.13	334.10	–453.81	Ruzinov 1975
1000.00	–140.10	347.82	–487.92	Ruzinov 1975

TABLE 5.90
$2\ TlNO_3 = Tl_2O + 2\ NO_2(g) + 1/2\ O_2(g)$

T K	ΔH° kJ	ΔS° J deg^{-1}	ΔG° kJ	K
298.15	385.01	398.94	266.068	2.411E – 047
400.00	378.96	381.61	226.320	2.774E – 030
500.00	343.68	304.63	191.362	1.016E – 020
600.00	336.78	292.04	161.558	8.589E – 015
700.00	330.54	282.40	132.857	1.217E – 010
800.00	324.88	274.83	105.010	1.390E – 007
900.00	351.10	305.55	76.107	3.824E – 005
1000.00	348.77	303.09	45.680	4.109E – 003

TABLE 5.91
Decomposition Pressures of $TlNO_3$

T K	P (O_2) bar	P (NO_2) bar	P (NO) bar	P (Tot) bar
298.15	$<10^{-15}$	$<10^{-15}$	$<10^{-15}$	$<10^{-13}$
400.00	2.59E – 011	7.38E – 013	3.43E – 011	6.09E – 011
500.00	1.03E – 007	5.64E – 009	1.35E – 007	2.39E – 007
600.00	1.93E – 005	1.40E – 006	2.53E – 006	6.40E – 005
700.00	7.74E – 004	6.62E – 005	1.01E – 003	1.78E – 003
800.00	1.19E – 002	1.13E – 003	1.55E – 002	2.85E – 002
900.00	0.103	1.09E – 002	0.134	0.247
1000.00	0.615	7.24E – 002	0.796	1.481

TABLE 5.92
Thermodynamic Data of $Th(NO_3)_4$

T K	ΔHf° kJ mol^{-1}	S° J mol^{-1} deg^{-1}	ΔGf° kJ mol^{-1}	Reference
298.15	–1446.00	335.00	–1543.88	Phillips et al. 1988
400.00	–1446.00	335.00	–1580.00	Phillips et al. 1988
500.00	–1446.00	335.00	–1613.50	Phillips et al. 1988
600.00	–1446.00	335.00	–1647.00	Phillips et al. 1988
700.00	–1446.00	335.00	–1680.50	Phillips et al. 1988

TABLE 5.93
$Th(NO_3)_4 = ThO_2 + 4\ NO_2(g) + O_2(g)$

T K	ΔH° kJ	ΔS° J deg^{-1}	ΔG° kJ	K
298.15	351.97	895.45	84.99	1.286E – 015
400.00	377.54	969.06	–10.084	2.075E + 001
500.00	404.58	1029.33	–110.087	3.174E + 011

TABLE 5.94
Thermodynamic Data of $Sn(NO_3)_2$

T K	ΔHf° kJ mol^{-1}	S° J mol^{-1} deg^{-1}	ΔGf° kJ mol^{-1}	Reference
298.15	–456.10	208.80	–518.35	Kubaschewski et al. 1993
400.00	–434.79	270.28	–542.90	Kubaschewski et al. 1993
500.00	–413.87	316.96	–572.35	Kubaschewski et al. 1993
600.00	–392.95	355.10	–606.01	Kubaschewski et al. 1993
700.00	–372.03	387.35	–643.18	Kubaschewski et al. 1993

TABLE 5.95
Thermodynamic Data of SnO

T K	ΔHf° kJ mol^{-1}	S° J mol^{-1} deg^{-1}	ΔGf° kJ mol^{-1}	Reference
298.15	–285.77	56.48	–302.61	Barin 1993
400.00	–280.72	71.04	–309.13	Barin 1993
500.00	–275.50	82.67	–316.84	Barin 1993
600.00	–270.11	92.50	–325.61	Barin 1993
700.00	–264.58	101.02	–335.20	Barin 1993

TABLE 5.96
Thermodynamic Data of $Zn(NO_3)_2$ (298.15 K)

	ΔHf° kJ mol^{-1}	S° J mol^{-1} deg^{-1}	Reference
$Zn(NO_3)_2$	–483.7	193.7	49, 133
ZnO	–348.2	43.64	133

REFERENCES

1. Kelley, K. K., *U.S. Bur. Mines Rept. Invest.*, 3776, 1944.
2. Ketelaar, J. A., *Z. Krist.*, 95, 383, 1936.
3. Ziegler, G. E., *Phys. Rev.*, 38, 1040, 1931.
4. Cotton, F. A. and Wilkinson, G., *Advanced Inorganic Chemistry*, 2nd ed., Interscience, New York, 1966, 349.
5. Addison, C. C. and Sutton, D., *Advances in Inorganic Chemistry*, Vol. 8, Cotton, F.A., Ed., Interscience, New York, 1967, 195.
6. Addison, C. C. and Logan, N., *Adv. in Inorganic Chemistry*, 6, 72, 1964.
7. Addison, C. C. and Logan, N., *Preparative Inorg. Reactions*, 1, 142, 1964.
8. Hardy, C. J. and Field, B. O., *J. Chem. Soc.*, 5130, 1963.
9. Chasan, D. E. and Norwitz, G., *Applications of Spectroscopy to the Analysis of Inorganic Nitrates: Phase II. Test Rept. T–69–3–1*, Dept. of the Army, Frankford Arsenal, March 1969.
10. Wait, S. C. and Janz, G. J., *Q. Revs.*, 17, 225, 1963.
11. Hardy, C. J. and Field, B. O., *J. Chem. Soc.*, 5130, 1963.
12. Addison, C. C. and Boorman, P. M., *J. Chem. Soc.*, A, 1434, 1966.
13. Centnerszwer, M. and Piekelny, W., *Bull. Acad. Polon.*, 9A, 389, 1933.
14. Protsenko, P. I. and Bordyushkova, E. A., *Russ. J. Inorg. Chem.*, 39, 1049, 1963.
15. Protsenko, P. I. and Brykova, N. A., *Russ. J. Inorg. Chem.*, 8, 1130, 1963.
16. Oza, T. M. and Jha, J. C., *J. Indian. Chem. Soc.*, 43, 759, 1966.
17. Protsenko, P. I. and Bordyushkova, E. A., *Russ. J. Inorg. Chem.*, 39, 1049, 1965.
18. Vratny, F., *J. Appl. Spectr.*, 13, 59, 1959.
19. Patil, K. C., Gosavi, R. K., and Rao, C. N. R., *Inorg. Chim. Acta*, 1, 155, 1967.
20. Oza, T. M., Jha, J. C., and Ezekiel, E. I., *J. Indian Chem. Soc.*, 45, 420, 1963.
21. Addison, C. C. and Coldrey, J. M., *J. Chem. Soc.*, 468, 1961.
22. Gordon, S. and Campbell, C., *Anal. Chem.*, 27, 1102, 1955.
23. Haigh, F. L., J. *Am. Chem. Soc*, . 27, 1102, 1955.
24. Addison, C. C. and Walker, A., *J. Chem. Soc.*, 1220, 1963.
25. Plekkotkin, K., *J. Appl. Chem. USSR*, 40, 2470, 1967.
26. Wagman D. D., Evans, W. H., Paker, V. B., Schumm, F. H., Halow, I., Bailey, S. M., Chuney, K. L., and Nutall, R. L., The NBS Tables of Thermodynamic Properties, *J. Phys. Chem. Ref. Data*, 11, Suppl. 2, 1982.
27. Straub, D. K., Sisler, H. H., and Ryschkewitch, G. E., *J. Inorg. Nucl. Chem.*, 24, 919, 1962.
28. Gmelin's *Handbuch der Anorganischen Chemie*, 8th ed., Verlag Chemie GmbH., Weinheim, Germany.
29. Addison, C. C. and Gatehouse, B. M., *J. Chem. Soc.*, 613, 1960.
30. Kleppa, O. J. and Menschel, S. V., *J. Phys. Chem.*, 67, 902, 1963.
31. Krause, H., *Apparatebau*, 42, 265, 1930.
32. Protsenko, P. I. and Brykova, N. A., *Russ. J. Inorg. Chem.*, 8, 1130. 1963.
33. Oza, T. M. and Oza, T. O., *J. Chem. Soc.*, 907, 1953.
34. Gordon, S. and Campbell, C., *Anal. Chem.*, 27, 1102, 1955.
35. Addison, C. C. and Coldrey, J. M., *J. Chem. Soc.*, 468, 1961.
36. Shirokov, A. V., *Zh. Prikl. Khim.*, 56, 1376, 1983.
37. Protsenko, P. I. and Bordyushkova, E. A., *Russ. J. Inorg. Chem.*, 10, 657, 1965.
38. Bagaratian, N. V., Verkhutov, E. V., Ilyin, M. K., Makarov, A. V., and Nikitin, O. T., *Adv. Mass Spectr.*, 7A, 578, 1978.
39. Rao, K. J. and Rao, C. N. R., *Materials Sci.*, 1, 238, 1966.
40. Bordyushkova, E. A., Protsenko, P. I., and Venerovskaya, L. N., *J. Appl. Chem.*, 40, 1386, 1967.
41. Ames, L. L., Ling-Fai Wen, J., and Margrave, J. L., *Inorg. Nucl. Letters*, 9, 1243, 1973.
42. Addison, C. C. and Chapman, D. J., *J. Chem. Soc.*, 540, 1964.
43. Chretien, A. and Longi, Y., *Compt. Rend.*, 217, 504, 1943.
44. Addison, C. C. and Sutton, D., *J. Chem. Soc.*, Suppl. No. 1, 5553, 1964.
45. Guntz, A. and Martin, M., *Bull. Soc. Chim. France*, 7, 313, 1910.
46. Addison, C .C. and Logan, N., *Adv. in Inorg. Chem. Radiochem.*, 6, 72, 1964.
47. Gmelin's *Handbuch der Anorganischen Chemie*, 8th ed., Verlag Chemie GmbH, Weinheim, Germany.

48. Addison, C. C. and Hathaway, B. J., *J. Chem. Soc.,* 3009, 1958.
49. Latimer, W. M., *The Oxidation States of the Elements and their Potentials in Aqueous Solutions*, 2nd ed., Prentice Hall, New York, 1952.
50. Dupre, A., *Compt. Rend.,* 86, 720, 1878.
51. Plekhotin, V. K., *J. Appl. Chem. USSR*, 40, 2470, 1967.
52. Field, B. O. and Hardy, C. J., *J. Chem. Soc.,* 4428, 1964.
53. Tuseev, N. I., Izmailovich, A. S., and Komissarova, L. N., *Vestnik Moskovskogo Universitete, Khimia,* 34, 453, 1979.
54. Schmeisser, M. and Koehler, G., *Angew. Chem.*, 77, 456, 1965.
55. Addison, C. C., Hathaway, B. J., and Logan, N., *J. Chem. Soc.,* 51, 1958.
56. Patil, K. C., Gosavi, R. K., and Rao, C. N. R., *Inorg. Chim. Acta,* 1, 155, 1967.
57. Baekeland, L., *J. Am. Chem. Soc.*, 26, 391, 1904.
58. Neumann, B. and Sonntag, A., *Z. Elektrochem.,* 39, 799, 1933.
59. Vratny, F. and Gugliotta, F., *J. Inorg. Nucl. Chem.,* 25, 1129, 1963.
60. Bodenstein, M., *Z. Phys. Chem.*, 100, 68, 1922.
61. Neumann, B. and Sonntag, A., *Z. Elektrochem.,* 41, 611, 1935.
62. Margulis, E. V., Shokarev, M. M., Savchenko, L. A., Beisekeva, L. I., and Vershinina, F. I., *Russ. J. Inorg. Chem.,* 17, 21, 1972.
63. Protsenko, P. I. and Bordyushkova, E. A., *Zh. Neorg. Khim.*, 10, 1215, 1965; *Zh. Fiz. Khim.*, 39, 1978, 1965.
64. Lee, A. K. and Johnson, E. F., *Inorg. Chem.*, 11, 782, 1972.
65. Krasvoskaya, V. A., Ganz, S. N., Gulyanov, Yu. M., and Sergeiko, L. A., *Zh. Prikl. Khim.*, 54, 991, 1981.
66. ibid. *Prikl. Khim.,* 56, 533, 1983.
67. Protsenko, P.I., Protsenko, A. V., and Razamouskaya, O. N., *Russ. J. Inorg. Chem.,* 10, 405, 1965.
68. Haigh, F. L., *J. Am. Chem. Soc.,* 34, 1137, 1912.
69. Janz, G. J., Lakshminarayanan, G. R., Lorenz, P. K., and Tomkins, R. P. T., *Molten Salts: Volume 1. Electrical Conductance, Density, and Viscosity Data,* NSRDS-NBS 15, 1968.
70. Francosini, P. and Sinistry, C., *Ricerca Sci. Re.*, A3, 411, 1963.
71. Latimer, W. M., *The Oxidation States of the Elements and their Potentials in Aqueous Solutions*, 2nd ed., Prentice-Hall, New York, 1952.
72. Shirokov, A. V., *Zh. Prikl. Khim.*, 56, 1376, 1983.
73. Guntz, A. and Martin, M., *Bull. Soc. Chim. France,* 7, 313, 1910.
74. Dehnicke, K. and Straehle, J., *Chem. Ber.,* 97, 2502, 1964.
75. De Bruijn, T. J. W., Ipekoglu, A. N., DeJong, W. A., VandenBerg, P. J., *Thermochim. Acta,* 45, 265, 293, 305, 1981.
76. Addison, C. C. and Gatehouse, B. M., *J. Chem. Soc.,* 3099, 1958.
77. Dehnicke, K. and Strähle, J., *Chem. Ber.*, 97, 2502, 1964.
78. Ray, P. C., *J. Chem. Soc.,* 523, 1904.
79. Oza, T. M., Jha, J. C., and Ezekiel, E. I., *J. Indian Chem. Soc.,* 45, 420, 1968.
80. Potts, R. A. and Alfred, A. L., *Inorg. Chem.*, 5, 1066, 1966.
81. Oza, T. M. and Oza, T. O., *J. Chem. Soc.,* 907, 1963.
82. Field, B. O. and Hardy, C. J., *J. Chem. Soc.,* 4428, 1964.
83. Addison, C. C., Johnson, B. F. G., Logan, N., and Wojcicki, *Proc. Chem. Soc.*, 306, 1961 .
84. Addison, C. C. and Logan, N., *Adv. Inorg. Radiochem.*, 6, 72, 1964.
85. Field, B. O. and Hardy, C. J., *J. Chem. Soc.,* 4428, 1964.
86. Manchot, W. and Waldmuller, A., *Ber. Deutsch. Chem. Ges.*, 59B, 2363, 1926.
87. Oza, J. M., *J. Indian Chem. Soc.,* 22, 173, 1945.
88. Oza, T. M. and Patel, S. A., *J. Indian Chem. Soc.,* 31, 519, 1954.
89. Freeman, E. S., *J. Am. Chem. Soc.,* 79, 838, 1957.
90. Kramer, C. A., Munir, Z. A., and Volponi, J., *Thermochim. Acta,* 55, 11, 1982.
91. Mustajoki, A., *Ann. Acad. Sci. Fenn.*, A VI, 99, 1966.
92. Newns, D. M. and Stavely, L. A. K., *Chem. Rev.*, 66, 267, 1967.
93. Stromme, K. O., *Acta Chem. Scand.*, 23, 1625, 1969.
94. Bordyushkova, E. A., Protsenko, P. I., and Venerovskaya, I. N., *J. Appl. Chem. USSR*, 40, 1386,1967.
95. Bartholemew, R. F., *J. Phys. Chem.*, 70, 3442, 1966.

96. Sirotkin, G. D., *Russ. J. Inorg. Chem.,* 4, 1180, 1959.
97. Weigel, F. and Trinkl, A., *Radiochim. Acta*, 19, 199, 1973.
98. Patil, K. C., Gosavi, R. K., and Rao, C. N. R., *Inorg. Chim. Acta,* 1, 155, 1967.
99. Walker, A. and Ferraro, J. R., *J. Chem. Phys.,* 43, 2689, 1965.
100. Wendlandt, W. W., *Anal. Chim. Acta*, 15, 435, 1956.
101. Wendlandt, W. W. and Bear, J. L., J. *Nucl. Chem*., 12, 276, 1960.
102. Protsenko, P. I. and Bordyushkova, E. A., *Russ. Inorg. Chem*., 10, 657, 1965.
103. Erdey, L., Liptay, G., and Gal, S., *Talanta,* 12, 883, 1965.
104. Mustajoki, A., Suom. Tideakat Toim. *Ann. Acad. Sci. Fenn*., A VI, No. 9, 1958.
105. Hardy, C. J. and Field, B. O., *J. Chem. Soc.,* 5130, 1963.
106. Protsenko, P. I. and Protsenko, A. V., Russ. *J. Phys. Chem*., 8, 1130, 1963.
107. Haigh, F. L., *J. Am. Chem. Soc.,* 34, 1137, 1912.
108. Mellor, J. W., *Comprehensive Treatise on Inorganic Chemistry*, Vol. V, Longmans Green, New York, 1924, 493.
109. Patil, K., Gosavi, R. K., and Rao, C. N. R., *Inorg. Chim. Acta,* 1, 155, 1967.
110. Centnerszwer, M. and Blumenthal, M., *Bull. Intern. Polon. Sci., Ser. A*, 470, 1936.
112. Vratny, F., J. *Appl. Spectry.*, 13, 59, 1959.
113. Szper, J. and Fizman, K., *Z. Anorg. Chem.,* 206, 257, 1932.
114. Freeman, E. S., *J. Phys. Chem*., 60, 1487, 1956.
115. Bond, B. D. and Jacobs, P. W. M., *J. Chem. Soc.,* A, 1265, 1966.
116. Bordyushkova, E. A., Protsenko, P. I., and Venerovskaya, L. N., *J. Appl. Chem. USSR*, 40, 1386, 1967.
117. Centnerszwer, M. and Piekelny, W., *Bull. Acad. Polon*., 9A, 389, 1933.
118. Protsenko, P. I. and Bordyushkova, E. A., Russ. *J. Phys. Chem*., 39, 1049, 1965.
119. Oza, T. M. and Patel, S. A., *J. Indian Chem. Soc.,* 31, 519, 1954.
120. Lumsden, J., *Thermodynamics of Molten Salt Mixtures*, Academic Press, New York, 1966.
121. Gordon, S. and Campbell, C., *Anal. Chem*., 27, 1102, 1955.
122. Addison, C. C. and Coldrey, J. M., *J. Chem. Soc.,* 468, 1961.
123. Gordon, S. and Campbell, C., *Anal. Chem.,* 27, 1102, 1955.
124. Protsenko, P. I. and Brykova, N. A., *Russ. J. Inorg. Chem.,* 8, 1130, 1963.
125. Kleppa, O. J. and McCarthy, F. G., *J. Chem. Engin. Data*, 8, 331, 1963.
126. Cubiciotti, D., *High Temp. Sci*., 2, 131, 1970.
127. Addison, C. C. and Simpson, W. B., *J. Chem. Soc.,* 598, 1965.
128. Field, B. O. and Hardy, C. J., *J. Chem. Soc.,* 5278, 1963.
129. Reihlen, H. and Hake, A., *Lieb. Ann*., 452, 47, 1927.
130. Addison, C. C., Lewis, J., and Thompson, R., *J. Chem. Soc.,* 2829, 1951.
131. Field, B. O. and Hardy, C. J., *Proc. Chem. Soc*., 76, 1962.
132. Bureau, J., *Ann. Chim*., 8, 1, 1957.
133. Wagman D. D., Evans, W. H., Parker, V. B., Schumm, R. H., Halow, I., Bailey, S. M., Churney, K. L., and Nuttall, R. L. The NBS Tables of Chemical and Thermodynamic Properties, *J. Phys. Chem. Ref. Data*, 11, Suppl. 2, 1982.
134. Mustajoki, A., Suom. Tiedeakat. Toim. *Ann. Akad. Sci. Fenn*., AVI, No. 5, 1957.
135. Rossini, F. D., Wagman, D. D., Evans, W. H., Levine, S., and Jaffe, I., *Selected Values of Thermodynamic Properties*. NBS Circ. 500, 1952.
136. Rainsborough, V. C. and Wetmore, F. E. W., *Austr. J. Chem*., 20, 1, 1967.
137. Janz, G. J. and Kelly, K. K., *J. Phys. Chem*., 67, 2848, 1963.
138. Kozlowski, T. R. and Bartholemew, B. F., *J. Electrochem. Soc.,* 114, 937, 1967.
139. Mellor, J. W., *Comprehensive Treatise of Inorganic Chemistry*, Vol IV, Longmans Green, New York, 1924, 93.
140. Kleppa, O. J. and Meschel, S. V., *J. Phys. Chem*., 67, 902, 1963.
141. Hovi, V., Arell, A., and Varteva, M., *Ann. Acad. Sci. Fenn*., AVI, No. 39, 1959.
142. Mustajoki, A., *Suom. Tiedeakat. Toim. Ann. Akad. Sci. Fenn.,* AVI, No. 7, 1957.
143. Rolla, M., Francosini, P., and Riccardi, R., *Dicussions Faraday Soc*., 32, 84, 1961.
144. Misciatelli, P., *Gazzette*, 60, 882, 1932.

The following references occur in thermodynamic tables:

CRC 94: *CRC Handbook of Physics and Chemistry*, 74th ed., CRC Press, Boca Raton, FL, 1994.

Knacke et al. 91: Knacke, O., Kubaschewski, O., and Hasselman, K., *Thermochemical Properties of Inorganic Substances*, 2nd ed., Springer Verlag, Berlin, 1991.

Fabricius 94: Fabricius, G., Liukkonen, S., Sundholm, G., Fusi-kasilisen Kemiu taulukoita Otatieto, Espoo, 1994.

Gurvich 90: Gurvich, L. V., *Termodinamicheski Svoistva*, several volumes.

Phillips 88: Phillips, S. L., Hale, F. V., Silvester, L. F., *Thermodynamic Tables for Nuclear Waste Isolation*, Sandia National Laboratory, Albuquerque, NM, 1988.

The following reference appears in some tables:

Kubaschewski, O., Alcock, C., and Spencer, P. J., *Materials Thermochemistry*, 6th ed., Pergamon Press, London, 1993.

6 Oxysalts of the Halogens

6.1 GENERAL

All of the halogens except fluorine, which is more electronegative than oxygen, form oxyanions in which the valence of the halogen varies, e.g., chlorite, chlorate, perchlorate.

The chemistry of specific classes of salts, e.g., chlorates, is discussed in the introduction to the particular halogen. Further information is available in textbooks on inorganic chemistry.

A book by Solymosi[1] deals with the chemistry of solid oxyhalogens in great detail. This book covers the literature through 1974, and includes kinetic studies. The author's previous monograph[2] was published the same year as part of a larger work on salts with oxyanions. A search of the literature through 1996 has shown a noticeable decrease in publications in this field since about 1985. Material from all of the above works has been included in this and succeeding chapters.

In contrast to other classes of salts with oxyanions, high-temperature thermodynamic data are largely lacking; e.g., HSC only lists data for four. These have been listed as well as ΔHf° and S° from Wagman et al.[3] In this chapter thermodynamic data for the three halogen gases have been listed (Tables 6.1 to 6.3)

6.2 PHASE TRANSITIONS

Since many oxyhalogen salts decompose in the solid state, melting points are available for relatively few compounds. In some cases the decomposition is slow enough that the melting point of a nearly pure salt can be measured. Values have been listed whenever possible, even if they are not very precise.

6.3 DENSITY

Except for $LiClO_4$, which is stable for some distance above its melting point, no molten salt densities seem to have been measured. Most ambient- temperature measurements are of fairly low accuracy. The most reasonable values have been listed for convenience. These are obtained either experimentally (usually pycnometrically, by displacement of a liquid such as CCl_4) or calculated from X-ray data. Solids frequently exhibit voids and/or occlude solvent so that reported values may scatter considerably. X-ray values are those for a perfect lattice and thus are only approximate guides to densities of "real" crystals. Whenever recent data are not available, the evaluation from *International Critical Tables*[4] has been accepted.

6.4 CHEMISTRY OF THE HALOGEN-OXYSALTS

Apparently the halogen-oxysalts are predominantly ionic, i.e., the anion constitutes a definite charged entity with covalent bonds between the halogen and oxygen atoms, but only electrostatic interactions between cations and anions. This is in contrast to the nitrates where some are ionic and some covalent. Because the outer electronic structure of the halogens is so similar, it is not surprising that the structures of corresponding ions are also similar, e.g. ClO_3^-, BrO_3^-, and IO_3^- are pyramidal. Table 6.4 is a listing of halogen oxyanions given by Wells.[5]

One of the most striking differences between the halogens is the large number of anhydrous iodine compounds, compared to those of chlorine and bromine. However, this difference may only be due to lack of interest exhibited by chemists in preparing these compounds. For example, although chlorites have been known for a long time, anhydrous bromites and iodites have been synthesized only relatively recently. Similarly, perbromates were said to be too unstable to exist, but several of them have now been synthesized, and the BrO_4^- ion seems about as stable as ClO_4^- and IO_4^-. This topic has been discussed in detail by Cox and Moore.[6] In nearly all respects the paralellism is complete: hypohalites (XO^-) are not known as pure anhydrous salts; the halites ClO_2^-, BrO_2^-, and IO_2^- are known, although relatively few bromites and iodites have been prepared; the halates ClO_3^- , BrO_3^-, and IO_3^- exist; the (meta) perhalates ClO_4^-, BrO_4^-, and IO_4^- are known. In each of these groups the structures are quite similar. Only the tendency of iodine in its +7 oxidation state to polymerize leads to a variety of iodates, a phenomenon apparently quite absent from chlorine and bromine.

In general, the stabilities of the anions for each oxidation state increase in the order Cl < Br < I. How great this difference is may depend on the property used to measure it. For example, both perchlorates and metaperiodates may be heated to several hundred degrees before decomposing, but this similarity is probably due to the kinetics of the decomposition, since perchlorates are thermodynamically unstable even at ambient temperature.

Since the chemistry of the individual classes of oxyhalogens, i.e., chlorine, bromine, and iodine, are independent of each other, they have been placed in separate chapters, each with their own references.

TABLE 6.1
Thermodynamic Data of Cl_2(g)

T K	ΔHf° kJ mol^{-1}	S° J mol^{-1} deg^{-1}	ΔGf° kJ mol^{-1}	Reference
298.15	0.00	223.08	–66.51	JANAF 1985
400.00	3.53	233.26	–89.77	JANAF 1985
500.00	7.11	241.24	–113.51	JANAF 1985
600.00	10.74	247.86	–137.97	JANAF 1985
700.00	14.41	235.52	–163.05	JANAF 1985
800.00	18.11	258.46	–188.65	JANAF 1985
900.00	21.83	262.84	–214.72	JANAF 1985
1000.00	25.56	266.77	–241.21	JANAF 1985
1100.00	29.31	270.34	–268.07	JANAF 1985
1200.00	33.07	273.61	–295.27	JANAF 1985

TABLE 6.2
Thermodynamic Data of Br_2(g)

T K	ΔHf° kJ mol^{-1}	S° J mol^{-1} deg^{-1}	ΔGf° kJ mol^{-1}	Reference
298.15	30.91	245.39	–42.25	Barin et al. 1977
400.00	34.62	256.10	–67.81	Barin et al. 1977
500.00	38.31	264.33	–93.85	Barin et al. 1977
600.00	42.03	271.11	–120.64	Barin et al. 1977
700.00	45.77	276.87	–148.04	Barin et al. 1977
800.00	49.52	281.87	–175.98	Barin et al. 1977
900.00	53.27	286.30	–204.40	Barin et al. 1977
1000.00	57.04	290.27	–233.23	Barin et al. 1977
1100.00	60.81	293.86	–262.44	Barin et al. 1977
1200.00	64.59	297.16	–291.99	Barin et al. 1977

TABLE 6.3
Thermodynamic Data of $I_2(g)$

T K	ΔHf° kJ mol^{-1}	S° J mol^{-1} deg^{-1}	ΔGf° kJ mol^{-1}	Reference
298.15	62.42	260.69	–15.30	Barin 1993
400.00	66.20	271.58	–42.43	Barin 1993
500.00	69.93	279.91	–70.02	Barin 1993
600.00	73.68	286.75	–98.37	Barin 1993
700.00	77.45	292.55	–127.34	Barin 1993
800.00	81.22	297.59	–156.85	Barin 1993
900.00	85.00	302.04	–186.84	Barin 1993
1000.00	88.79	306.03	–217.25	Barin 1993
1100.00	92.58	309.65	–248.03	Barin 1993
1200.00	96.38	312.96	–279.17	Barin 1993

TABLE 6.4
Bonding of the Halogen Oxyanions

Number of σ-Electron Pairs	Type of Hybrid	Number of Lone Pairs	Bond Arrangement	Examples
4	sp^3	0	Tetrahedral	ClO_4,IO_4
		1	Trigonal bi-pyramidal	ClO_3,BrO_3,IO_3
5	sp^3d_z	0	Trigonal bi-pyramidal	IO_5 (?)
6	sp^3d_y	0	Octrahedral	IO_6

REFERENCES

1. Solymosi, F., *Structure and Stability of Salts of Halogen Oxyacids in the Solid Phase*, John Wiley and Sons, London, New York, Sydney, Toronto, 1977.
2. Stern, K. H., High temperature properties and decomposition of inorganic salts. Part 4, Oxysalts of the halogens, *J. Phys. Chem. Ref. Data,* 3, 481, 1974.
3. Wagman, D. D., Evans, W. H., Parker, V. B., Schumm, R. H., Halow, I., Bailey, S. M., Churney, K. L., and Nuttall, R. L., The NBS tables of chemical thermodynamic properties, *J. Phys. Chem. Ref. Data,* 11, Suppl. 2, 1982.
4. *International Critical Tables*, Vol. I., McGraw Hill New York, 1926.
5. Wells, A. F., *Structural Inorganic Chemistry,* 3rd ed., Academic Press, New York, 1962.
6. Cox, M. M., *J. Phys. Chem.*, 74, 627, 1970.

7 Oxysalts of Chlorine

7.1 GENERAL

There are only three oxyanions that form anhydrous salts: chlorites (ClO_2^-), chlorates (ClO_3^-), and perchlorates (ClO_4^-). Their stability increases with the increasing oxidation state of the halogen. There is no clear evidence for the existence of stoichiometric, anhydrous hypochlorites (ClO^-). Even for the few reported salts (NaClO, $Ca(ClO)_2$) one finds reports of "smell of chlorine," indicating continuous decomposition at ambient temperature.

The known and doubtful salts are shown in Figures 7.1 to 7.3. Although more elements form perchlorates than chlorites and chlorates, the chlorates seem underrepresented. Any metal that forms a chlorite should also form a chlorate. One would also expect the lanthanides to form all three classes of salts.

The quantitative high-temperature chemistry of chlorine oxysalts is rather underdeveloped. There are very few thermodynamic data for these compounds above 298 K. Even when they exist, they must be applied cautiously, since there may be kinetic rather than thermodynamic factors that determine decomposition behavior. Although the thermal decomposition of a few compounds has been studied very carefully (e.g., the $KClO_4$ literature extends back for more than a hundred years because of the compound's use in explosives), the bulk of the available information is qualitative or semiquantitative. In recent years this has changed somewhat with increasing use of automated techniques such as DTA and TGA. Many of the reactions are complex, with mechanisms frequently controversial and not completely worked out. Decomposition products may depend on experimental conditions; e.g., salts are frequently prepared by dehydration of their hydrates, and residual water may affect the course of the decomposition.

As pointed out above, perchlorates are thermodynamically unstable even at ambient temperature so that calculations are not useful in predicting the actual behavior of the salt. Thermodynamic properties of only a few salts have been listed by HSC above 298 K, and these are reproduced in this work, but no calculations have been done for reactions. Thermodynamic data at 298.15 K are given[1] under each element.

7.2 CLASSES OF OXYCHLORINE SALTS

7.2.1 CHLORITES

The best-studied chlorite, $AgClO_2$, is probably not typical of chlorites generally, since the short Ag–Cl distance is suggestive of $AgClO_2$ molecules. The O–Cl–O angle in the ClO_2 ion is 110°, with a Cl–O interatomic distance of 1.5 Å.

When heated, most chlorites disproportionate according to

$$3\ ClO_2^- = 2\ ClO_3^- + Cl^-$$

although under very rapid heating only O_2 and chlorides may form explosively, probably because the chlorate then decomposes simultaneously.

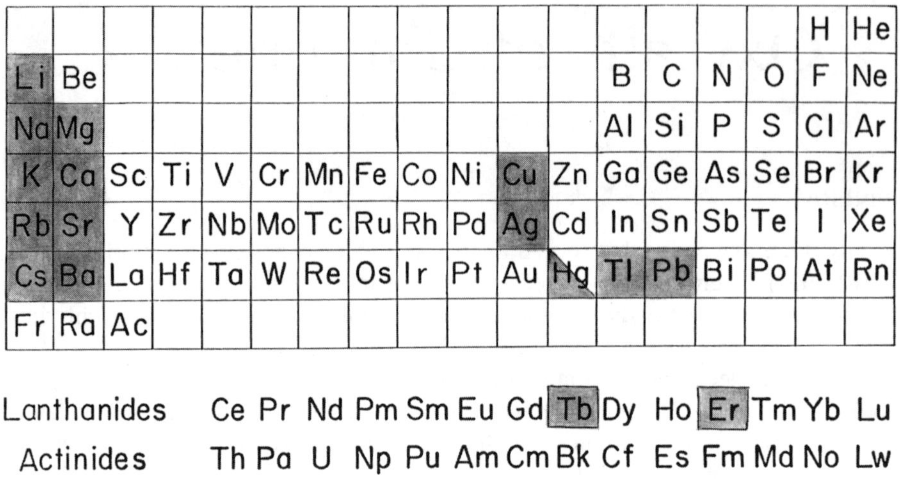

FIGURE 7.1 The known chlorites.

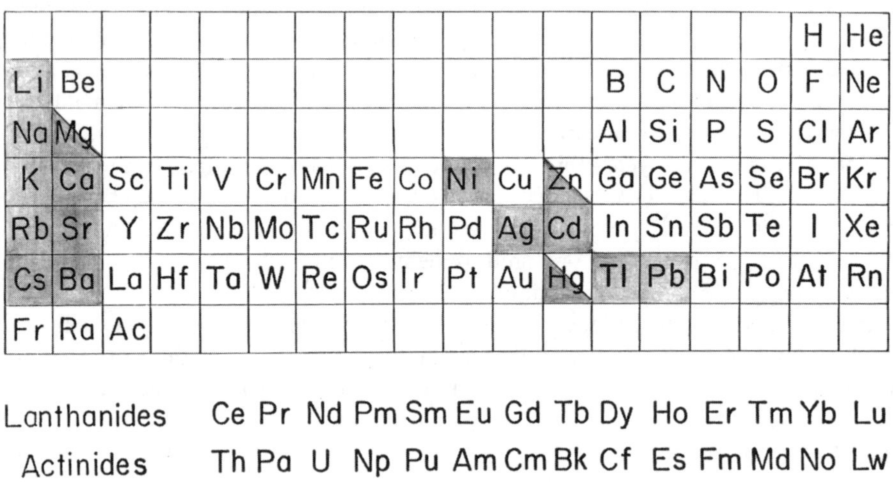

FIGURE 7.2 The known chlorates.

7.2.2 Chlorates

The chlorate ion is a rather flat pyramid with Cl at the apex and the O–Cl–O bond angle of 110°. The Cl–O distance is near 1.46 Å. A picture of the ion, constructed by Weigel et al.[2] from known data, is shown in Figure 7.4.

Most chlorates undergo two decomposition reactions which usually occur simultaneously:

$$4ClO_3^- = 3\ ClO_4^- + Cl^-$$

$$ClO_3^- = Cl^- + 3/2\ O_2$$

The subsequent fate of the perchlorate depends on the temperature. At relatively low temperatures it is stable and thus remains. At temperatures high enough for it to decompose, the final products are chloride and oxygen. When the metal exists in more than one oxidation state, the oxide may form. The chlorates of the rare earths, scandium, and yttrium form oxychlorides.

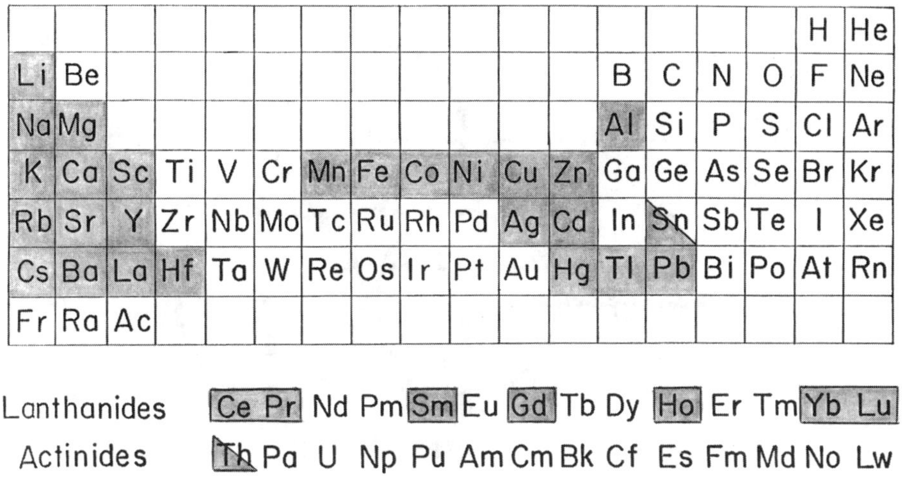

FIGURE 7.3 The known perchlorates.

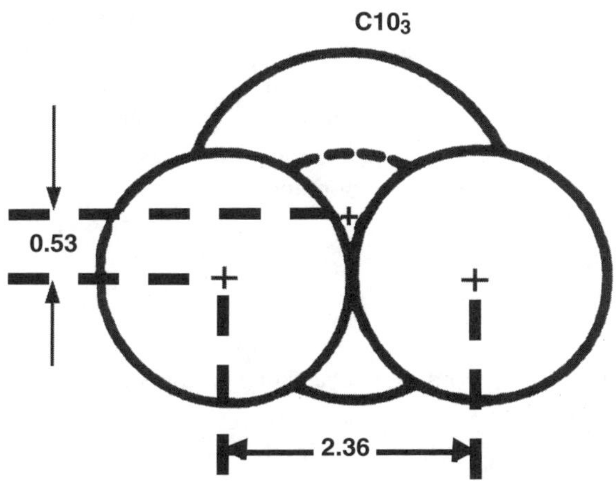

FIGURE 7.4 Structure of the chlorate ion. (From Weigel, D., Imelik, B., and Prettre, M., *Bull. Soc. Chim. France*, 1427, 1962. With permission.)

7.2.3 Perchlorates

The perchlorate ion (ClO_4) is tetrahedral, with a Cl–O distance near 1.4Å (Figure 7.5).[2] The perchlorates decompose by one of two reactions: salts of the more electropositive metals go to the chloride

$$MClO_4 = MCl + 2\ O_2$$

the others, e.g., Cd, Zn, go to the oxide:

$$MClO_4 = 1/2\ M_2O + 1/2\ Cl_2 + 7/2\ O_2$$

Only $Mg(ClO_4)_2$ simultaneously decomposes by both reactions.

Markowitz[2a] has discussed perchlorate decomposition in terms of the thermodynamics of the above two reactions. Since $\mathbf{\Delta}S°$ for both reactions is highly positive (because gaseous products are

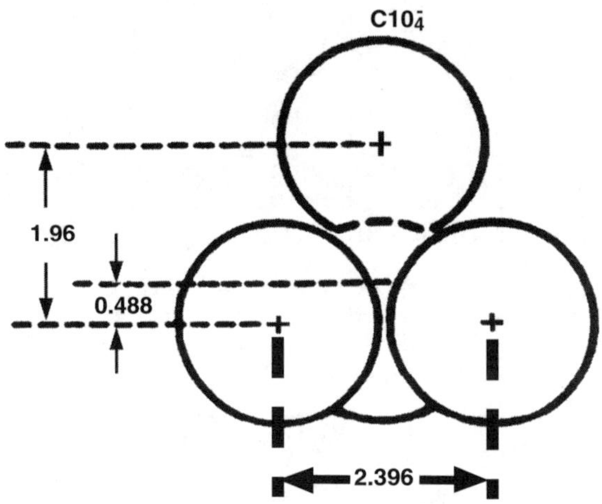

FIGURE 7.5 Structure of the perchlorate ion. (From Weigel, D., Imelik, B., and Prettre, M., *Bull. Soc. Chim. France*, 1427, 1962. With permission.)

formed) the determining factor is ΔH°. Since ΔH° is not very temperature dependent, values at 298 K can be used to predict the reaction. In effect, when the difference per equivalent in ΔHf° (chloride-oxide) is negative, the product will be the chloride. Most perchlorates form either the chloride or the oxide. $Mg(ClO_4)_2$ and some transition metal perchlorates form both because the enthalpy difference is not very large. However, for both kinds of reaction ΔH is negative, so it is only a question of which reaction is more spontaneous.

7.3 INDIVIDUAL SALTS

7.3.1 Aluminum

The anhydrous salt, $Al(ClO_4)_3$, decomposes according to the stoichiometry

$$2\ AlClO_4)_3 = Al_2O_3 + 3\ Cl_2O + 9\ O_2$$

Between 240 and 290°C decomposition kinetics fit the Erofeev equation, according to which the fraction decomposed is given by

$$\alpha = 1 - \exp(-kt^n)$$

where the value of n depends on the shape of the nuclei. From the temperature dependence of k, an Arrhenius activation energy of 82.8 kJ was calculated.[4]

7.3.2 Barium

7.3.2.1 $Ba(ClO_2)_2$

The salt is rather unstable, but its precise mode of decomposition depends on the rate at which it is heated.[5,6] If it is heated slowly or at a constant temperature in the range 130 to 180°C, it disproportionates according to

$$3\ Ba(ClO_2)_2 = 2\ Ba(ClO_3)_2 + BaCl_2$$

In this temperature range, $Ba(ClO_3)_2$ is stable. If the salt is heated so rapidly that the reaction has not been completed, $Ba(ClO_2)_2$ decomposes explosively at 200°C according to

$$Ba(ClO_2)_2 = BaCl_2 + 2\ O_2$$

7.3.2.2 $Ba(ClO_3)_2$

Two studies of the decomposition disagree somewhat.[7,8] There is agreement that both the perchlorate and the chloride are formed:

$$4\ Ba(ClO_3)_2 = 3\ Ba(ClO_4)_2 + BaCl_2$$

$$BaClO_3)_2 = BaCl_2 + 3\ O_2$$

Both agree that the decomposition does not become appreciable until about 380°C. (The disproportionation does not lead to a weight loss and can only be detected by analysis.) At 396°C the salt melts and decomposition then becomes rapid since it occurs in the liquid. At this temperature the second reaction predominates, whereas slow decomposition is the major one below 396°C.[8] $Ba(ClO_2)_2$ may be an intermediate in the decomposition, but it is too unstable to detect.

7.3.2.3 $Ba(ClO_4)_2$

The salt exhibits two solid-state transitions: a rhombic-cubic one at 557 K, and a cubic to γ one at 633 K.[9]

The density of the salt is 3.681 gcm^{-3} at 298 K.[10]

Above 435°C the salt decomposes according to the overall reaction[11–13]

$$Ba(ClO_4)_2 = BaCl_2 + 4\ O_2$$

However, the decomposition mechanism appears to be quite complex and has been the subject of several studies. Acheson and Jacobs[11] found that α = fraction decomposed vs. time plots showed a distinct break at $\alpha = 0.52$, corresponding to the formation of $Ba(ClO_2)_2$. This discontinuity can be prevented by pumping off the oxygen. Addition of $BaCl_2$ to the perchlorate before decomposition (a) removes the induction period and (b) reduces the value of α at which the discontinuity occurs. Solymosi and Braun[12] proposed that the rapid reaction occurs in the melt, the slower rate being caused by the freezing of the reaction mixture, whereas Acheson and Jacobs[11] proposed that (a) the perchlorate decomposes by successive removal of oxygen atoms, and (b) $Ba(ClO_2)_2$ undergoes disproportionation as it is formed. The discrepancies between the two studies were then resolved by a newer study.[14] The main features are that O atoms are successively removed from the perchlorate ion; in an isothermal experiment these stages overlap; the complete removal of perchlorate is not achieved by $\alpha = 0.52$, nor is there any accumulation of chlorate. The authors also emphasize the differences between experiments carried out isothermally and under rising temperature regimes, such as DTA and TGA. For a more complete explanation the original literature should be consulted.

The detailed kinetics of the decomposition mechanism have also been reexamined by Bohidar and Mohanty[15] who also studied the effect of radiation. The decomposition proceeds through four stages: initial rapid gas evolution, induction, a short acceleratory stage, and a long decay stage. The kinetics follow the Prout-Tompkins equation, according to which individual product nuclei catalyze neighboring molecules, which is clearly more feasible in the molten state. The activation energy corresponds to the Cl–O bond energy.

Thermodynamic data at 298.15 K are given in Table 7.1.

7.3.3 Cadmium

7.3.3.1 $Cd(ClO_3)_2$

The anhydrous salt has been prepared,[16] but its high-temperature properties have not yet been determined.

7.3.3.2 $Cd(ClO_4)_2$

Decomposition of the anhydrous salt has been determined in the range 377 to 403°C.[17] At the lower end of the range the decomposition is quite slow, approximately 10% of the material decomposing in two hours. In contrast, Hering and Leray[18] report decomposition in the range 150 to 200°C. This would be expected in the presence of traces of water. Most of the decomposition goes according to

$$Cd(ClO_4)_2 = CdCl_2 + 4\ O_2$$

and about 3% to the oxide:

$$Cd(ClO_4)_2 = CdO + Cl_2 + 7/2\ O_2$$

There are no thermodynamic data for $Cd(ClO_3)_2$ at 298.15 K. An older NBS compilation gives –227.7 $kJmol^{-1}$,[19] but this has been dropped from the newer publication.[1]

7.3.4 Calcium

7.3.4.1 $Ca(ClO)_2$

Although several authors have described various materials containing this compound,[20] these have usually been commercial bleaching powders containing various amounts of $CaCl_2$, $Ca(ClO_3)_2$, $CaCO_3$, and H_2O. The most nearly pure anhydrous salt seems to have been prepared by Urano.[21] The compound is quite stable at 45°C, losing weight only slightly (<5%) over several hundred hours. Above 150°C it decomposes violently with the evolution of O_2, leaving $CaCl_2$.

7.3.4.2 $Ca(ClO_2)_2$

This salt is a white, crystalline powder, probably pseudocubic in structure.[22] A nearly (93%) pure preparation has been decomposed at 180 to 200°C. At this temperature the salt decomposes according to[23]

$$3\ Ca(ClO_2)_2 = 2\ Ca(ClO_3)_2 + CaCl_2$$

The compound has been reported to explode due to mechanical shock.

7.3.4.3 $Ca(ClO_3)_2$

The high-temperature properties of this salt appear not to have been studied since about 1900. According to the older literature,[20] the salt is very hygroscopic.

7.3.4.4 $Ca(ClO_4)_2$

The thermal decomposition of this salt has been studied by several researchers.[24,25] The anhydrous salt exhibits polymorphic transitions at 615 and 679 K. It begins to decompose at 666 K, mostly by the reaction

$$Ca(ClO_4)_2 = CaCl_2 + O_2$$

At 681 K a liquid phase forms, probably containing perchlorate and chloride.

The densities of $Ca(ClO_2)_2$ and $Ca(ClO_4)_2$ are shown in Table 7.2, thermodynamic data in Table 7.3.

7.3.5 Cesium

7.3.5.1 $CsClO_2$

In the temperature range 190 to 210°C the salt disproportionates according to Levi and Cippolone[29]

$$3\ CsClO_2 = 2\ CsClO_3 + CsCl$$

In this temperature range the chlorate is stable.

7.3.5.2 $CsClO_3$

The salt is stable to the melting point and at least a little beyond it.[30] As expected from the large cationic size, $CsClO_3$ is the most stable of the chlorates. Like the other alkali metal chlorates, the salt decomposes by the parallel paths

$$CsClO_3 = CsCl + 3/2\ O_2 \quad \text{(a)}$$

$$4\ CsClO_3 = 3\ CsClO_4 + CsCl \quad \text{(b)}$$

Since the perchlorate is stable near the melting point of the chlorate, both CsCl and $CsClO_4$ accumulate. No detailed study of the relative rates of the two reactions seems to have been carried out, but indications are that the rate of reaction (b) is about three times as fast as that of (a) at 400°C.[30] At higher temperatures $CsClO_4$ also decomposes.

7.3.5.3 $CsClO_4$

The temperature of the orthorhombic-cubic transition is somewhat dependent on the heating rate. Cabane and Benard[31] think the transition occurs reversibly at 489 K. The melting point is estimated because the salt begins to decompose in the solid state. As is usual for the perchlorates, overall decomposition leads to the chloride and O_2, with $CsClO_3$ as an unstable intermediate:

$$CsClO_4 = CsClO_3 + 1/2\ O_2$$

$$CsClO_3 = CsCl + 3/2\ O_2$$

$$CsClO_4 = CsCl + 2\ O_2$$

Decomposition kinetics has been studied by Cabane and Benard[31] in the range 385 to 467°C and by Solymosi[32] between 528 and 558°C, both using weight loss measurements. In addition, Cordes and Smith[33] studied the first 3% of the solid decomposition mass spectrometrically between 330 and 500°C. This study has the advantage of avoiding melting, which occurs as the melting point is lowered by the accumulation of reaction products. $CsClO_4$, in contrast to the other alkali metal perchlorates, is quite volatile, approximately 50% of the sample subliming above 500°C in 1 to 2 hours.[33] The solid residue contains approximately ten times as much $CsClO_3$ as CsCl. Cabane and Benard[31] found an Arrhenius activation energy of 117 kJ, approximately equal to the activation energy for diffusion, and argued that the rate-determining step involved the diffusion of the perchlorate ion. Cordes and Smith[33] obtained 187 ± 8.8 kJ over the same temperature range. Their mechanism, which is virtually the same for all the alkali metal perchlorates is discussed under $KClO_4$. At the higher temperature,

Solymosi found the activation energy to be near 233 kJ, so there may be a change of mechanism as the temperature increases.

Densities of $CsClO_4$ are given in Table 7.4, phase transitions of $CsClO_3$ and $CsClO_4$ in Table 7.5, thermodynamic data at 298 K in Table 7.6, and high-temperature thermodynamic data of $CsClO_4$ in Table 7.7.

7.3.6 Cobalt

The preparation of the pure anhydrous salt, $Co(ClO_4)_2$, has been reported.[34] Its density is 3.2 gcm^{-3}, and it is stable in a vacuum to 90°C.

The decomposition kinetics have been studied[35] between 150 and 185°C. The major reaction is

$$Co(ClO_4)_2 = CoCl_2 + 2\ O_2$$

with a lesser contribution from

$$Co(ClO_4)_2 = CoO + Cl_2 + 3.5\ O_2$$

7.3.7 Copper

7.3.7.1 $Cu(ClO_2)_2$

The nearly pure anhydrous salt, but with a slightly low chlorine content, has been prepared.[29] It is very unstable, transforming into a green powder even in a closed vessel and exploding on being heated.

7.3.7.2 $CuClO_4$

The preparation of this salt was first described by Gomberg,[36] but no further work seems to have been reported since then.

7.3.7.3 $Cu(ClO_4)_2$

Studies of this salt are in serious disagreement. Hathaway[37] reported the preparation of the anhydrous salt by fractional vacuum sublimation of $Cu(ClO_4)_{2-x}(NO_3)_x$. An approximate melting point of 230 to 240°C was found. On the other hand, Vilcu and Georgescu[35] report the melting point of the (supposedly) anhydrous salt begins near 75°C, simultaneously with decomposition, which they studied in the range 80 to 110°C. It involves the simultaneous reactions

$$Cu(ClO_4)_2 = CuCl_2 + 4\ O_2$$

$$Cu(ClO_4)_2 = CuO + Cl_2 + 3.5\ O_2$$

The ratio of the two reactions is determined by the gas-phase composition through the equilibrium

$$CuO + Cl_2 = CuCl_2 + 1/2\ O_2$$

Particularly above 110°C the oxide is favored. It is possible that some water was present (the starting material was a hydrate) and the salt melted in its hydration water, but not enough information is given in the paper to know whether this the case. In contrast, Hathaway[37] reports that $Cu(ClO_4)_2$ decomposes at 130°C to a basic perchlorate.

7.3.8 Iron

7.3.8.1 $Fe(ClO_4)$

The thermal decomposition of a "dried" (at 110°C) hydrate in the range 90 to 135°C proceeds entirely to the oxide[35]

$$2\ Fe(ClO_4) = Fe_2O_3 + 3\ Cl_2 + 10.5\ O_2$$

7.3.9 Lead

7.3.9.1 $Pb(ClO_2)_2$

The salt is pseudotetragonal.[38] It disproportionates in the range 70 to 95°C according to the reaction

$$3\ Pb(ClO_2)_2 = 2\ Pb(ClO_3) + PbCl_2$$

If the compound is heated rapidly, this conversion is apparently not complete and the chlorite decomposes explosively between 100 and 125°C.[39] A complete analysis was not carried out, but apparently both O_2 and Cl_2 evolve.

7.3.9.2 $Pb(ClO_3)_2$

Decomposition is complex.[40] Decomposition rates were measured between 194° and 216°C in terms of the gas evolved. Four stages could be distinguished: (1) initial rapid decomposition, (2) slow reaction, (3) extremely rapid decomposition in which approximately two thirds of the material react, (4) a short, slow decay period. The major part of the residue is PbO_2, with some $PbCl_2$. No perchlorate is formed. The reactions, therefore, probably are

$$Pb(ClO_3)_2 = PbCl_2 + 3\ O_2$$

$$Pb(ClO_3)_2 = PbO_2 + 2\ ClO_2\ (Cl_2 + 2\ O_2)$$

7.3.9.3 $Pb(ClO_4)_2$

The salt is very hygroscopic, but can be prepared by vacuum treatment of its pyridine adduct.[41] At the melting point, decomposition is slow, but becomes more rapid above 300°C. Decomposition goes to both the chloride and to the oxide:[42]

$$Pb(ClO_4)_2 = PbCl_2 + 4\ O_2$$

$$Pb(ClO_4)_2 = PbO + Cl_2 + 7/2\ O_2$$

The first of these reactions is the major one.[43] Kinetics were studied in the range 320 to 355°C.[43] The melting point is estimated as 546 K.[41] Densities of $Pb(ClO_3)_2$ and $Pb(ClO_4)_2$ are listed in Table 7.8, $\Delta Hf°$ values of $Pb(ClO_4)_2$ and $PbCl_2$ in Table 7.9. Entropies are not known.

7.3.10 Lithium

7.3.10.1 LiClO

Preparation of the anhydrous salt has been described in numerous patents, e.g., Robson and Soule[44] and Soule,[45] because of its importance as a bleaching agent, but not even the range of its thermal stability seems to be known.

7.3.10.2 $LiClO_2$

When heated in the range 200 to 210°C, the salt disproportionates rather violently:[46]

$$3\ LiClO_2 = 2\ LiClO_3 + LiCl$$

7.3.10.3 $LiClO_3$

The phase transitions of this salt are in some doubt. Based on the measurements of Bruehl, three crystalline forms (two transitions) have been listed in a critical compilation.[47] A careful reexamination by DTA[30] has yielded only a single solid-state transition. The high-temperature form was found to supercool easily, by as much as 50°C, which may account for the earlier results. Liquid $LiClO_3$ is stable over a longer temperature range than other chlorates. Up to 300°C[48] no weight loss was detected. At 330°C[48] the salt decomposes slowly. The major reactions are

$$LiClO_3 = 3/4\ LiClO_4 + 1/4\ LiCl$$

$$LiClO_3 = LiCl + 3/2\ O_2$$

Both of these reactions, occurring in the ratio Rate 1/ Rate 2 = 5.3, are catalyzed by LiCl. At the temperature of the experiment, $LiClO_3$ decomposes quite slowly. Approximately 4% disappears through oxide formation:

$$LiClO_3 = 1/2\ Li_2O + 1/2\ Cl_2 + 5/4\ O_2$$

At 370°C decomposition becomes quite rapid. At this and higher temperatures decomposition of $LiClO_4$ is appreciable, so that the kinetics become quite complicated.

7.3.10.4 $LiClO_4$

The decomposition of this salt has been the subject of several studies[49,50] that agree on the major features. In contrast to the other alkali metal perchlorates, $LiClO_4$ is stable in the solid state and in the liquid up to approximately 390°C. The overall reaction is

$$LiClO_4 = LiCl + 2\ O_2$$

Very small amounts of $LiClO_3$ form as an intermediate, but this salt is very unstable at temperatures at which $LiClO_4$ decomposes. Very small quantities of Cl_2 and Li_2O have also been detected. Markowitz and Boryta[51] studied by TGA the decomposition kinetics of both pure $LiClO_4$ and $LiClO_4$-LiCl mixtures below 420°C, the highest temperature at which the reaction mixture was not heated appreciably by the heat released in the exothermic reaction. The α vs. time curve of pure $LiClO_4$ exhibits an induction period and is sigmoidal, indicating that it is autocatalytic. For that range of the decomposition corresponding to the attainment of a saturated solution of LiCl in $LiClO_4$, the kinetics fit the Prout-Tompkins equation originally derived for branching reaction nuclei in solids. When the decomposition is studied for $LiClO_4$ initially saturated with LiCl, the autocatalytic behavior is absent and kinetics are first order. Evidently, LiCl functions as a catalyst. Solymosi[50] studied the kinetics of pure $LiClO_4$ decomposition by measuring gas evolution between 411 and 430°C and reached similar conclusions. For the autocatalytic phase both studies obtained 222 kJ for the activation energy, consistent with $LiClO_4 = LiClO_3 + O$, i.e., with the breaking of a Cl–O bond as the rate-determining step. For first-order kinetics, however, there is no agreement. Solymosi finds 172 kJ and Markowitz gives 259 kJ. There is no obvious reason for this discrepancy.

Densities of $LiClO_3$ and $LiClO_4$ are listed in Table 7.10, phase transitions in Table 7.11, and high-temperature thermodynamic data of $LiClO_4$ in Table 7.12.

7.3.11 Magnesium

7.3.11.1 $Mg(ClO_2)_2$

Preparation of this salt has been reported,[56] but no evidence was presented that it is completely anhydrous.

7.3.11.2 $Mg(ClO_3)_2$

Although the infrared spectrum of this salt has been described,[57] there exists no evidence that the anhydrous salt has been prepared.

7.3.11.3 $Mg(ClO_4)_2$

Decomposition of this salt has been studied by several investigators[11,43,58–60] who obtained different results, indicating that the reaction is complex. The two major reactions that have been proposed are

$$Mg(ClO_4)_2 = MgCl_2 + 4\ O_2 \qquad (a)$$

$$Mg(ClO_4)_2 = MgO + Cl_2 + 1/2\ O_2 \qquad (b)$$

As pointed out by Markowitz,[113] $\Delta Gf°$ of $MgCl_2$ and MgO are comparable (they are equal at 427°C) so that they may interconvert through the equilibrium

$$MgCl_2 + 1/2\ O_2 = MgO + Cl_2 \qquad (c)$$

This equilibrium is achievable in practice and thus product distribution will be determined by the gas composition at the reaction interface and the rate of equilibration of reaction (c). These predictions are borne out by a number of experimental studies. For example, Chudinova[61] found reaction (a) to be dominant at the lower end of the temperature range 360 to 420°C, and (b) at the higher end in a system in which the condensed phase remained in contact with the gas. Very similar results were obtained by Solymosi[43] in the same temperature range. In contrast, Acheson and Jacobs[11] found MgO to be the predominant product when the salt was heated under vacuum in the range 390 to 440°C. This suggests that reaction (b) occurs directly since reaction (c) could not occur in a vacuum. Above 600°C the product distribution appeared to depend only on the internal gaseous environment and to be independent of the gaseous atmosphere above the solid phase.

A more recent, very careful experimental reexamination of this problem was carried out by Lemesheva and Krivtsov.[62] In this work, both the gas and solid phase were examined under isothermal conditions in the range 360 to 440°C, and also under increasing temperature. They found the first stage to be reaction (a). The $MgCl_2$ formed then reacts with the perchlorate:

$$MgCl_2 + Mg(ClO_4)_2 = 2\ MgO + 2Cl_2 + 3\ O_2$$

Thus the overall reaction will depend on experimental conditions, e.g., whether gas is continuously removed.

The density of $Mg(ClO_4)_2$ has been reported as 2.21 gcm^{-3}.[62a] Thermodynamic data of $Mg(ClO_4)_2$ and $MgCl_2$ are given in Table 7.13.

7.3.12 Manganese

The decomposition of a $Mn(ClO_4)_2$ sample of unknown water content (vacuum dried at 150°C) was studied in the range 125 to 155°C.[63] The main reaction is

$$Mn(ClO_4)_2 = MnCl_2 + 4\ O_2$$

with a smaller fraction decomposing to the oxide:

$$Mn(ClO_4)_2 = MnO + Cl_2 + 3.5\ O_2$$

7.3.13 Mercury

7.3.13.1 $Hg_2(ClO_2)_2$

The salt is so unstable that its existence must be regarded as marginal. Several attempts to prepare it[64,65] resulted in the preparation of somewhat basic, yellow crystals which turned red in air and decomposed explosively on slight heating.

7.3.13.2 $Hg(ClO_2)_2$

The salt has been prepared.[38,66] It consists of red crystals whose composition indicates some contamination by HgO. It is fairly unstable, tending to decompose even at ambient temperatures.

7.3.13.3 $Hg(ClO_3)_2$

A salt of approximately this composition was prepared and studied by several investigators in the 19th century. This work has been summarized by Gmelin. However, it is questionable whether the pure salt has ever been prepared. No work seems to have been done since 1910.

7.3.13.4 $Hg_2(ClO_4)_2$

A salt of this composition was obtained by dehydration of the 6-hydrate.[67] It melts at 210°C, with the evolution of chlorine oxides.

7.3.14 Nickel

7.3.14.1 $Ni(ClO_3)_2$

There is no detailed description of the anhydrous salt, although spectral measurements of a vacuum-dried salt were reported by Rocchiccioli.[68]

7.3.14.2 $Ni(ClO_4)_2$

Attempts to prepare the salt by vacuum dehydration of the 2-hydrate were unsuccessful.[69] A light-green, not quite pure anhydrous salt was prepared by the incomplete reaction of $HClO_4$ with $Ni(CF_3COO)_2$ in trifluoroacetic acid.[70] Borgne and Weigel[71] reported the preparation of the anhydrous salt by the slow, stepwise dehydration of the hexahydrate. Final heating was carried out at 180°C. Its structure was determined by X-ray diffraction, and a density of 2.985 gcm^{-3} calculated. This result conflicts with a more recently reported preparation[72] in which the density (probably determined pycnometrically) is given as 3.4 gcm^{-3}. The salt was said to be stable up to 130°C in a vacuum.

The decomposition of a salt of unknown water content was studied by Vilcu and Georgescu[63] between 130 and 170°C. They found the major reaction to be

$$Ni(ClO_4)_2 = NiCl_2 + 4\ O_2$$

with a minor simultaneous reaction forming the oxide

$$Ni(ClO_4)_2 = NiO + Cl_2 + 3.5\ O_2$$

7.3.15 Potassium

7.3.15.1 $KClO_2$

The salt is quite hygroscopic. When sealed into glass tubes and heated at 180 to 200°C, 98% of it disproportionates, as do all the other alkali metal chlorites, according to Levi and Garrini[73]

$$3\ KClO_2 = 2\ KClO_3 + KCl$$

7.3.15.2 $KClO_3$

Although the molten salt is stable for some distance above its melting point,[74] no density data for the liquid are available. A mass-spectrometric study[75] shows no $KClO_3$ molecules in the vapor. The thermal decomposition of the salt has been studied by many investigators, frequently as part of the decomposition of $KClO_4$ in which it occurs as an intermediate (see below). The earlier literature has been summarized by Rudloff and Freeman[76] who studied the kinetics in detail. Below 500°C, where $KClO_4$ is stable, the decomposition products are KCl, O_2, and $KClO_4$, consistent with the frequently cited overall stoichiometry[74] in terms of two reactions:

$$KClO_3 = KCl + 3/2\ O_2 \tag{a}$$

$$4\ KClO_3 = 3\ KClO_4 + KCl \tag{b}$$

Clearly, reaction (b) cannot correspond to a simple mechanism. However, Rudloff and Freeman[76] found that below 500°C $KClO_3$ loses only one third of the weight expected from reaction (a), and equivalent amounts of $KClO_4$ and KCl are formed. In this range the kinetics are second order. On this basis they propose the reaction sequence

$$\begin{array}{c} 2KClO_3 \rightarrow KClO_4 + (KClO_2) \\ (KClO_2) \rightarrow (KClO) + 1/2\ O_2 \rightarrow KCl + O_2 \\ \hline 2KClO_3 \rightarrow KClO_4 + KCl + O_2 \end{array}$$

$KClO_2$ is unstable above 200°C and cannot be detected.[74] It can therefore only be present as an unstable intermediate. The rate-determining step is thus an oxygen atom transfer between two chlorate ions:

$$2\ ClO_3^- \rightarrow ClO_4^- + ClO_2^-$$

The frequently observed deceleration of the reaction by KCl is attributed to dilution. Above 500°C the decomposition becomes more complicated because $KClO_4$ then also decomposes and $KClO_3$ appears as an intermediate reaction product. Moreover, KCl now catalyzes the reaction. Explanations for this phenomenon, e.g., in terms of polarization centers, are quite speculative.

7.3.15.3 $KClO_4$

Because of its use in pyrotechnics and propellants, thermal decomposition of this salt has been studied for a long time. Simchen[77] has given references back to 1816. High-temperature thermodynamic functions are listed in the JANAF tables. We have given those listed by HSC.

The course of the decomposition is complicated by simultaneously occurring phase changes resulting from a lowering of the melting point by the accumulating reaction products, KCl and $KClO_3$, and a subsequent freezing of the melt as the material becomes predominantly KCl. For this reason it is also impossible to determine a precise melting point, although it appears to be near 580°C.[78]

Tracer experiments have established that O_2 and Cl^- are formed[79,80] irreversibly, but that ClO_3^- is transformed to ClO_4^-. Harvey et al.[81,82] established that in the range 556 to 580°C the decomposition of $KClO_4$ is first order. As reaction products accumulate, the solid melts and the rate increases 50-fold.

The kinetics of the liquid-state reaction are also first order. For both the solid and liquid the activation enthalpies and entropies are the same, $\Delta S = 8$ J deg^{-1}, indicating that the rate-determining step is the breaking of a Cl–O bond, with the formation of atomic oxygen

$$KClO_4 \rightarrow KClO_3 + O \quad \Delta H° = 290 \text{ kJ}$$

The activated complex does not differ much from the ClO_4^- ion. These kinetic results have been confirmed by Solymosi.[32]

Reactions subsequent to the rate-determining step are presumably those described under $KClO_3$ above. Since KCl and O_2 are formed irreversibly, the final result is consistent with the stoichiometry of the overall reaction

$$KClO_4 = KCl + O_2$$

Two studies at lower temperatures have obtained much lower activation energies. Cabane and Benard[83] measured both decomposition rates and ClO_4^- diffusion between 375 and 480°C. Since for both of these processes the activation energy was the same for several alkali metal perchlorates, 126 kJ for $KClO_4$, they argued that the rate-determining step must be the same, i.e., diffusion of ClO_4^-.

A much more detailed experimental and theoretical study of the decomposition kinetics over approximately the same temperature range was carried out by Cordes and Smith,[84] using only the first 3% of the solid-state decomposition in their analysis. For $KClO_4$, an extrapolation of their data on an Arrhenius plot fits smoothly into the high-temperature data of Harvey et al.,[81,82] i.e., at lower temperatures the plot is linear but exhibits a curvature corresponding to a higher apparent activation energy at higher temperatures. For $KClO_4$ the low temperature data are represented by

$$k = A \exp(-E/RT)$$

where $A = 10^7 s^{-1}$ and E = 210 kJ. They argue from this, that at least at low temperatures, the rate-determining step cannot be $ClO_4^- \rightarrow ClO_3^- + O$. Instead, they carried out a detailed analysis in terms of Cordes' theory[85] of pre-exponential factors and a molecular model in which internal motions are specified, which leads to a pre-exponential factor in good agreement with experiment. In this model the activated complex is assumed to consist of two neighboring, freely rotating ClO_4^- ions joined by a peroxide linkage ($(ClO_3)–O–(ClO_3^-)$). The rate-determining step then consists in the formation of O_2 and two ClO_3^- ions.

To account for the higher activation energies at higher temperatures, it is necessary to postulate either increased freedom of motion of the ClO_3^- ion in the lattice, or a change in the mechanism proposed by Harvey et al.[81,82] A similar rise in the activation energy is observed for $NaClO_4$.

Since 1968 no further work has been reported on this problem.

Densities of $KClO_3$ and $KClO_4$ are listed in Table 7.14, phase transitions in Table 7.15, and high-temperature thermodynamic data in Tables 7.16 and 7.17 for $KClO_3$ and $KClO_4$, respectively.

7.3.16 Rare Earths

7.3.16.1 Chlorites

Nearly anhydrous $Tb(ClO_2)_3$ and $Er(ClO_2)_3$ have been prepared, but none of their properties have been reported. Chlorites of several other rare earths (La, Ce, Pr, Nd, Sm) have been prepared as hydrates.[86]

7.3.16.2 Chlorates

Only hydrates of the chlorates of the general formula $Ln(ClO_3)_3.5\ H_2O$ of the following rare earths have been prepared: La, Nd, Gd, Dy, Er, Th, Y, Lu,[87]

When these salts are heated, or when they are just kept under vacuum for a long time at ambient temperature, they decompose with the evolution of O_2 and Cl_2. The kinetics of the decomposition have been reported,[88] but in view of the simultaneous evolution of H_2O, Cl_2, and O_2, the interpretation of the data probably cannot result in an unambiguous mechanism.

7.3.16.3 Perchlorates

The preparation of the anhydrous salts (by vacuum-drying in the presence of P_2O_5 with the temperature gradually rising to 155°C) has been reported.[89] Whether or not the anhydrous salts can be prepared in this way may depend on experimental conditions, such as the heating rate. For example, Bel'kova and Aleksenko[90] and Bel'kova, Aleksenko, and Serebrennikov[91] reported brief plateaus in the TGA curves of the hydrated perchlorates of La, Ce, Pr, and Sn at weights corresponding to the anhydrous salts, whereas Steinberg and Shidlovsky[92] reported finding only inflections for these salts and $Nd(ClO_4)_3$. Except for Nd, these findings are consistent since the latter authors used a faster heating rate.

Bel'kova, Aleksenko, and Serebrennikov[93] reported an isothermal decomposition study of several rare-earth perchlorates in which weight loss was measured under flowing nitrogen. The stoichiometry of all the salts could be represented by

$$2\ M(ClO_4)_3 = MOCl + MCl_3 + Cl_2O + 11\ O_2$$

Between 240 and 370°C the kinetics followed the Erofeev equation in which the fraction decomposed is represented by

$$\alpha = 1 - \exp(-kt^n)$$

The value of n is determined by the properties of the nucleus. The Arrhenius activation energy was determined from the temperature dependence of the rate constant k. The results are shown in Table 7.18. In any case there is still a question as to whether the kinetics were studied for truly anhydrous salts.

7.3.17 Rubidium

7.3.17.1 $RbClO_2$

When heated in a closed container at 200 to 210°C the salt disproportionates according to Levi and Garrini[73]

$$3\ RbClO_2 = 2\ RbClO_3 + RbCl$$

Similar results have been obtained by Solymosi and Bansagi.[94]

7.3.17.2 $RbClO_3$

Knowledge of high-temperature properties is largely based on a study by Markowitz, Boryta, and Stewart.[74] In contrast to the perchlorate (see below) the salt is quite stable for 20 to 30°C above its melting point (342°C). Decomposition becomes rapid at 480°C. The salt disappears by two paths:

$$RbClO_3 = RbCl + 3/2\ O_2 \qquad (a)$$

$$4\ RbClO_3 = 3\ RbClO_4 + RbCl \qquad (b)$$

Reaction (b) is the major one by a factor of eight. Only near 600°C does the reaction proceed further with decomposition of the perchlorate. Chlorites and hypochlorites seem to be absent as intermediates.

7.3.17.3 $RbClO_4$

The salt decomposes during fusion[60] and therefore the melting point can only be determined approximately. The decomposition is exothermic and yields only RbCl and O_2 as the final products. The high-temperature kinetic data can be fitted by various rate equations, all of which yield an Arrhenius activation energy near 293 K,[95] consistent with the loss of one oxygen as the rate-determining step:

$$RbClO_4 = RbClO_3 + O$$

Since $RbClO_3$ is quite unstable in the decomposition range of the perchlorate, decomposition proceeds to the chloride. In addition to the decomposition reactions

$$RbClO_4 = RbClO_3 + O$$

$$RbClO_3 = RbCl + 3/2\ O_2$$

the salts also undergo fusion or solution in the reaction melt, so the physical state of the system depends on both the temperature and the extent of decomposition.

At lower temperatures (330 to 480°C) a much lower activation energy, 192 kJ, was observed for the initial 3% of the solid-state decomposition.[84] This, and the observed pre-exponential factor, were interpreted in terms of the abstraction of O_2 from two neighboring ClO_4 ions, (ClO_3=O=O=(ClO_3). The theory is discussed in more detail under $KClO_4$.

Densities of $RbClO_3$ and $RbClO_4$ are listed in Table 7.19, transitions in Table 7.20, and thermodynamic data at 298.15 K in Table 7.21.

7.3.18 Scandium

The thermal decomposition of the anhydrous salt, $Sc(ClO_4)_2$, proceeds according to

$$3\ Sc(ClO_4)_3 = ScOCl + Sc_2O_3 + 4\ ClO_2 + 14\ O_2$$

in the range 240 to 370°C the kinetics follow the Erofeev equation (see under Rare Earths) with an Arrhenius activation energy of 87.4 kJ.[91]

7.3.19 Silver

7.3.19.1 $AgClO_2$

The anhydrous salt exists as shiny, yellow platelets, orthorhombic in structure.[96,97] The compound is rather unstable and disproportionates, beginning at 83°C:

$$3\ AgClO_2 = 2\ AgClO_3 + AgCl$$

The reaction is first order with an activation energy of 109 kJ.

Above 100°C the thermal decomposition of the salt becomes increasingly important. The precise nature of the decomposition is not quite clear, but the major gaseous product is ClO_2, and the solid residue contains both silver and silver oxide. Above 145°C the reaction becomes explosive. A detailed kinetic analysis of the disproportionation reaction and decomposition has been published.[98]

The density is 4.30 gcm^{-3}.[99]

7.3.19.2 $AgClO_3$

The salt undergoes a crystalline transition from tetragonal to cubic at 412 K; ΔH for the transition is 2.49 kJ mol^{-1}. The temperature dependence of the lattice parameters and thermal expansion coefficient has been carefully studied.[100]

The decomposition has been measured in the range 316 to 339°C,[101] i.e., of the liquid salt. However, the accumulation of reaction products increases the melting point, so the mass solidifies in this temperature range after about 80% of it has decomposed. The gaseous product is mainly O_2, with 2 to 3% Cl_2. The reaction is therefore

$$2\ AgClO_3 = 2\ AgCl + 3\ O_2$$

The activation energy is 239 kJ, consistent with the rupture of a Cl–O bond as the rate-determining step:

$$AgClO_3 = AgClO_2 + O$$

Approximately 10% of the chlorate disproportionates:

$$4AgClO_3 = 3\ AgClO_4 + AgCl$$

These reactions were also found by Udupa,[102] except that he found decomposition begins at a much lower temperature.

7.3.19.3 $AgClO_4$

There are some contradictions in the literature concerning the high-temperature behavior of this salt. Several authors agree that $AgClO_4$ undergoes a transition from rhombic to cubic in the range 155 to 159°C,[9,59,103] but there is some confusion regarding the melting point and thermal decomposition. For example, Gordon and Campbell[59] report a slight decomposition in the solid state at 277°C, with violent bubbling of the melt at 473°C. However, Carnelly and O'Shea[105] report a melting point of 476°C. Markowitz has suggested that in Gordon and Campbell's experiment, the AgCl, which is formed as decomposition product, lowers the melting point. This suggestion was confirmed experimentally by Solymosi[101] who reports a melting point of 476°C. This author also studied the decomposition kinetics between 415 and 445°C. The only reaction is

$$AgClO_4 = AgCl + 2\ O_2$$

The absence of $AgClO_3$ is not surprising, since this salt decomposes at a much lower temperature. The reaction is probably autocatalytic.

Densities of the silver salts are listed in Table 7.22, transitions in Table 7.23, and thermodynamic data at 298.15 K in Table 7.24. Thermodynamic data for $AgClO_3$ are listed in Table 7.25.

7.3.20 Sodium

7.3.20.1 $NaClO_2$

Thermal decomposition has been studied by several researchers who agree on the main features of the reaction.[6,22,106,107]

In an open container, reaction begins as low as 140°C and is certainly rapid above 160°C. The major reaction is disproportionation:

$$3\ NaClO_2 = 2\ NaClO_3 + NaCl$$

The fraction of $NaClO_2$ which disappears by this path as compared with the direct decomposition

$$NaClO_2 = NaCl + O_2$$

varies from 75 to 99%, depending on the experimental details. The lower value was reported for rapid heating.[107] All others report 97 to 99%. The kinetics are first order, with an Arrhenius activation energy of 227 kJ, which is less than the Cl–O bond energy.

7.3.20.2 $NaClO_3$

A very thorough crystallographic study has been reported by Abrahams and Bernstein.[104]

The thermal expansion coefficient of the solid was measured by Sharma,[108] using an interferometric method, and by Deshpande and Mudholker,[109] using X-ray diffraction data. At ambient temperature both sets of data agree within 5%, but near 500 K, Sharma's values are much higher. Deshpande and Mudholker suggested that the macroscopic expansion coefficient is made up of a sum of terms:

$$\alpha(\text{macr.}) = \alpha(\text{lattice}) + \alpha(\text{Schottky}) - \alpha(\text{readjustment})$$

so that α(macr.) may differ, depending on the relative contributions of the defects and the readjustment of the small single crystals packed together in a mosaic structure with a small amount of misorientation. Sharma's data are within 3% of those given by Kelley,[110] so the interferometric measurements may reflect the behavior to be expected from "real" $NaClO_3$, whereas the X-ray results refer to an ideal lattice.

$NaClO_3$ is stable to well above its melting point. Three thermal decomposition studies of liquid $NaClO_3$ have been published[40,74,111] that agree in their main features. Decomposition starts near 400°C. No Cl_2 and ClO_2 are formed. The major reaction is

$$NaClO_3 = NaCl + 3/2\ O_2 \quad \text{(a)}$$

However, $NaClO_4$ is also found to be a product. The amount of this formed product probably depends on reaction conditions. The overall stoichiometry can be written as

$$4\ NaClO_3 = 3\ NaClO_4 + NaCl \quad \text{(b)}$$

but this is not a likely mechanism. Osada et al.[111] suggested that the atomic oxygen formed as the initial product in reaction (a) oxidizes unreacted $NaClO_3$:

$$NaClO_3 + O = NaClO_4 \quad \text{(c)}$$

A proper stoichiometric combination of (a) and (c), and allowing for $O + O = O_2$, yields reaction (b). Perchlorate formation becomes increasingly important with increasing temperature, from a few percent at the lower end to perhaps equal amounts with NaCl near 450°C. $NaClO_4$ is stable to near 530°C. The activation energy of 285 kJ[40] is consistent with the breaking of a Cl–O bond as the rate- determining step of reaction (a).

The relation between events occurring on the surface and the bulk substructure close to the melting point has been studied in single crystals.[112] As the melting point is approached, dislocations in both regions increase.

7.3.20.3 $NaClO_4$

In contrast to $NaClO_3$, the perchlorate begins to decompose well below the melting point. Hence the melting point can only be determined approximately. However, since the temperature at which this decomposition becomes appreciable is higher than the stability region of the chlorate, very little $NaClO_3$ is found in the decomposition of $NaClO_4$, even though the initial reaction is probably

$$NaClO_4 = NaClO_3 + 1/2\ O_2$$

The initial stages of the solid-state decomposition have been studied by Cordes and Smith[84] between 330 and 440°C. An Arrhenius plot was linear to 390°C. Above this temperature, the slope of the plot increased. This result, as well as the low preexponential factor, was accounted for by a detailed molecular model (discussed under $KClO_4$) in which O_2 is abstracted from two neighboring perchlorate ions:

$$(ClO_3)\text{–O–O–}(ClO_3) = 2\ ClO_3 + O_2$$

The increase in activational energy with rising temperature is accounted for by increased rotational freedom of the chlorate ion.

In contrast, Cabane and Benard[31] observed an activation energy of only 117 kJ between 390 and 428°C, equal to that for diffusion of ClO_3 through the lattice. They argued that this implies that ion diffusion is the rate- determining step.

In the liquid state (495 to 515°C) the reaction is catalyzed by NaCl, but the activation energy is near 250 kJ which is more consistent with the breaking of a Cl–O bond as the rate-determining step:

$$NaClO_4 = NaClO_3 + O$$

$NaClO_3$ decomposes to NaCl as rapidly as it is formed. The overall decomposition therefore is

$$NaClO_4 = NaCl + 2\ O_2$$

although a small part of the perchlorate disappears by the reaction[113]

$$2\ NaClO_4 = Na_2\ O + Cl_2 + 1/2\ O_2$$

Above 525°C decomposition becomes quite rapid.

Transitions of $NaClO_3$ and $NaClO_4$ are shown in Table 7.26, and high-temperature thermodynamic data of $NaClO_3$ and $NaClO_4$ in Tables 7.27 and 7.28.

7.3.21 Strontium

7.3.21.1 $Sr(ClO_2)_2$

The preparation of the salt has been described.[114,115] Heating in the range 180 to 200°C leads to virtually complete disproportionation according to Levi and Garrini[116]

$$2Sr(ClO_2)_2 = 2\ Sr(ClO_3)_2 + SrCl_2$$

7.3.21.2 $Sr(ClO_3)_2$

The salt is of orthorhombic-bipyramidal structure.[117] It can be heated to at least 120°C without decomposing.[118]

7.3.21.3 $Sr(ClO_4)_2$

Only a brief decomposition has been carried out.[119] Oxygen evolution begins near 415°C. During the decomposition the salt appears to melt at 450°C, but it is then no longer pure. The product is almost entirely $SrCl_2$.

Densities of the strontium salts are listed in Table 7.29 and thermodynamic data in Table 7.30.

7.3.22 Thallium

7.3.22.1 $TlClO_2$

Little is known about this salt. It explodes from mechanical shock and thermally decomposes above 70°C.[120]

7.3.22.2 $TlClO_3$

Decomposition becomes noticeable near 150°C. The kinetics have been studied from this temperature up to 170°C.[8] The overall reaction can be represented by

$$7\ TlClO_3 = 3\ Tl_2O_3 + TlCl + 6\ ClO_2$$

but the mechanism is obviously more complex. Possible reactions by which Tl_2O_3 and TlCl might be produced are

$$TlClO_3 = TlCl + 3/2\ O_2$$

$$2\ TlClO_3 = Tl_2O_3 + ClO_2 + 1/2\ Cl_2 + 1/2\ O_2$$

Tl_2O may also be formed and oxidized to Tl_2O_3:

$$2\ TlClO_3 = Tl_2O + 2\ ClO_2 + 1/2O_2$$

$$Tl_2O + O_2 = Tl_2O_3$$

7.3.22.3 $TlClO_4$

The phase transitions are well established, but little work has been done on the decomposition. DTA measurements[121] indicate that it occurs near 400°C. Both TlCl and Tl_2O_3 are formed.

Densities of the thallium salts are listed in Table 7.31 and transitions in Table 7.32.

7.3.23 Thorium

7.3.23.1 $Th(ClO_4)_4$

The evidence for the existence of the anhydrous salt is not conclusive. Some crystals were reported to have formed in the reaction between thoria and perchloric acid,[122] but no analysis was carried out. The salt would certainly be very hygroscopic. Attempts to dehydrate the hydrate $Th(ClO_4)_4.4\ H_2O$ resulted in the formation of the anhydrous oxysalt $ThO(ClO_4)_2$.[123]

7.3.24 Tin

7.3.24.1 Perchlorates

Neither the stannous nor the stannic salts seem to have been prepared in the pure anhydrous state, although there are indications that they could be. Thus Gomberg[124] reported the preparation of several anhydrous perchlorates, including stannous and stannic perchlorates, by reacting the metal with an ethereal solution of chlorine tetroxide, but the products were not characterized. Schmidt[125] reported an acetonitrile adduct $Sn(ClO_4)_2.2\ CH_3CN$, but did not attempt to remove the acetonitrile.

7.3.25 Yttrium

7.3.25.1 $Y(ClO_4)_2$

Thermal decomposition of the anhydrous salt follows the overall stoichiometry

$$2\ Y(ClO_4)_3 = YOCl + YCl_3 + Cl_2O + 11\ O_2$$

In the range 240 to 370°C the kinetics follow the Erofeev equation (see Rare Earths) with an Arrhenius activation energy of 116.3 kJ.[6]

7.3.26 Zinc

7.3.26.1 $Zn(ClO_3)_2$

The preparation of several divalent chlorides, including $Zn(ClO_3)_2$, was reported by Amiel,[126] but the salts were not characterized or analyzed. It is therefore doubtful that the anhydrous salt has been prepared,

7.3.26.2 $Zn(ClO_4)_2$

The anhydrous salt has been prepared in a two-stage process whereby the water of hydration is first replaced with dioxane which can then be removed under vacuum without decomposing the salt.[127] The salt is quite stable up to 130°C. Chudinova has reported the decomposition to occur above 190°C.[127] The kinetics were studied by Solymosi[43] in the range 280 to 310°C.

The salt decomposes by two parallel reactions:

$$Zn(ClO_4)_2 = ZnCl_2 + 4\ O_2$$

$$Zn(ClO_4)_2 = ZnO + Cl_2 + 7/2\ O_2$$

which take place to nearly the same extent. Both reactions appear to be first order.

Rosolovskii and Krivtsov[128] have used their measurements of the enthalpy of solution of $Zn(ClO_4)_2$ in 0.9M $HClO_4$ to calculate ΔHf° for this salt. Recalculating their values with newer data gives –247.86 ± 2.18 kJ mol^{-1}.

TABLE 7.1
Thermodynamic Data (298.15 K)

Compound	ΔHf° kJ mol^{-1}	S° J mol^{-1} deg^{-1}	Reference
$Ba(ClO_2)_2$	–680.3	197	1
$Ba(ClO_3)_2$	–771.5	—	1
$Ba(ClO_4)_2$	–800.0	—	1
$BaCl_2$	–858.6	123.68	1

TABLE 7.2
Densities of Ca-Cl salts

Phase	T(K)	d (gcm^{-3})	Reference
$Ca(ClO_2)_2$	298	2.71	26
	293	2.65	27
$Ca(ClO_4)_2$	298	2.651	28

TABLE 7.3
Thermodynamic Data (298.15 K)

Compound	ΔHf° kJ mol^{-1}	S° J mol^{-1} deg^{-1}	Reference
$Ca(ClO_2)_2$	–678.2	—	1
$Ca(ClO_4)_2$	–736.79	—	1
$CaCl_2$	–795.8	104.6	1

TABLE 7.4
Densities of Cs-Cl Salts

Phase	T(K)	d (gcm^{-3})	Reference
$CsClO_3$	Ambient	3.626	26
$CsClO_4$ (orthorhombic)	289	3.327	27, 28
$CsClO_4$ (cubic)	>498	3.01	29

TABLE 7.5
Transitions of Cs-Cl Salts

Phase Change	T(K)	ΔH kJ/mol	ΔS J/(mol*K)	Reference
$CsClO_3$, $\alpha \rightarrow \beta$	578	—	—	26
$\beta \rightarrow$ liquid	615	—	—	26
$CsClO_4$(orthorhombic) $\rightarrow$ cubic	498	8.4	—	30
$\beta \rightarrow$ liquid	850(est)	—	—	30

TABLE 7.6
Thermodynamic Data (298.15 K)

Compound	ΔHf° kJ mol^{-1}	S° J mol^{-1} deg^{-1}	Reference
$CsClO_3$	–411.7	151.6	1
$CsClO_4$	–443.09	175.06	1
CsCl	–443.04	101.17	1

Note: High temp. data for $CsClO_4$ (Table 7.7) differ at 298 K.

TABLE 7.7
Thermodynamic Data of $CsClO_4$

T K	ΔHf° kJ mol^{-1}	S° J mol^{-1} deg^{-1}	ΔGf° kJ mol^{-1}	Reference
298.15	–437.23	175.27	–489.48	Knacke et al. 1991
400.00	–424.89	210.66	–509.16	Knacke et al. 1991
500.00	–410.17	243.42	–531.88	Knacke et al. 1991
600.00	–385.04	290.54	–559.36	Knacke et al. 1991
700.00	–367.42	317.69	–589.81	Knacke et al. 1991
800.00	–349.81	341.21	–622.78	Knacke et al. 1991
900.00	–332.19	361.96	–657.96	Knacke et al. 1991
1000.00	–314.58	380.52	–695.10	Knacke et al. 1991

TABLE 7.8
Density of Pb-Cl Salts

Phase	T(K)	d (gcm^{-3})	Reference
$Pb(ClO_3)_2$	393	3.89	41
$Pb(ClO4)_2$	Ambient	3.86	42

TABLE 7.9
Thermodynamic Data (298.15 K)

Compound	ΔHf° kJ mol^{-1}	S° J mol^{-1} deg^{-1}	Reference
$Pb(ClO_4)_2$	–218.4	—	43
$PbCl_2$	–326.31	—	1

TABLE 7.10
Densities of Li-Cl Salts

Phase	T(K)	d (gcm^{-3})	Reference
$LiClO_3$	Ambient	2.631	52
	410	2.0832	53
	421	2.0756	53
	430	2.0681	53
	440	2.0605	53
	450	2.0529	53
$LiClO_4$	530	2.013	53
	540	2.007	53
	550	2.000	53
	560	1.994	53
	570	1.988	53
	580	1.982	53
	590	1.976	53
	600	1.970	53
	610	1.964	53
	620	1.958	53
	630	1.951	53
	640	1.945	53
	650	1.939	53

TABLE 7.11
Transitions of Li-Cl Salts

Phase Change	T(K)	ΔH kJ/mol	ΔS J/(mol*K)	Reference
$LiClO_3$, c,II → c, I	384	—	—	52
c,I → liquid	402	—	—	52
$LiClO_4$, c → liquid	520	20.9	40	54, 55

TABLE 7.12
Thermodynamic Data of $LiClO_4$

T K	ΔHf° kJ mol⁻¹	S° J mol⁻¹ deg⁻¹	ΔGf° kJ mol⁻¹	Reference
298.15	–380.74	125.52	–418.17	Barin 1989
400.00	–368.69	160.12	–432.74	Barin 1989
500.00	–354.95	190.72	–450.31	Barin 1989
600.00	–309.71	277.32	–476.10	Barin 1989
700.00	–293.60	302.16	–505.11	Barin 1989
800.00	–277.49	323.67	–536.42	Barin 1989
900.00	–261.38	342.64	–569.76	Barin 1989
1000.00	–245.28	359.61	–604.89	Barin 1989

TABLE 7.13
Thermodynamic Data (298.15 K)

Compound	ΔHf° kJ mol⁻¹	S° J mol⁻¹ deg⁻¹	Reference
$Mg(ClO_4)_2$	–568.90	—	1
$MgCl_2$	–641.32	89.62	1

TABLE 7.14
Density of K-Cl Salts

Phase	T(K)	d(gcm⁻³)	Reference
$KClO_3$	293	2.338	130
$KClO_4$	284	2.524	131
	291	2.527	104
	298	2.530	132
	613	2.181	103

TABLE 7.15
Transitions of K-Cl Salts

Phase Change	T(K)	ΔH kJ/mol	ΔS J/(mol*K)	Reference
$KClO_3$, c → liquid	357	—	—	129
$KClO_4$ c, rhombic → c,cubic	572.7	13.8	24.0	19
c, cubic → liquid	853 (est.)	—	—	81

TABLE 7.16
Thermodynamic Data of $KClO_3$

T K	ΔHf° kJ mol^{-1}	S° J mol^{-1} deg^{-1}	ΔGf° kJ mol^{-1}	Reference
298.15	–389.11	143.09	–431.76	Knacke et al. 1991
400.00	–378.17	174.57	–448.00	Knacke et al. 1991
500.00	–366.36	200.89	–466.80	Knacke et al. 1991
600.00	–356.72	223.91	–488.07	Knacke et al. 1991
700.00	–340.35	244.51	–511.51	Knacke et al. 1991
800.00	–326.29	263.28	–536.91	Knacke et al. 1991
900.00	–311.57	280.60	–564.11	Knacke et al. 1991
1000.00	–296.21	296.79	–592.99	Knacke et al. 1991

TABLE 7.17
Thermodynamic Data of $KClO_4$

T K	ΔHf° kJ mol^{-1}	S° J mol^{-1} deg^{-1}	ΔGf° kJ mol^{-1}	Reference
298.15	–430.12	151.04	–475.15	Knacke et al. 1991
400.00	–417.18	188.17	–492.45	Knacke et al. 1991
500.00	–402.50	220.87	–512.94	Knacke et al. 1991
600.00	–372.77	273.98	–537.16	Knacke et al. 1991
700.00	–355.79	300.13	–565.88	Knacke et al. 1991
800.00	–337.95	323.94	–597.10	Knacke et al. 1991
900.00	–319.32	345.87	–630.61	Knacke et al. 1991
1000.00	–299.96	366.27	–666.23	Knacke et al. 1991

TABLE 7.18
Activation Energies for the Decomposition of Rare Earth Perchlorates

Salt	E_a (kJ)
$La(ClO_4)_3$	142.7
$Sm(ClO_4)_3$	120.5
$Gd(ClO_4)_3$	119.2
$Ho(ClO_4)_3$	108.8
$Yb(ClO_4)_3$	105.9
$Lu(ClO_4)_3$	105.0

TABLE 7.19
Densities of Rb-Cl Salts

Phase	T(K)	d(gcm^{-3})	Reference
$RbClO_3$	Ambient	3.184	60
$RbClO_4$	Ambient	3.04	50

TABLE 7.20
Transitions of Rb-Cl Salts

Phase Change	T(K)	ΔH kJ/mol	ΔS J/(mol*K)	Reference
$RbClO_3$, α → β	596	—	—	48
$RbClO_4$, rhombohedral → cubic	557	12.6	—	60
cubic → liquid	597	—	—	60

TABLE 7.21
Thermodynamic Data (298.15 K)

Compound	ΔHf° kJ mol^{-1}	S° J mol^{-1} deg^{-1}	Reference
$RbClO_3$	–402.9	151.9	1
$RbClO_4$	–437.23	161.1	1
RbCl	–435.35	95.90	1

TABLE 7.22
Density of Ag-Cl Salts

Phase	T(K)	d(gcm^{-3})	Reference
$AgClO_2$	493	4.67	97
$AgClO_3$	493–496	4.43–4.46	
$AgClO_4$	433	3.96 (cubic,calc.)	133

TABLE 7.23
Transitions of Ag-Cl Salts

Phase Change	T(K)	ΔH kJ/mol	ΔS J/(mol*K)	Reference
$AgClO_3$, c → l	504	—	—	19
$AgClO_4$, rhombic →cubic	412	—	—	59
cubic → l	759	—	—	105

TABLE 7.24
Thermodynamic Data (298.15 K)

Compound	ΔHf° kJ mol^{-1}	S° J mol^{-1} deg^{-1}	Reference
$AgClO_2$	8.79	134.56	1
$AgClO_3$	–30.29	142.	1
$AgClO_4$	–31.13	38.8 (est.)	1
AgCl	–127.068	96.2	1

TABLE 7.25
Thermodynamic Data of $AgClO_3$

T K	ΔHf° kJ mol^{-1}	S° J mol^{-1} deg^{-1}	ΔGf° kJ mol^{-1}	Reference
298.15	30.29	141.84	72.58	Barin 1989
400.00	19.43	173.10	88.67	Barin 1989
500.00	7.50	199.65	107.33	Barin 1989
600.00	5.68	223.66	128.51	Barin 1989
700.00	20.15	245.93	152.00	Barin 1989
800.00	35.89	266.93	177.66	Barin 1989
900.00	52.91	286.97	205.36	Barin 1989
1000.00	71.22	306.25	235.02	Barin 1989

TABLE 7.26
Transitions of Na-Cl Salts

Phase Change	T(K)	ΔH kJ/mol	ΔS J/(mol*K)	Reference
$NaClO_3$, c → l	528	22.6	42.7	19
$NaClO_4$, c, rhombic → cubic	582	2.5	—	60
c → l	741	—	—	19

TABLE 7.27
Thermodynamic Data of $NaClO_3$

T K	ΔHf° kJ mol^{-1}	S° J mol^{-1} deg^{-1}	ΔGf° kJ mol^{-1}	Reference
298.15	–357.73	126.36	–395.41	Knacke et al. 1991
400.00	–346.66	158.19	–409.94	Knacke et al. 1991
500.00	–334.22	185.88	–427.16	Knacke et al. 1991
600.00	–298.29	252.99	–450.09	Knacke et al. 1991
700.00	–284.99	273.50	–476.44	Knacke et al. 1991
800.00	–271.68	291.27	–504.70	Knacke et al. 1991
900.00	–258.38	306.94	–534.62	Knacke et al. 1991
1000.00	–245.07	320.96	–566.03	Knacke et al. 1991
1100.00	–231.77	333.64	–598.77	Knacke et al. 1991
1200.00	–281.46	345.22	–632.72	Knacke et al. 1991

TABLE 7.28
Thermodynamic Data of $NaClO_4$

T K	ΔHf° kJ mol–1	S° J mol^{-1} deg^{-1}	ΔGF° kJ mol^{-1}	Reference
298.15	–382.75	142.26	–425.17	Barin 1989
400.00	–370.01	178.84	–441.55	Barin 1989
500.00	–355.65	210.82	–461.06	Barin 1989
600.00	–326.14	263.18	–484.05	Barin 1989
700.00	–309.69	288.52	–511.65	Barin 1989
800.00	–292.47	311.50	–541.67	Barin 1989
900.00	–274.56	332.58	–573.89	Barin 1989
1000.00	–256.00	352.13	–608.14	Barin 1989
1100.00	–236.83	370.41	–644.27	Barin 1989
1200.00	–217.05	387.61	–682.18	Barin 1989

TABLE 7.29
Density of Sr-Cl Salts

Phase	T(K)	d (gcm^{-3})	Reference
$Sr(ClO_2)_2$	Ambient	3.23	27
$Sr(ClO_3)_2$	Ambient	3.152	90
$Sr(ClO_4)_2$	298	3.00	28, 119

TABLE 7.30
Thermodynamic Data (298.15 K)

Compound	ΔHf° kJ mol^{-1}	S° J mol^{-1} deg^{-1}	Reference
$Sr(ClO_4)_2$	–762.79	—	1
$SrCl_2$	–828.9	114.85	1

TABLE 7.31
Density of Tl-Cl Salts

Phase	T(K)	d	Reference
$TlClO_3$	Ambient	4.58	134
$TlClO_4$	Ambient	4.959 (rhombic)	135
	553	4.58 (cubic)	103

TABLE 7.32
Transitions of Tl-Cl Salts

Phase Change	T(K)	ΔH kJ/mol	ΔS J/(mol*K)	Reference
$TlClO_4$ c, rhombic	539	—	—	136,137
cubic → l	778	—	—	136

REFERENCES

1. Wagman, D. D., Evans, W. H., Parker, V. B., Schumm, R. H., Halow, I., Bailey, S. M., Churney, K L., and Nuttall, R. L., The NBS tables of chemical thermodynamic properties, *J. Phys. Chem. Ref. Data,* 11, Suppl. 2, 1982.
2. Weigel, D., Imelik, B., and Prettre, M., *Bull. Soc. Chim. France,* 1427, 1962.
2a. Markowitz, M. M., *J. Inorg. Chem.*, 25, 407, 1963.
3. Lumme, P. and Lumme, H., *Suomen Kemi.*, 35, 129, 1962 .
4. Belkova, M. M., Alekseenko, I. A., and Serebrennikov, V. V., *Russian J. Inorg. Chem.*, 10, 747, 1965.
5. Levi, G. R. and Garrini, E., *Gazz. Chim. Ital.*, 90, 447, 1956.
6. Solymosi, F. and Bansagi, T., *Acta Chim. Acad. Sci. Hung.*, 56, 251, 1968.
7. Lumme, P. and Lumme, H., *Suomen Kemi.*, 35, 129, 1962.
8. Solymosi, F. and Bansagi, T., *Acta Chim. Acad. Sci. Hung.,* 56, 337, 1968.
9. Vorlaender, D. and Kaascht, E., *Berichte,* 56, 1157, 1923.
10. Smith, G. F. and Koch, E. G., *Z. Anorg. Chem.,* 223, 17, 1935.
11. Acheson, R. J. and Jacobs, P. M. W., *J. Phys. Chem.*, 74, 281, 1970.
12. Solymosi, F. and Braun, G., *Acta Chim. Acad. Sci. Hung.,* 52, 1, 1967.
13. Zinov'ev, A. A., Chudinova, L. I., and Smolina, L. P., *Russian J. Inorg. Chem.*, 1, 163, 1956.
14. Jacobs, P. W. M., Solymosi, F., and Rasko, J., *Combustion and Flame*, 17, 125, 1971.
15. Bohidar, N. and Mohanty, S. R., *Radiochim. Acta,* 27, 19, 1980.
16. Rolla, M., *Gazz. Chim. Ital.*, 69, 779, 1939.
17. Solymosi, F., *Acta Chim. Acad. Sci. Hung.*, 57, 35, 1968.
18. Hering, H., and Leray, A., *Bull. Soc. Chim. France,* 6, 1034, 1939.
19. Rossini, F. D., Wagman, D. D., Evans, W. H., Levine, S., and Jaffe, I., *Selected Values of Chemical Thermodynamic Properties*, NBS Circular 500, U.S. Govt. Printing Office, Washington, D.C., 1952.
20. See Gmelin, *Handbook of Inorganic Chemistry, Volume on Calcium.*
21. Urano, S., *Trans. Electrochem. Soc.*, 49, 65, 1926.
22. Levi, G. R. and Garrini, E., *Gazz. Chim. Ital.*, 90, 447, 1956.
23. Levi, G. R. and Cipollone, C., *Gazz. Chim. Ital.*, 53, 200, 1923.
24. Berezkina, L. G., Borisova, S. I., and Tamm, N. S., *Kin. i Kataliz.*, 13, 1168, 1972.
25. Tsolovich, N. and Antich, M., Russian *J. Phys. Chem.*, 58, 1645, 1984.
26. Levi, G. R. and Scherillo, A., *Atti. Linc.,* 311, 212, 1922.
27. Garrini, E. R. and Rigant, V., *Gazz. Chim. Ital.*, 90, 321, 1960.
28. Smith, G. F. and Koch, E. G., *Z. Anorg, Chem.* 223, 17, 1935.
29. Levi, G. R. and Cippolone, C., *Gazz. Chim. Ital.*, 200, 1923.
30. Markowitz, M. M., Boryta, D. A., and Stewart, H. J., *J. Chem. Eng. Data,* 9, 573, 1964.
31. Cabane, J. and Benard, J., *Bull. Soc. Chim.,* 36, 1961.
32. Solymosi, F., *Acta Chim. Acad. Sci. Hung.*, 57, 11, 1968.
33. Cordes, H. F. and Smith, F. R., *J. Phys. Chem.*, 72, 2189, 1968.
34. Elson, R. E. and Stuckey, J. E., *J. Inorg. Nucl. Chem.,* 35, 1029, 1973.
35. Vilcu, R. and Georgescu, N., *Rev. Roumaine Chim.,* 17, 1791, 1972.
36. Gomberg, M., *J. Am. Chem. Soc.,* 45, 398, 1923.
37. Hathaway, B. J., *Proc. Chem. Soc.,* 344, 1958.
38. Levi, G. R., *Selecta Chimica,* 13, 1091, 1953.
39. Solymosi, F. and Bansagi, T., *Acta Chim. Acad. Sci. Hung.*, 56, 251, 1968.
40. Solymosi, F. and Bansagi, T., *Acta Chim. Acad. Sci. Hung.*, 56, 337, 1968.
41. Chudinova, L. I., *Russ. J. Inorg. Chem.,* 12, 1778, 1967.
42. Willard, H. H. and Kassner, J. L., *J. Am. Chem. Soc.,* 52, 2391, 1930.
43. Solymosi, F., *Acta Chim. Acad. Sci. Hung.*, 57, 35, 1968.
44. Robson, H. L. and Soule, E. C., *Chem. Abst.,* 41, 2215e, 1947.
45. Soule, E. E., *Chem. Abst.*, 41, 2215c, 1947.
46. Levi, G. R. and Garrini, E., *Gazz. Chim. Ital.*, 90, 447, 1956.
47. Rossini, F. D., Wagman, D. D., Evans, W. H., Levine, S., and Jaffe, I., *Selected Values of Chemical Thermodynamic Properties*, NBS Circular 500, U.S. Govt. Printing Office, Washington, D.C., 1952.
48. Markowitz, M. M., Boryta, D. A., and Stewart, H., *J. Phys. Chem.*, 68, 2282, 1964.

49. Markowitz, M. M. and Boryta, D. A., *J. Phys. Chem.*, 65, 1419, 1961.
50. Solymosi, F., *Acta Chim. Acad. Sci. Hung.*, 57, 11, 1968.
51. Markowitz, M. M. and Boryta, D. A., *J. Phys. Chem.*, 65, 1419, 1961.
52. Markowitz, M. M., Boryta, D. A., and Stewart, H., *J. Chem. Eng. Data,* 9, 573, 1964.
53. Janz, G. J., Dampier, F. W., Lakshminarayanan, G. R., Lorenz, P. K ., and Tomkins, R. P. T., NSRDS-NBS 15, 1968.
54. Markowitz, M. M. and Boryta, D. A. M., *J. Phys. Chem.*, 69, 1114, 1965.
55. Markowitz, M. M. and Boryta, D. A. M., *J. Phys. Chem.*, 65, 1419, 1961.
56. Riolo, C. B., *Ann. Chim. (Rome),* 43, 564, 1953.
57. Rocchichioli, C., *Compt. Rend.,* 242, 2922, 1956.
58. Chudinova, L. I., *Izv. Vyssch. Uchen. Zavedenni Chim. i Chim. Tech.*, V, 357, 1962.
59. Gordon, S. and Campbell, C., *Anal. Chem.,* 27, 1102, 1955.
60. Levi, G. R. and Scherillo, A., *Att. Linc.,* 311, 216, 1922.
61. Chudinova, L. I., *Russ. J. Inorg. Chem.,* 10, 707, 1965.
62. Lemesheva, D. G. and Krivtsov, *Zh. Neorg. Khim.*, 30, 2216, 1985.
62a. Moles, E. and Roguero, C., *Anal. Espan.*, 31, 181, 1933.
63. Vilcu, R. and Georgescu, N., *Rev. Roumaine Chim.*, 17, 1791, 1972.
64. Bruni, G. and Levi, G., *Gazz. Chim. Ital.*, 53, 245, 1923.
65. Levi, G. and Bisi, C., *Gazz. Chim. Ital.*, 86, 392, 1956.
66. Barbieri-Melardi, E., *Ann. Chim. (Rome)*, 44, 20, 1954.
67. Fujioka, G. S. and Cady, G. H., *J. Am. Chem. Soc.,* 79, 2541, 1957.
68. Rocchiccioli, C., *Compt. Rend.,* 242, 2922, 1956.
69. Zinov'ev, A. A. and Naumova, V. L., *Russ. J. Inorg. Chem.,* 4, 910, 1959.
70. Fujioka, G. S. and Cady, G. H., *J. Am. Chem. Soc.,* 79, 2451, 1957.
71. Borgne, G. and Weigel, D., *Bull. Soc. Chim. France,* 3081, 1972.
72. Elson, R. E. and Stuckey, J. E., *J. Inorg. Nucl. Chem.*, 35, 1029, 1973.
73. Levi. G. R. and Garrini, E., *Gazz. Chim. Ital.*, 90, 447, 1956.
74. Markowitz, M. M., Boryta, D. A., and Stewart, H., *J. Phys. Chem.*, 68, 2282, 1964.
75. Buechler, A., Meschi, D. A., Mohazzebi, P., and Searcy, A. W., *J. Chem. Phys.,* 64, 4800, 1976.
76. Rudloff, W. K. and Freeman, E. S., *J. Phys. Chem.*, 73, 1209, 1969.
77. Simchen, A. E., *J. Phys. Chem.*, 65, 1093, 1961.
78. Markowitz, M. M. and Boryta, D. A., *J. Phys. Chem.*, 69, 1114, 1965.
79. Stern, K. H. and Bufalini, M., *J. Phys. Chem.*, 64, 1781, 1960.
80. Van der Bosch, A. and Aten, A. H. W., *J. Am. Chem. Soc.,* 75, 3835, 1953.
81. Harvey, A. E., Edmison, M. T., Jones, E.D., Seybert, R. A., and Catto, K. A., *J. Am. Chem. Soc.,* 76, 3270, 1954.
82. Harvey, A. E., Wassink, C. J., Rodgers, T. A., and Stern, K. H., *Ann. N.Y. Acad. Sci.,* 79, 971, 1960.
83. Cabane, J. and Benard, J., *Bull. Soc. Chim.,* 36, 1961.
84. Cordes, H. F. and Smith, S. R., *J. Phys. Chem.*, 72, 2189, 1968.
85. Cordes, H. F., *J. Phys. Chem.*, 72, 2185, 1968.
86. Bisi, C. C. and Clerici, A., *Chim. Ital.,* 93, 1438, 1963.
87. Yakunina, G. M. and Serebrennikov, V. V., *Russian J. Inorg. Chem.*, 15, 879, 1970.
88. ibid. *Russ. J. Inorg. Chem.,* 48, 1619, 1974.
89. Slavkina, R. I., Sorokina, G. E., and Serebrennikov, V. V., *Tr. Tomskogo Gos. Univ. Ser. Khim.*, 157, 135, 1963.
90. Bel'kova, M. M. and Aleksenko, L. A., *Russ. J. Inorg. Chem.,* 10, 747, 1965.
91. Bel'kova, M. M., Aleksenko, L. A., and Serebrennikov, V. V., *Russ. J. Inorg. Chem.,* 40, 1363, 1966.
92. Steinberg, M. and Shidlovsky, I., *J. Inorg. Nucl. Chem.*, 26, 887, 1964.
93. Bel'kova, M. M., Aleksenko, L. A., and Serebrennikov, V. V., *Russ. J. Inorg. Chem.,* 40, 1363, 1966.
94. Solymosi, F. and Bansagi, T., *Acta Chim. Acad. Sci. Hung.*, 56, 328, 1968.
95. Solymosi, F., *Acta Chim. Acad. Sci. Hung.*, 57, 11, 1968.
96. Cooper, J. and Marsh, R. E., *Acta Cryst.,* 14, 202, 1961.
97. Curti, R., Riganti, V., and Locchi, S., *Acta Cryst.*, 10, 687 1957.
98. Solymosi, F., *Z. Phys. Chem.,* (NF) 57, 1, 1968.

99. Solymosi, F., *Structure and Stability of Salts of Halogen Oxyacids in the Solid Phase*, John Wiley and Sons, London, New York, 1977, 15.
100. Deshpande, V. and Suryanarayama, S. V., *J. Matls. Sci.,* 17, 2968, 1982.
101. Solymosi, F., *Z.f. Phys. Chem. (NF),* 57, 1, 1968.
102. Udupa, M. R., *Indian J. Chem.*, 18, 355, 1979.
103. Hermann, K. and Ilge, W., *Z. Kryst.*, 75, 41, 1931.
104. Abrahams, S. C. and Bernstein, J. L., *Acta Cryst.* B33, 3601, 1977.
105. Carnelly, T. and O'Shea, K., *J. Chem. Soc.,* 45, 409, 1884.
106. Ishi, G., *Kogyo Kagaku Zasshi,* 65, 1011, 1965.
107. Riolo, C. B., *Gazz. Chim. Ital.*, 84, 884, 1954.
108. Sharma, S. S., *Proc. Indian Acad. Sci.,* 31A, 83, 1950.
109. Deshpande, V. T. and Mudholker, V. M., *Acta Cryst.*, 13, 483, 1960.
110. Kelley, K. K., *U.S. Bur. Mines Bull.*, 584, 1960.
111. Osada, H., Kusumoto, K., and Mukaj, K., *Kogyo Kayagu Kyokaishi*, 27, 91, 1966.
112. Begg, I. D., Halfpenny, P. J., Hooper, R. M., Norang, R. S., Roberts, K. J., and Sherwood, J. N., *Proc. Royal Soc. London,* A 386, 431, 1983.
113. Markowitz, M. M., *Inorg. Nucl. Chem.*, 25, 407, 1963.
114. Pierron, P., *Bull. Soc. Chim. France,* 8, 660, 1941.
115. Riolo, C. B., *Ann. Chim. (Rome),* 43, 564, 1953.
116. Levi, G. R. and Garrini, E., *Gazz. Chim. Ital.*, 90, 447, 1956.
117. Traube, H., *Z. Krist.,* 23, 135, 1894.
118. Philibert, P., *Thesis, Clermont-Ferrand,* 1958.
119. Chudinova, L. I., *Russ. J. Inorg. Chem.,* 7, 219, 1962.
120. Levi, G. R., *Atti. Lincei,* 311, 216, 1922.
121. Petru, F. and Kutek, F., Z. *Chem.*, 6, 426, 1926 .
122. Caven, R. M. and Bryce, G., *J. Chem. Soc.,* 514, 1934.
123. Murthy, P. R. and Patel, C. C., *Indian J. Chem.*, 3, 134, 165.
124. Gomberg, M., *J. Am. Chem. Soc.*, 45, 398, 1923.
125. Schmidt, H., *Z. Anorg. Allgem. Chem.*, 271, 305, 1953.
126. Amiel, J., *Compt. Rend.,* 198, 10333, 1954.
127. Chudinova, L. I., *Russ. J. Inorg. Chem.,* 10, 707, 1965.
128. Rosolovskii, V. Ya. and Krivtsov, N. V., *Russ. J. Inorg. Chem.,* 13, 18, 1968 129. Carpenter, C. D., *Chem. Metall. Eng.,* 24, 569, 1921.
130. Cohen, E. and Cohen-de Meester, W. A. T., *K. Akad. Wetenschappen Amsterdam Proc.,* 35, 1255, 1932.
131. Muethmann, W., *Z. Krist.,* 22, 543, 1894.
132. Wulff, P. and Heigl, A., *Z. Phys. Chem.,* A. 153, 187, 1931.
133. Braekken, H. and Harang, L., *Z. Krist.,* 75, 538, 1930.
134. *International Critical Tables*, Vol. I, McGraw-Hill, New York, 1926.
135. Wulff, P. and Schaller, D., *Z. Krist.,* 87, 43, 1934.
136. Kanellakopolos, B., *J. Inorg. Nucl. Chem.,* 28, 813, 1966.
137. Petru, F. and Kutek, F., *Z. Chem.*, 6, 426, 1966.

8 Oxysalts of Bromine

8.1 GENERAL

Less is known about the anhydrous bromine-containing oxysalts than about the corresponding compounds of chlorine and iodine. This is true both with respect to the total number of such salts, as well as the information available on each salt. Since 1974, when the last review of these compounds was published,[1] the situation has changed only slightly, with the number of known bromites increasing from 2 to 3, bromates from 15 to 17, and perbromates from 3 to 8. There are still no thermodynamic data above 298 K. Consequently, this chapter is fairly brief, includes no high-temperature equilibrium calculations, and updates qualitative and semi-quantitative information using material published since 1972.

There are only three oxyanions which form anhydrous salts: bromites (BrO_2^-), bromates (BrO_3^-), and perbromates (BrO_4^-). There is no evidence for the existence of stoichiometric, anhydrous hypobromites (BrO^-), nor is there any literature concerning a search for them.

The known and doubtful oxybromine salts are shown in Figures 8.1 to 8.3.

8.2 CLASSES OF OXYBROMINE SALTS

8.2.1 Bromites

So far, only the bromites of barium, lithium, and strontium have been prepared,[2] by heating the corresponding bromate and bromide together. The method should be applicable to other bromites as well. So far, very little is known of these salts, except that heating produces the bromide

$$MBrO_2 = MBr + O_2$$

Other methods that have also been used for the preparation of bromites are (a) disproportionation of hypobromites[3–5] and (b) radiolysis of bromates,[6,7] and decomposition of bromates.[8,9] The former depends on the use of a hydrate as the starting material and is thus outside the scope of this work; similarly for (b). The cited references will allow the interested reader access to the literature.

8.2.2 Bromates

These constitute the largest group of bromine oxysalts. The BrO_3^- ion is a flattened pyramid similar to ClO_3^- and IO_3^-. A model derived by Weigel et al.[10] from literature data is shown in Figure 8.4. The quadrupole spectra of bromates[11] show that the electronic structure of bromine is affected by the metal cation of the salt. The cation is nearer to one of the oxygen atoms than to bromine, e.g., in $NaBrO_3$, Na–Br 4.1Å, Na–O 2.38Å. The Br–O distance of 1.78Å is consistent with a virtually pure single-bond character.

As pointed out by Bancroft and Gesser,[4,12] bromates may decompose by two competing reactions:

$$MBrO_3 = MBr + 3/2\ O_2$$

$$MBrO_3 = 1/2\ M_2O + 1/2\ Br_2 + 5/4\ O_2$$

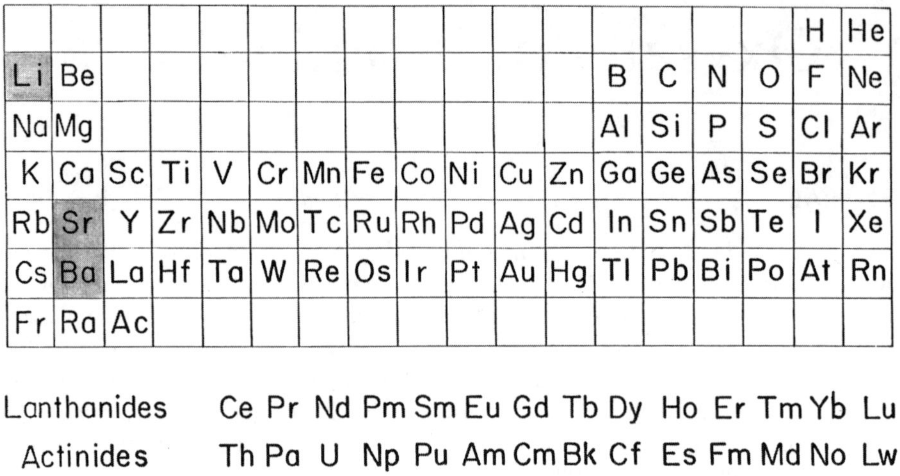

FIGURE 8.1 The known bromites.

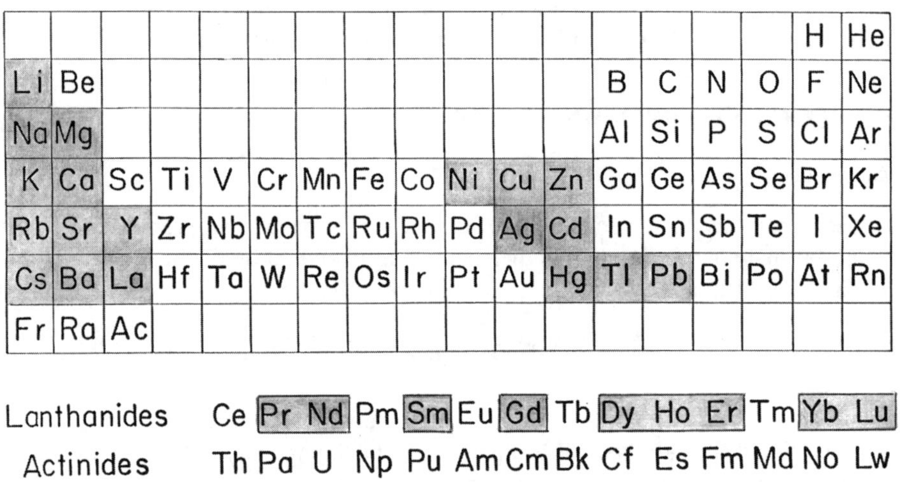

FIGURE 8.2 The known bromates.

If thermodynamics were the sole determinant as to which reaction occurs, the argument developed by Markowitz[13] for predicting the decomposition products of the perchlorates could be used, i.e., if [ΔGf°(MBr) – ΔGf°] > 0, the oxide should form. Bancroft and Gesser[4] applied this method to the results of their TGA experiments on several bromates. Their results are shown in Table 8.1.

The rule works reasonably well, particularly when the Gibbs energy difference is large, the most notable exception being $Cd(BrO_3)_2$. However, it cannot be expected to be exact in predicting the composition of product mixtures since this is frequently affected by kinetic factors which are particularly difficult to analyze for reactions involving solids. In this connection it is of interest to examine the activation energies for the decomposition reactions in Table 8.1. Bancroft and Gesser[4] applied a method developed by Horowitz and Metzger[14] which requires that the rate constant obeys an Arrhenius equation. If the reaction is first order, the resulting expression is

$$\ln\left[\ln\frac{W_o - W_f}{W - W_f}\right] = \frac{E^{*}\Theta}{RT_s^2}$$

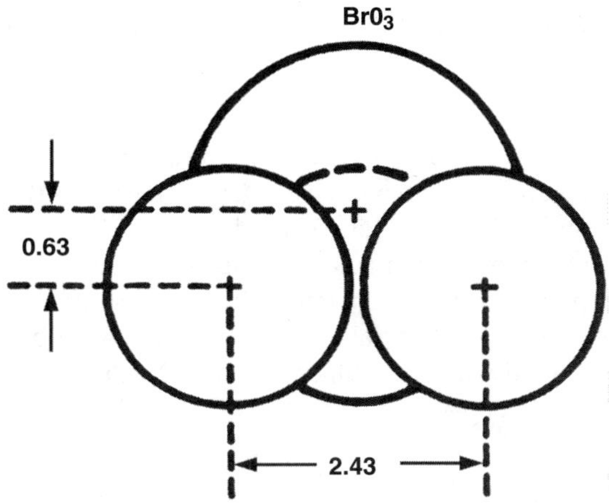

FIGURE 8.3 Structure of the bromate ion. (From Weigel, D., Imelik, B., and Prettre, M., *Bull. Soc. Chim. France*, 1427, 1962. With permission.)

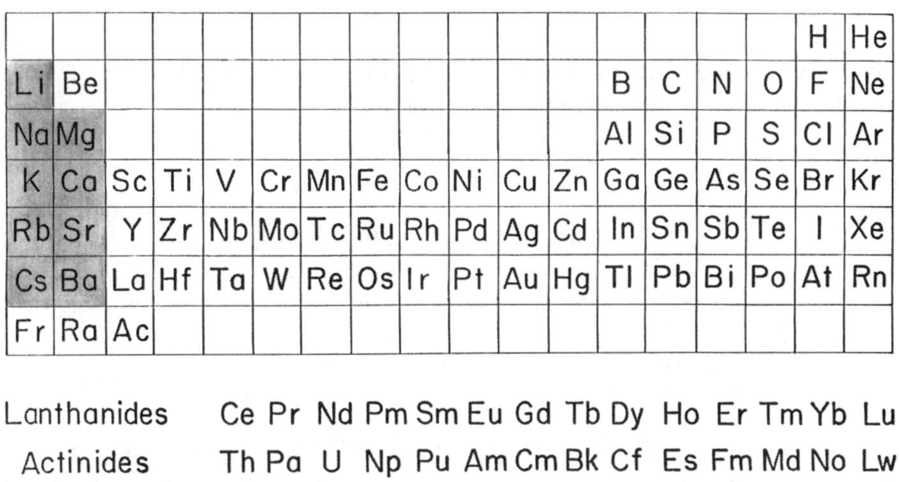

FIGURE 8.4 The known perbromates.

where W_o and W_f are the initial and final weights, respectively; T_s is the temperature at which $[(W - W_f) - (W_o - W_f)] = 1/e$ and $\eta = T - T_s$. Most of the bromates obeyed this equation over at least the first half of the decomposition.

For $RbBrO_3$ and $CsBrO_3$ the higher value was obtained with a higher heating rate. In addition, differences in constitutional factors, such as particle size, and defects in the crystal structure, may affect the activation energy in some unknown way.[15] Large anomalies are quite apparent for $LiBrO_3$ and $Ca(BrO_3)_2$, but it is not known whether this is because the reaction is not first order, or for some other reason. It is also noticeable that $E^{\neq}$ values vary considerably even when salts which decompose only to the bromide are considered. This is true even within the same group in the periodic table, such as alkali metal and alkali earth bromates, where great differences in the mechanism and/or bond energies within the BrO^- would not be expected. Clearly, Table 8.1 is only an argument for further kinetic studies, particularly isothermal ones. For convenience, information on decomposition of bromates, collected by Solymosi,[16] has been given in Table 8.2. From this it appears that the stabilities of the alkali metal bromates decrease in the order Li > Na > K > Cs > Rb.

The kinetics consist of rapid nucleation, followed by a contracting sphere model.[17]

Trivalent bromates generally exist only as hydrates, although some of them, La, Pr, Nd, Sm, Er, are reported to have been dehydrated to the anhydrous compound without decomposition.[18]

8.2.3 Perbromates

In contrast to the well-known +7 oxidation state of chlorine and iodine exhibited by the perchlorates and periodates, attempts to prepare perbromates were, until rather recently, unsuccessful. References to the earlier work are given by Bancroft and Gesser.[4]

The synthesis of $HBrO_4$ and rubidium and potassium salts was accomplished, using oxidation of bromate by XeF_2 or (preferably) molecular fluorine in aqueous solution.[19] Spectral studies[20] show that the perbromate ion is tetrahedral in both the solid-state and aqueous solutions. The thermodynamic properties and thermal decomposition of individual salts are discussed under each element. A more general article on various properties of the perbromates was published by Herrell and Gayer.[21]

8.3 INDIVIDUAL BROMINE — OXYSALTS

8.3.1 Barium

8.3.1.1 $Ba(BrO_2)_2$

The salt has been prepared[22] by the thermal decomposition of $Ba(BrO_3)_2$ at 280°C under an O_2 atmosphere. Decomposition probably proceeds according to

$$Ba(BrO_3)_2 = BaBr_2 + 3\ O_2$$

$$2\ Ba(BrO_3)_2 + BaBr_2 = 3\ Ba(BrO_2)_2$$

$BaBr_2$ is separated by extraction with anhydrous ethanol. Above 795°C the bromite decomposes rapidly.

8.3.1.2 $Ba(BrO_3)_2$

Two crystalline forms of $Ba(BrO_3)_2$ have been reported.[23] The low-temperature form was produced by rapid heating of the monohydrate, whereas isothermal heating of the hydrate between 105 and 120°C produced the high- temperature form. The transformation I → II was detected by DTA at 160°C.

Several groups have studied the thermal decomposition of the salt.[23,24] They agree that the final product is $BaBr_2$, i.e., the overall reaction is

$$Ba(BrO_3)_2 = BaBr_2 + 3\ O_2$$

but both $Ba(BrO_2)_2$ (see above) and $Ba(BrO_4)_2$ have been suggested as intermediates, the latter on the basis of a great increase in the rate after one third of the oxygen had disappeared. However, no such increase was noted by Bancroft and Gesser.[4] Dupuis and Rocchiccioli[25] observed differences in the spectra of $Ba(BrO_3)_2$ heated below and above 159°C. They argued against the formation of bromite at the higher temperature and hypothesized the existence of two crystalline forms of the bromite. It seems, however, that the work of Tanguy et al.[22] provides firmer evidence for the formation of bromite as an intermediate. The kinetics of the overall decomposition were measured by Lumme and Lumme[22a] using TGA. They found decomposition occurring in the range 290 to 380°C in air, and 265 to 315°C in vacuum. The Arrhenius activation energy of 209 kJ calculated from their data agrees well with the 218 kJ given by Bancroft and Gesser.[12]

The decomposition kinetics were also studied by Mohanty and co-workers,[26–28] who were interested in elucidating the effect of radiation and doping. The effect of both is to introduce defects into the crystal structure. In the range 219 to 315°C the decomposition of normal $Ba(BrO_3)_2$ can be divided into five regions: (1) initial gas evolution, up to $\alpha = 0.06$; (2) a short induction period; (3) a slow linear reaction followed by a sigmoidal zone consisting of (4) acceleratory and (5) decay stages. The decomposition begins entirely in the solid state:

$$Ba(BrO_3)_2 = BaBr_2 + 3\ O_2$$

and the initial stage (1) is likely due to physical desorption and removal of occluded gases. The induction period is attributed to the formation of a small number of nuclei at lattice imperfections. The linear stage is attributable to surface decomposition caused by these nuclei. These nuclei slowly penetrate the interior of the crystal. The following sigmoidal stage has been attributed to the formation of new nuclei and the formation of cracks at $\alpha = 0.2$, with the reaction proceeding down these cracks. Irradiation produces various kinds of defects which affect different stages of the decomposition differently.

8.3.1.3 $Ba(BrO_4)_2$

The decomposition of this salt was studied by Russian researchers[29] who prepared it by the reaction of $HBrO_4$ with BaO. The salt may not have been entirely anhydrous. Studies by DTA and TGA showed that decomposition to the bromate occurred at 462 K, which then decomposed further at 548 K.

Thermodynamic data of $Ba(BrO_3)_2$ and $BaBr_2$ are given in Table 8.3.

8.3.2 Cadmium

8.3.2.1 $Cd(BrO_3)_2$

The salt can be prepared by step-wise dehydration of the dihydrate. The salt is then stable up to 250°C, above which it decomposes, largely to the oxide.[12]

8.3.3 Calcium

8.3.3.1 $Ca(BrO_3)_2$

Not much work has been done on this salt. It decomposes in the solid state according to

$$Ca(BrO_3)_2 = CaBr_2 + 3\ O_2$$

in the range 260 to 340°C.[12] In vacuum the decomposition begins at 220°C. The kinetics follows the same mechanism[30] as was described for $Ba(BrO_3)_2$.

8.3.3.2 $Ca(BrO_4)_2$

The salt decomposes according to the following scheme:

$$Ca(BrO_4)_2 = Ca(BrO_3)_2 + O_2$$

$$Ca(BrO_3)_2 = CaBr_2 + 3\ O_2$$

The first decomposition occurs at 464 K, the second at 532 K.

Thermodynamic data are shown in Table 8.4.

8.3.4 Cesium

8.3.4.1 $CsBrO_3$

A specific gravity at 16°C of 4.109 was reported.[33] Thermal decomposition proceeds to the bromide:

$$CsBrO_3 = CsBr + 3/2\ O_2$$

The kinetics at 673 K were reported[32] to follow Prout-Tompkins kinetics, passing through an initial period of gas evolution, short acceleratory regime, and long decay.

8.3.4.2 $CsBrO_4$

The decomposition proceeds according to

$$CsBrO_4 = CsBrO_3 + 3/2\ O_2$$

$$CsBrO_3 = CsBr + 3/2\ O_2$$

in the range 473 to 592 K.[29]

Thermodynamic data are shown in Table 8.5.

8.3.5 Copper

8.3.5.1 $Cu(BrO_3)_2$

The anhydrous salt has not yet been definitely prepared. The hydrate decomposes at 180 K, with the loss of oxygen and possibly bromine, and the formation of Cu_2O and a basic bromate or oxybromate.[34]

8.3.6 Lead

8.3.6.1 $Pb(BrO_3)_2$

Very little work has been done on this salt since two people were killed[35] while grinding the material which had been prepared from $KBrO_3$ and lead acetate, although it was later established that the compound thus formed is diacetatoplumbo bromate.[36] $Pb(BrO_3)_2.H_2O$ is reported to decompose near 100°C into PbO_2, $PbBr_2$, and Br_2.

The only thermodynamic value is $\Delta Gf° = -16.30$ kJ at 298.15 K.[56]

8.3.7 Lithium

8.3.7.1 $LiBrO_2$

Preparation of this salt was carried out by heating $LiBrO_3$ and LiBr between 190 and 220°C:

$$2\ LiBrO_3 + LiBr = 3\ LiBrO_2$$

The salt is very hygroscopic. Near 225°C it decomposes while melting:

$$LiBrO_2 = LiBr + O_2$$

8.3.7.2 $LiBrO_3$

The salt decomposes by two competing reactions:[2,12]

$$LiBrO_3 = LiBr + 3/2\ O_2\ (a)$$

$$LiBrO_3 = 1/2\ Li_2O + 1/2\ Br_2 + 5/4\ O_2\ (b)$$

95% of the decomposition proceeds by reaction (a). Curiously, the oxide is formed predominantly in the initial stages; it then stops and the major part of the decomposition goes to LiBr. The products of (b) do not react further so that LiBr cannot be formed by the reaction 1/2 Li_2O + $1/2Br_2$ = LiBr + 1/4 O_2. The decomposition temperature is in doubt. Bancroft and Gesser[12] report it to occur as low as 214°C, whereas Simmons and Waldeck[37] report the melting point to lie between 248° and 260°C. The former results were obtained at very low pressures and this may account for the difference.

8.3.7.3 $LiBrO_4$

Decomposition follows the sequence

$$LiBrO_4 = LiBrO_3 + 1/2\ O_2$$

$$LiBrO_3 = LiBr + 3/2\ O_2$$

The decomposition temperature of $LiBrO_4$, 573 K, coincides with its melting.

Densities are given in Table 8.6, thermodynamic data in Table 8.7.

8.3.8 Magnesium

8.3.8.1 $Mg(BrO_3)_2$

No thermodynamic data are available. The only thermal decomposition study is that of Bancroft and Gesser[12] who found that decomposition, beginning near 250°C, goes almost entirely to the oxide:

$$Mg(BrO_3)_2 = MgO + Br_2 + 5/2\ O_2$$

8.3.8.2 $Mg(BrO_4)_2$

Decomposition to $Mg(BrO_3)_2$ occurs at 472 K[29] and, in contrast to the barium salt, proceeds largely to the oxide.

8.3.9 Mercury

8.3.9.1 $HgBrO_3$

The preparation of this salt is mentioned by Mellor[38] and Gmelin.[39] Its thermal decomposition apparently leads to the formation of metallic mercury, but the decomposition does not seem to have been studied in detail.

8.3.9.2 $Hg(BrO_3)_2$

The anhydrous salt decomposes slightly when heated to near 125°C and then deflagrates near 155°C, with $HgBr_2$ as the major product.[40]

8.3.10 Nickel

8.3.10.1 $Ni(BrO_3)_2$

The salt begins to decompose near 160°C, the product being largely NiO, with some $NiBr_2$ formed as a secondary product.[12]

8.3.11 Potassium

8.3.11.1 $KBrO_3$

There appears to be no reliable melting point for this salt since its decomposition begins in the solid state. Japanese workers[41] report an endothermic DTA peak at 698 K which they identify with the melting point. However, decomposition has been reported as low as 615 K;[42] somewhat higher decomposition temperatures were reported by Duval[43] and Bancroft and Gesser,[12] but this obviously depends on the sensitivity with which weight loss can be measured.

Jach[42] studied decomposition kinetics from 615 K up. The overall decomposition is entirely

$$KBrO_3 = KBr + 3/2\ O_2$$

Jach found somewhat different kinetics over the ranges 615 to 640 K and 652 to 685 K, with a transition region between the two ranges. Over both the reaction appeared to be first order, but the Arrhenius activation energies in the low and high regions were 260 and 221 kJ, respectively. Over the higher range melting evidently occurred, with the formation of a eutectic at the interface between KBr and $KBrO_4$ and bulk melting near the top of the range. Over both ranges the reaction begins with rapid nucleation of the entire surface, followed by a contracting reaction zone. However, the first-order kinetics do not fit this mechanism, and Jach suggested that extensive cracking of crystals, which is observed during decomposition, is responsible for this. The chemical mechanism of the decomposition has not yet been studied.

8.3.11.2 $KBrO_4$

The salt decomposes in two stages:[29]

$$KBrO_4 = KBrO_3 + 1/2\ O_2$$

$$KBrO_3 = KBr + 3/2\ O_2$$

The decomposition begins at 275°C, a temperature at which $KBrO_3$ is still stable.

The density of $KBrO_3$ is 3.27 gcm^{-3} at 290 K.[33]

Thermodynamic data at 298.15 K are given in Table 8.8.

8.3.12 Rare Earths

8.3.12.1 Bromates

The anhydrous bromates of Y, La, Pr, Nd, Sm, Gd, Dy, Ho, Er, Yb, and Lu have been prepared by treating the oxides with $HBrO_3$.[44,45] The salts decompose near 170°C to the oxybromides:[45]

$$M(BrO_3)_3 = MOBr + Br_2 + 4\ O_2$$

and finally to the oxides. Rate constants and activation energies for the decomposition have been reported.[46]

8.3.13 Rubidium

8.3.13.1 $RbBrO_3$

A specific gravity value at 16°C of 3.680 has been reported.[47]

The salt decomposes entirely to the bromide:

$$RbBrO_3 = RbBr + 3/2\ O_2$$

beginning at 360°C.[29] This value is clearly inconsistent with the melting point of 430°C reported by Buell and McCroskey[48] for a carefully purified sample. The very low pressures in Bancroft and Gesser's apparatus would favor the decomposition.

8.3.13.2 $RbBrO_4$

The salt decomposes in the solid state near 285°C to $RbBrO_3$.[29]

An ambient temperature density of 3.68 gcm^{-3} has been reported.[33]

Thermodynamic data are listed in Table 8.9.

8.3.14 Silver

8.3.14.1 $AgBrO_3$

The salt decomposes to the bromide:

$$AgBrO_3 = AgBr + 3/2\ O_2$$

Decomposition starts near 200°C and is complete near 300°C.[12a] In the presence of dust, decomposition is explosive.[49]

A density value of 5.206 gcm^{-3} at 293 K has been reported.[47] Thermodynamic data are given in Table 8.10.

8.3.15 Sodium

8.3.15.1 $NaBrO_3$

The linear thermal expansion coefficient has been measured by Sharma[50] between 36.5 and 285.1°C, and by Ganesan[51] between –149 and +171.5°C. Although both authors agree well in their results for $NaClO_3$, they disagree considerably for $NaBrO_3$. Near 50°C both sets of data agree fairly well, but Sharma obtained a much smaller temperature coefficient for α. There is no obvious reason for the difference.

$NaBrO_3$ decomposes entirely to the bromide:

$$NaBrO_3 = NaBr + 3/2\ O_2$$

At low pressures decomposition begins near 320°C.[12] The kinetics were studied by Jach[42] in the range 323 to 430°C. Surface nucleation by NaBr is very rapid, after which a contracting envelope mechanism is followed. An Arrhenius plot exhibits a marked discontinuity at 330°C. Above this temperature the plot is linear, below it the rate drops abruptly to rates much lower than expected from the extrapolation of the high-temperature plot. In the high-temperature region, Jach postulates a liquid eutectic at the $NaBrO_3$-NaBr interface. The effect of particle size and dislocation is discussed, but no detailed mechanism is presented.

A density value of 3.339 gcm^{-3} at 290 K has been reported.[33]

8.3.15.2 $NaBrO_4$

The salt was prepared from NaOH and $HBrO_4$. Its decomposition occurred at 266 ± 10°C.[29]

Thermodynamic data are given in Table 8.11.

8.3.16 Strontium

8.3.16.1 $Sr(BrO_3)_2$

Very little information is available for this salt. According to the older literature[52] decomposition begins near 240°C and results mainly in the formation of $SrBr_2$, as well as some SrO. The more recent results of Bancroft and Gesser[12] show that the decomposition is entirely to the bromide:

$$Sr(BrO_3)_2 = SrBr_2 + 3\ O_2$$

beginning near 265°C. Similar results were also obtained earlier by Hackspill and Winterer.[3]

8.3.16.2 $Sr(BrO_4)_2$

The decomposition of this salt has been studied by TGA and DTA,[53] beginning with a hydrate. At 473 K the salt decomposes to the bromate:

$$Sr(BrO_4)_2 = Sr(BrO_3)_2 + O_2$$

which then decomposes at 532 K:

$$Sr(BrO_3)_2 = SrBr_2\ 3\ O_2$$

8.3.17 Thallium

8.3.17.1 $TlBrO_3$

Decomposition kinetics have been studied in considerable detail.[54,55] Decomposition becomes appreciable near 125°C and results largely in the formation of black thallic oxide:

$$4\ TlBrO_3 = 2\ Tl_2O_3 + Br_2 + 3\ O_2$$

A small amount of TlBr also seems to be formed.

Kinetically, the reaction appears to occur in three stages: (a) an initial phase of up to 3 to 4% decomposition which fits a first-order rate equation, the activation energy for this stage is 150 kJ; (b) an acceleratory stage fitted by the Prout-Tompkins equation, the activation energy for this stage, terminating near 35% decomposition, is 209 kJ; (c) the decay stage is well fitted by the contracting sphere model with an activation energy of 150 kJ. A tentative mechanism proposed to account for these observations identifies the initial phase with rapid decomposition at the subgrain boundaries, followed by progression of the reaction into the subgrains.

No thermodynamic data are available.

8.3.18 Zinc

The thermal decomposition of $Zn(BrO_3)_2$ was studied by Bancroft and Gesser.[12] Decomposition begins near 150°C and proceeds nearly entirely to the oxide:

$$Zn(BrO_3)_2 = ZnO + Br_2 + 5/2\ O_2$$

No thermodynamic data are available.

TABLE 8.1
Gibbs Energy Differences [ΔGf° (Br⁻)–ΔGf° (O⁻²)]$_{equ.}$ (kJ)

		T(K)		
Salt	%Br⁻	300	1000	E≠
$LiBrO_3$	90	–57.7	–47.3	1159
$NaBrO_3$	100	–151	–137	352
$KBrO_3$	100	–217	–196	494
$RbBrO_3$	100	–233		205–226
$CsBrO_3$	100			100–138
$Mg(BrO_3)_2$	5	+70.7	+94.6	84
$Ca(BrO_3)_2$	100	–50.2	–21	1205
$Sr(BrO_3)_2$	100	–143	–110	278
$Ba(BrO_3)_2$	100	–205	–162	218
$Zn(BrO_3)_2$	1	+12	+21	193
$Cd(BrO_3)_2$	2	–70.7	–36	159
$AgBrO_3$	100	–80.3		188
$Ni(BrO_3)_2$	7	+3	+35	193

TABLE 8.2
Characteristic Data on the Thermal Stability of Bromates[4,16,46]

Salt	Temperature Range of Decomposition (°C)
$LiBrO_3$	236–256
$NaBrO_3$	298–368
$KBrO_3$	366–395
$RbBrO_3$	328–416
$CsBrO_3$	265–388
$Mg(BrO_3)_2$	245–329
$Ca(BrO_3)_2$	245–270
$Sr(BrO_3)_2$	248–275
$Ba(BrO_3)_2$	267–328
$Ni(BrO_3)_2$	168–221
$Zn(BrO_3)_2$	160–205
$Cd(BrO_3)_2$	240–310
$La(BrO_3)_3$	210–231
$Pr(BrO_3)_3$	197–235
$Sm(BrO_3)_3$	214–242
$Dy(BrO_3)_3$	210–232
$Ho(BrO_3)_3$	205–232
$Er(BrO_3)$	214–235
$Yb(BrO_3)_3$	212–231
$Lu(BrO_3)_2$	225–252

TABLE 8.3
Thermodynamic Data (298.15 K)

Compound	ΔHf° kJ mol^{-1}	S° J mol^{-1} deg^{-1}	Reference
$Ba(BrO_3)_2$	–718.18	247.	56
$BaBr_2$	–757.3	146.	56

TABLE 8.4
Thermodynamic Data (298.15 K)

Compound	ΔHf° kJ mol^{-1}	S° J mol^{-1} deg^{-1}	Reference
$Ca(BrO_3)_2$	–685.8	—	56
$CaBr_2$	–682.8	130	56

TABLE 8.5
Thermodynamic Data (298.15 K)

Compound	ΔHf° kJ mol^{-1}	S° J mol^{-1} deg^{-1}	Reference
$CsBrO_3$	–375.81	163.6	56
CsBr	–405.8	113.05	56

TABLE 8.6
Density of $LiBrO_3$

Phase	T(K)	d	Reference
$LiBrO_3$	Ambient	3.62 (pyk)	
	Ambient	3.76 (X-ray)	57

TABLE 8.7
Thermodynamic Data (298.15 K)

Compound	ΔHf° kJ mol^{-1}	S° J mol^{-1} deg^{-1}	Reference
$LiBrO_3$	–346.98	—	56
LiBr	–351.213	74.27	56
Li_2O	–597.94	37.57	56

TABLE 8.8
Thermodynamic Data (298.15 K)

Compound	ΔHf° kJ mol^{-1}	S° J mol^{-1} deg^{-1}	Reference
$KBrO_4$	–287.86	170.08	56
$KBrO_3$	–360.24	149.16	56
KBr	–397.798	95.90	56

TABLE 8.9
Thermodynamic Data (298.15 K)

Compound	ΔHf° kJ mol^{-1}	S° J mol^{-1} deg^{-1}	Reference
$RbBrO_3$	–367.27	161.1	56
RbBr	–394.6	109.96	56

TABLE 8.10
Thermodynamic Data (298.15 K)

Compound	ΔHf° kJ mol^{-1}	S° J mol^{-1} deg^{-1}	Reference
$AgBrO_3$	–10.5	151.9	56
AgBr	–100.37	107.1	56

TABLE 8.11
Thermodynamic Data (298.15 K)

Compound	ΔHf° kJ mol^{-1}	S° J mol^{-1} deg^{-1}	Reference
$NaBrO_3$	–334.09	128.9	56
NaBr	–361.0	86.82	56

REFERENCES

1. Stern, K. H. High Temperature Properties and Decomposition of Inorganic Salts, Part 4, Oxysalts of the Halogens, *J. Phys. Chem. Ref. Data,* 3, 481, 1974.
2. Hagenmueller, P. and Tanguy, B., *Compt. Rend.,* 260, 3974, 1965.
3. Hackspill, A. and Winterer, L., *Compt. Rend.,* 191, 665, 1930.
4. Bancroft, G. M. and Gesser, H.D., *J. Inorg. Nucl. Chem.,* 27, 1545, 1965.
5. Hugus, Z. Z., *J. Am. Chem. Soc.,* 74, 9076, 1952.
6. Nyholm, R. S., *Proc. Chem. Soc.,* 273, 1961.
7. Urch, D. S., *J. Inorg. Chem.*, 25, 771, 1963.
8. Ferreira, R., *An. Acad. Brasiliera de Ciencias,* 38, 407, 1966.
9. Pitzer, K. S., *J. Am. Chem. Soc.,* 70, 2140, 1948.
10. Weigel, D., Imelik, B., and Prettre, M. *Bull. Soc. Chim. France,* 1427, 1962.
11. Shinomura, K., Kushida, T., and Inoue, N., *J. Chem. Phys.,* 22, 1944, 1954.
12. Bancroft, G. M. and Gesser, H. D., *J. Inorg. Nucl. Chem.,* 27, 1537, 1965.
13. Markowitz, M. M., *J. Inorg. Nucl. Chem.,* 25, 407, 1963.
14. Horowitz, H. H. and Metzger, G., *Anal. Chem.,* 35, 1464, 1963.
15. Stern, K. H. and Weise, E. L., *NSRDS-NBS,* 30, 1969.
16. Solymosi, F., *Structure and Stability of Salts of Halogen Oxyacids in the Solid Phase*, John Wiley & Sons, New York, 1977, Chapter 7.
17. See Chapter 1, this volume.
18. Phillips, B. R., and Taylor, D., *J. Chem. Soc.,* 5583, 1963.
19. Appleman, E. H., *J. Am. Chem. Soc*., 90, 1900, 1968.
20. Appleman, E. H., *Inorg. Chem*., 223, 1969.
21. Herrell, A. Y. and Gayer, K. H., *J. Chem. Ed*., 49, 583, 1972.
22. Tanguy, B., Frit, B., Terrell, G., and Hagenmuller, P., *Compt. Rend.,* 264C, 301, 1967.
22a. Lumme, P. and Lumme, H., *Suomenkemistilehti,* 35, 129, 1962.
23. Dupuis, T. and Rocchiccioli-Deltcheff, C., *Anal. Chem*., 44, 165, 1969.
24. Hackspill, A. and Winterer, L., *Compt. Rend.,* 191, 663, 1930.
25. Dupuis, T. and Rocchiccioli, C., *Compt. Rend.,* 266B, 397, 1968.
26. Bhattamisra, S. D. and Mohanty, S. R., *J. Inorg. Nucl. Chem.,* 39, 2103, 1977.
27. Mohanty, S. R. and Satapathy, M., *Radioch. Radioanal. Letters,* 37, 345, 1979.
28. Jena, B., Mohanty, S. R., and Satapathy, M., *Indian J. Chem*., 19A, 1139, 1980.
29. Isupov, V. K., Gavrilov, V. V., and Kirin, I. S. *Russ. J. Inorg. Chem.,* 22, 1406, 1977.
30. Gavande, A. M. and Ray, M. N. *Indian J. Chem*., 20A, 326, 1981.
31. Buchanan, J. Y., *Proc. Edinburgh Soc.,* 31, 642, 1910/11.
32. Mohanty, S.R. and Padhee, G.R., *Radiochem. Radioanal. Lett.*, 42, 31, 1980.
33. *International Critical Tables*, McGraw-Hill, New York, 1926.
34. Rocchiccioli, C., *Thesis*, Paris, 1960.
35. Victor, Z., *Z. f. Angew. Chem*., 40, 841, 1927.
36. Guenzel, E. and Marcus, E., *Z. f. Angew. Chem*., 38, 929, 1925.
37. Simmons, J. P. and Waldeck, W. F., *J. Am. Chem. Soc.,* 53, 1725, 1931.
38. Mellor, J. W., A Comprehensive Treatise of Inorganic and Theoretical Chemistry, Longmans, Green, New York.
39. *Gmelin's Handbuch der Anorganischen Chemie*, 8th ed., Verlag Chemie, Weinheim, Germany.
40. Smith, G. F., *J. Am. Chem. Soc.,* 46, 1577, 1924.
41. Furuichi, R., Ishii, T., Yamanaka, Z., and Shimokawabe, M., *Thermochim. Acta,* 51, 199, 1981.
42. Jach, J. J., *J. Phys. Chem. Solids,* 24, 63, 1963.
43. Duval, C., *Anal. Chem*., 15, 223, 1956.
44. Petru, F. and Dusek, B., *Coll. Czech. Chem. Commun*., 33, 2397, 1968.
45. Mayer, I. and Glasner, Y., *J. Inorg. Nucl. Chem.,* 33, 2397, 1968.
46. Yakunina, G. M., Alekseenko, L. A., and Serebrennikov, V. V., *Zh. Fiz. Khim*., 44, 60, 1970.
47. Buchanan, J.Y., *Proc. Edinburgh Soc*., 31, 642, 1910/11.
48. Buell, H. D. and McCroskey, C. R., *J. Am. Chem. Soc.,* 43, 2031, 1921.
49. Shidlovskii, A. A., Voskresenskaya, A. A., and Shitikov, E. S., *Russ. J. Phys. Chem*., 41, 380, 1967.

50. Sharma, S. S., *Proc. Indian Acad. Sci.*, 31A, 83, 1950.
51. Ganesan, S., *J. Indian Inst. Sci.*, 41, 9, 1951.
52. ref. 38, vols. I, II.
53. Isupov, V. K., Gavrilov, V. V., and Kirin, I., *Russ. J. Inorg. Chem.,* 22, 1406, 1977.
54. Simpson, J., Taylor, D., Fanshawe, R. S., Norbury, J. M., and Watson, W. J., *J. Chem. Soc.*, 3323, 1958.
55. Udupa, M. R., *Indian J. Chem.*, 19A, 1014, 1980.
56. Chapter 7, reference 1.
57. Burns, J. H., *Acta Cryst.*, 15, 89, 1967.

9 Oxysalts of Iodine

9.1 GENERAL

The chief difference between the oxyanions of iodine and those of chlorine and bromine is the large number of periodates (see below). Otherwise, differences between the respective anions are largely accounted for by the larger size of iodine and its more metallic character.

Although thermodynamic data at 298.15 K, e.g., Wagman et al.[1] are available, very little above this temperature is in the standard compilations. In earlier work [2] equilibrium constants for decomposition reactions involving alkali metal and alkaline earth salts were calculated from a variety of sources. In the absence of newer data these were retained in this chapter.

The known oxysalts of iodine are shown in Figures 9.1 to 9.5.

9.2 CLASSES OF OXYIODINE SALTS

9.2.1 IODITES

Until recently no anhydrous iodites were known. In 1965 Harmelin and Duval[3] reported a general method for the preparation of iodites: the solid-state reduction of iodates by metallic lead. Since the reaction occurs without loss of weight, it was postulated to occur according to

$$MIO_3 + Pb = MIO_2 + PbO$$

The iodites of Na, K, Ca, Ba, and Pb were prepared in this way at 420 to 425°C and the existence of IO_2^- determined from its infrared spectrum. However, the pure compounds were not isolated and no further synthetic work seems to have been done since 1965. Since no further information is available, these compounds are not described under the respective elements.

9.2.2 IODATES

The iodates all contain the pyramidal IO_3^- ion in which the iodine atom is bonded covalently to the three oxygens. (See Figure 9.2.) As shown in Figure 9.1, all the alkali metals (except Fr) and alkaline earths (except Be) form anhydrous iodates. Beryllium salts are frequently difficult to obtain in the anhydrous state because of the high charge density of Be^{+2}. Dasent and Waddington[4] studied the infrared spectra of several iodates. They distinguish between ionic iodates in which the crystals contain discrete, pyramidal IO_3 groups, e.g., $NaIO_3$, and iodate compounds ($–O–IO_2$) in which the metal is bound covalently to oxygen. Examples of the latter type are $Pb(IO_3)_2$, and both mercuric and mercurous iodate.

In general, iodates are thermally rather stable. Many of them can be heated to several hundred °C without appreciable decomposition. When they do decompose, it is frequently in the solid state or on melting. Thus there are virtually no properties of liquid iodates known. The decomposition reactions fall into three groups:

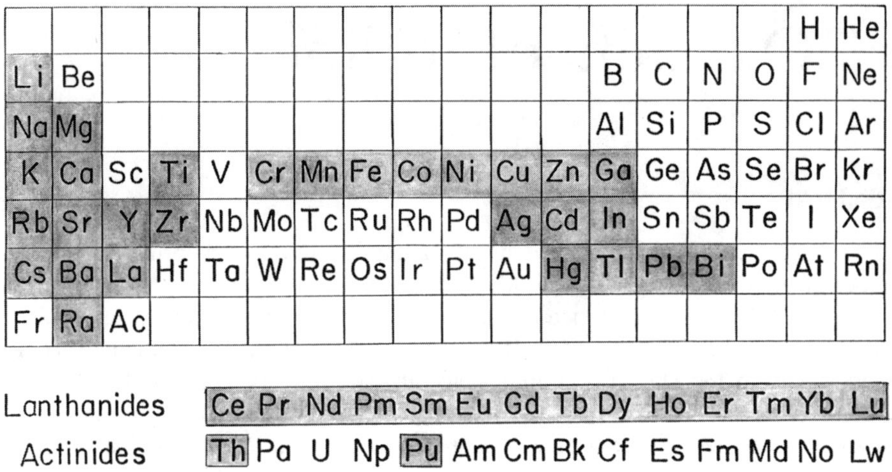

FIGURE 9.1 The known iodates.

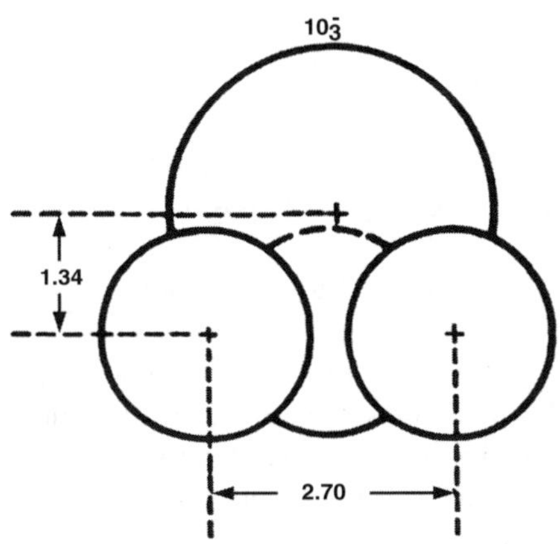

FIGURE 9.2 Structure of the IO_3 ion. (From Weigel, D., Imelik, B., and Prettre, M., *Bull. Soc. Chim. France*, 1427, 1962, With permission.)

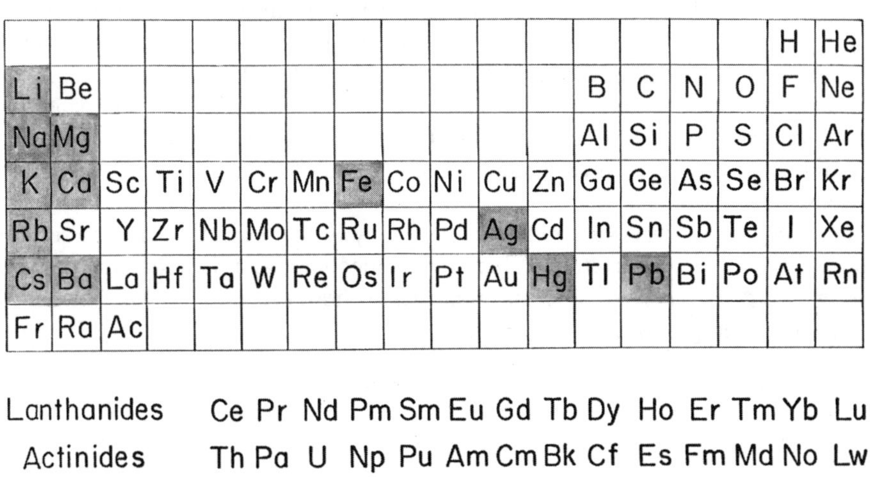

FIGURE 9.3 The known orthoperiodates.

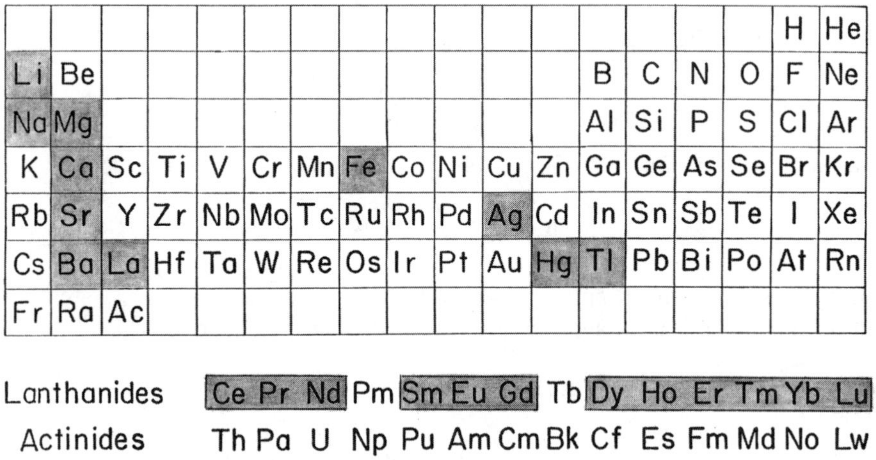

FIGURE 9.4 The known mesoperiodates.

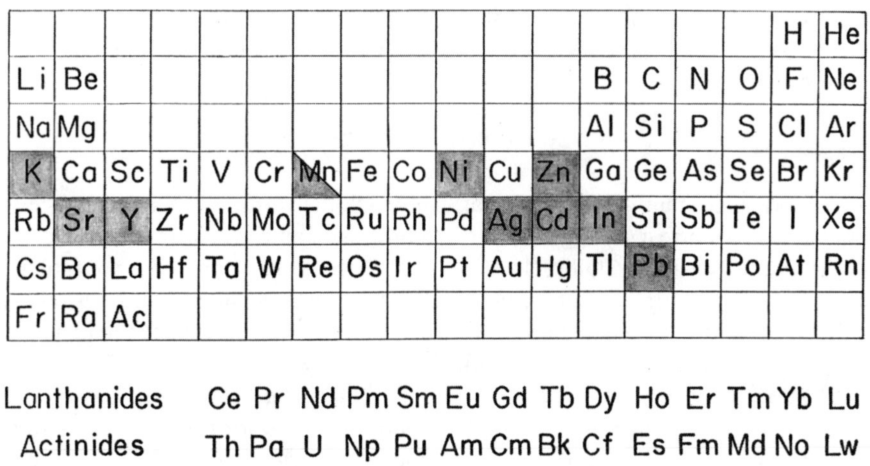

FIGURE 9.5 The known pyroperiodates.

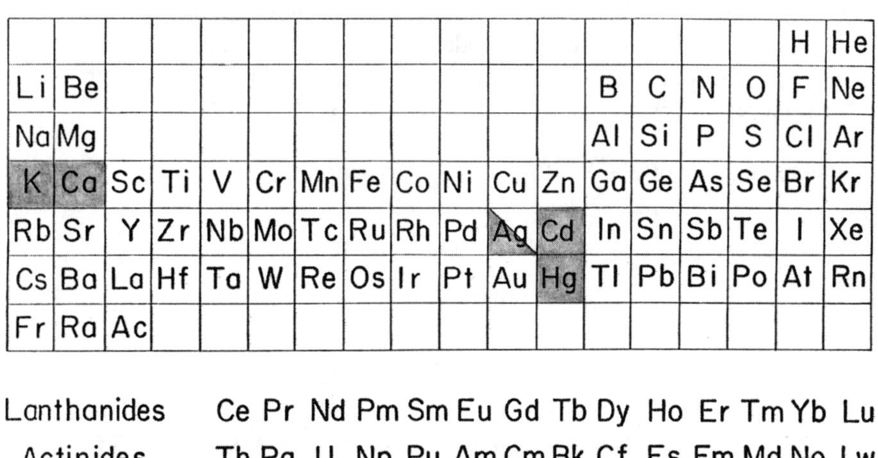

FIGURE 9.6 The known metaperiodates.

a. Iodates of the alkali metals (except Li), Ag, Co, and probably both Hg(I) and Hg(II) decompose according to

$$IO_3^- = I^- = 3/2\ O_2$$

b. Iodates of Cu, Cd, Zn, Ni, and probably Mg, Ti(I), and Pb go to the oxide

$$4\ IO_3^- = 2\ O^{2-} + I_2 + 5\ O_2$$

In addition, iodates of Na and K appear to go partly by this route.

c. Iodates of Li, Ca, Ba, Sr, and the rare earths form the orthoperiodate

$$5\ IO_3^- = IO^{5-} + 9/2\ O_2 + 2\ I_2$$

Reactions of type (a) for Na, K, Rb, and Cs can attain equilibrium, and those constants were calculated. Calculations are in good agreement with experimentally determined values. No calculations could be carried out for type (b) reactions. Only for $Mg(IO_3)_2$ have enthalpy increments above 298 K been measured, but in this case S°(298 K) is not known. There are no experimental studies. For reactions of type (c) only the calculation for Ca could be done since no S°(298 K) values for the periodates of Li, Ba, and Sr are available.

9.2.3 Periodates

The number of iodine-containing acids and salts in which the formal valence of iodine is +7 is far greater than that for the corresponding chlorine and bromine compounds. This behavior is generally consistent with the increasing metallic character of the heavier elements in a particular group.

Although various authors do not agree in all details on the number of periodic acids and periodates that exist, or their nomenclature, we list for the convenience of the reader the acids and salts which are generally recognized (see Table 9.1).

Structural information on the solid periodates is rather scarce. The IO_4^- ion is known to be tetrahedral[5,6] and IO^{5-} is octahedral.[6,7] The I–O bonding in these ions has been described [6], but the cation-anion distance has not been studied. Of the other periodates, only the structure of $K_4I_2O_9$ has been studied. The oxygen-bridged structure of the ion is described under potassium.

In Figures 9.3 to 9.6 the known anhydrous periodates have been listed. It is likely that the figures merely reflect work that has been done, rather than a limitation on the periodates that *could* exist. There are undoubtedly many more periodates that could be synthesized.

The various periodates decompose according to several different schemes, generally in the solid state; only the metaperiodates seem to have a melting point.

a. The metaperiodates of Na, K, Rb, and Mg decompose to the iodate:

$$IO_4^- = IO_3^- + 1/2\ O_2$$

b. The orthoperiodates of Li, Na, Ca, Sr, and Ba go to the oxide:

$$2\ IO_6^{5-} = 5\ O^{2-} + I_2 + 7/2\ O_2$$

c. Decomposition of the mesoperidates has not been studied except for the rare earths and Y:

$$2\ IO_5^{3-} = 3\ O^{2-} + 7/2\ O_2$$

d. Of the pyroperiodates, only cadmium salt has been studied:

$$2\ I_2O_9^{2-} = 4\ IO_4^{2-} + O_2$$

In this reaction a salt of hexavalent iodine is formed. Very little work on salts containing the IO_4^- ion has been reported yet. Decomposition of diorthosalts has not yet been studied.

9.3 INDIVIDUAL SALTS

9.3.1 Aluminum

9.3.1.1 $Al(IO_3)_3$

Anhydrous $Al(IO_3)_3$ can be prepared by the dehydration of its octahydrate. It then decomposes to the oxide in two stages in the range 460 to 560°C.[8] No further information was provided by the authors.

9.3.2 Barium

9.3.2.1 $Ba(IO_3)_2$

The salt is stable to near 580°C.[9–11] Above this temperature it undergoes the Rammelsberg reaction:

$$5\ Ba(IO_3)_2 = Ba_5(IO_6)_2 + 9\ O_2 + 4\ I_2$$

The *gef* for the iodate was calculated from the 298.15 K values in Wagman et al.,[1] and the enthalpy increments determined calorimetrically by David et al.[12] Unfortunately, equilibrium constants for the above reaction cannot be calculated, because no $S°_{298.15}$ is available for barium paraperiodate.

The density at 293 K is 5.23 gcm^{-3}.[13]

9.3.2.2 $Ba(IO_4)_2$

This salt decomposes to the iodate and O_2 in the range 150 to 250°C.[14,15] The iodate then decomposes further at 400 to 500°C:

$$Ba(IO_3)_2 = BaI_2 + 3\ O_3$$

9.3.2.3 $Ba_5(IO_6)_2$

This salt is stable to near 950°C.[9,11,16] Above this temperature it decomposes according to

$$Ba_5(IO_6)_2 = 5\ BaO + I_2 + 7/2\ O_2$$

This reaction is reversible.[16] The temperature at which decomposition occurs depends therefore on the gas phase; for example, under vacuum, decomposition becomes noticeable at as low as 500°C. The lack of S° values for this salt prevents any equilibrium calculations for the above reaction, as well as the *gef* calculation for the paraperiodate. The enthalpy and entropy increments are based on the calorimetric enthalpy increments of David, Mathurin, and Thevenot.[12]

Thermodynamic data for these salts are given in Tables 9.2 and 9.3.

9.3.3 Bismuth

Although the existence of $Bi(IO_3)_3$ has been reported,[17,18] the pure anhydrous salt has probably not yet been prepared.

9.3.4 CADMIUM

9.3.4.1 $Cd(IO_3)_2$

The salt is stable to ≈560°C. Above this temperature it decomposes according to Varhely and Kekedy[11]

$$2\ Cd(IO_3)_2 = 2\ CdO + I_2 + 5\ O_2$$

A ΔGf° value of 377.1 kJ gas been reported.[1]

9.3.4.2 $Cd_3(IO_5)_2$

Varhely and Kekedy[11] report that this salt forms by the dehydration of its monohydrate near 210°C. It is then stable to near 525°C. Above this temperature it decomposes to the oxide.

9.3.4.3 $Cd_5(IO_6)_2$

Preparation of this compound has been reported,[17] but its properties have not yet been reported.

9.3.4.4 $Cd_2I_2O_9$

The anhydrous compound has been prepared and its decomposition studied.[17] At temperatures as low as 110°C decomposition leads to the formation of hexavalent cadmium iodate

$$2\ Cd_2I_2O_9 = 4\ CdIO_4 + O_2$$

At a higher temperature the latter decomposes to the oxide.

The density of $Cd(IO_3)_2$ is 6.48 gcm^{-3}.[13]

9.3.5 CALCIUM

9.3.5.1 $Ca(IO_3)_2$, $Ca_5(IO_6)_2$

It is generally agreed[20,21] that the decomposition of $Ca(IO_3)_2$ occurs in two successive steps:

$$5\ Ca(IO_3)_2 = Ca_5(IO_6)_2 + 4\ I_2 + 9\ O_2 \quad \text{(a)}$$

$$Ca_5(IO_6)_2 = 5\ CaO + I_2 + 7/2\ O_2 \quad \text{(b)}$$

Since both reactions produce gaseous products, the stability range would be expected to vary with the gas-phase pressure and composition. Reaction (a) appears to begin near 550°C and to be complete near 770°C.[21] However, the reaction appears to be reversible and according to Bousquet and Remy[20] is the sum of the reactions

$$5\ Ca(IO_3)_2 = 5\ CaI_2 + 15\ O_2$$

$$5\ CaI_2 + 5/2\ O_2 = 5\ CaO + 5\ I_2$$

$$5\ CaO + I_2 + 7/2\ O_2 = Ca_5(IO_6)_2$$

Thus, in vacuum, heating of $Ca(IO_3)_2$ would produce CaI_2. In air, the first reaction is not displaced appreciably to the right below 510°C.[22] For this reason the calculation of the equilibrium constant for reaction (a) is somewhat misleading since it suggests appreciable decomposition of $Ca(IO_3)_2$

in air above 325°C. However, the equilibrium can probably not be established unless some CaI_2 is formed. Similarly, $Ca_5(IO_6)_2$ seems to be stable to a somewhat higher temperature, probably at least 725°C, than indicated by the calculation for reaction (b), but the reasons for this are not clear.

9.3.5.2 CaI_2O_9

This salt has been reported[11] as an intermediate in the dehydration of its 9-hydrate and to be stable to near 550°C. Above this temperature it decomposes, first to the hexavalent IO_4 ion and subsequently to the orthoperiodate.[14] The overall reaction is

$$5\ CaI_2O_9 = 2\ Ca_5(IO_6)_2 + 3\ I_2 + 5/2\ O_2$$

Thermodynamic data at 298.15 K are given in Table 9.4, and thermodynamic data for the decomposition of Ca $(IO_3)_2$ and $Ca_5(IO_3)_2$ are in Tables 9.5 and 9.6, respectively.

9.3.6 Cesium

9.3.6.1 $CsIO_3$

The ambient temperature density of $CsIO_3$ is uncertain. Pycnometrically determined values listed by Bousquet and David[23] vary from 4.83 to 4.94 gcm^{-3}, compared to 4.99 calculated from X-ray data. This suggests that crystals with varying void content were used.

Decomposition occurs on melting with oxygen evolution:[24,25]

$$CsIO_3 = CsI + 3/2\ O_2$$

In order to calculate the equilibrium constant for this reaction, S°(298.15 K) of $CsIO_3$ was estimated and enthalpy and entropy increments were calculated from the calometric data. The corresponding increments for CsI were obtained from the estimated C_p values of Kelley.[26] The resulting equilibrium constants indicate decomposition well below the melting point, but the rate in the solid state may be too slow to be detectable. Bousquet and Remy[27] calculated the equilibrium pressure of O_2 using thermodynamic values which differ little from those used here. They also measured oxygen pressures by both static and dynamic methods between 533 and 740 K and obtained good agreement with their own calculation. Their values of log K are approximately 0.5 more negative than those listed here over the entire temperature range. Most of this difference is accounted for by the different S° values selected by Bousquet and Remy.[27]

$CsIO_3$ and CsI form a eutectic at 781 K. Above this temperature decomposition leads to the formation of a liquid phase, but the oxygen pressure will remain independent of composition as long as a solid phase is present.

9.3.6.2 $CsIO_4$

The salt exists in three crystalline modifications.[28] Forms II and III are probably tetragonal and orthorhombic, respectively. The structure of I is less certain. The transformation I → II exhibits a large hysteresis. The transformation listed was obtained with rising temperature. On decreasing the temperature, it occurs about 30°C lower. The density of what is probably II is 4.25 gcm^{-3}.[29] Decomposition is reported[30] to go to the iodate near 500°C:

$$CsIO_4 = CsIO_3 + 1/2\ O_2$$

but no quantitative data on the reaction are available.

Thermodynamic data at 298.15 K are given in Table 9.7, and data for the decomposition of $CsIO_3$ are in Table 9.8.

9.3.7 Chromium

9.3.7.1 $Cr(IO_3)_3$

The preparation of the anhydrous salt has been reported,[31] but no information about its properties was given except that it decomposes with the formation of a volatile compound near 420°C.

9.3.8 Cobalt

9.3.8.1 $Co(IO_3)_2$

The blue-violet salt appears to be stable to at least 220°C[32] but decomposes in the range 415 to 425° to give Co_3O_4 according to the reaction[10,11]

$$3\ Co(IO_3)_2 = Co_3O_4 + 3\ I_2 + 7\ O_2$$

A density value of 5.138 gcm^{-3} has been given by Duval.[32]

9.3.9 Copper

9.3.9.1 $Cu(IO_3)_2$

The salt exists in three crystalline modifications with different densities:[33]

α light yellow green, monoclinic	5.275
β light green, triclinic	4.91
γ dark yellow, monoclinic	5.328

At 450°C it decomposes to the oxide:[10,34]

$$Cu(IO_3)_2 = CuO + I_2 + 5/2\ O_2$$

An activation energy of 325 kJ has been reported for this reaction in air from TGA data.[34]
No thermodynamic data have been listed in Wagman et al.,[1] but older values are given in Table 9.9.

9.3.10 Indium

9.3.10.1 $In(IO_3)_3$

The anhydrous white salt was first prepared by Mathers and Schluederberg[35] and the existence of the compound confirmed.[36] On heating, the crystals turn brown and evolve iodine vapor.[35] Decomposition at much higher temperature, 530 to 600°C, has been reported.[8]

9.3.10.2 $InIO_5$

Preparation of the yellowish crystalline compound was reported,[36] but none of its properties were examined.

9.3.11 Iron

9.3.11.1 $Fe(IO_3)_3$

Preparation of the anhydrous salt was reported by Endredy[37] with a decomposition temperature in air of 130°C. More recent studies[31,33,38] found $Fe(IO_3)_3$ to be much more stable. It began to decompose slowly only at 370°C, and rapidly at 470°C.

9.3.11.2 $Fe_5(IO_6)_2$

Reported by Kimmins[39] to be a brick-red crystalline salt, the iron content was slightly below the theoretical value.

9.3.11.3 $Fe(IO_4)_3$

Preparation was reported by both Kimmins[39] and Endredy.[37] The dry salt can be heated in air up to 125°C. Above this temperature it turns first brown and then black, while I_2 is evolved. The residue is Fe_2O_3.

9.3.11.4 $Fe_5(IO_6)_3$

Preparation of the salt was reported,[40] but its properties were not studied.

9.3.12 Lead

9.3.12.1 $Pb(IO_3)_2$

The salt is stable up to 400°C. Above this temperature it decomposes with the evolution of I_2 and O_2.[41] A density of 6.50 gcm^{-3} for $Pb(IO_3)_2$ at ambient temperature has been reported.[45]

9.3.12.2 $Pb(IO_3)_4$

The anhydrous salt was prepared by dehydration of its dihydrate between 100 and 120°C,[42] but no further work seems to have been reported.

9.3.12.3 $Pb(IO_4)_2$

The preparation of this salt in a slightly impure state was described by Kimmins.[39]

9.3.12.4 $Pb_3(IO_5)_2$

The preparation of this salt was described by Willard and Thompson.[43] Decomposition begins in the range 430 to 450°C. Nasanen and Uggla[44] report an intermediate 8 $PbO.PbI_2$, whereas Varhely and Kekedy show only a plateau on their TGA curve corresponding to PbO.

Thermodynamic data for $Pb(IO_3)_2$ at 298.15 K are shown in Table 9.10.

9.3.13 Lithium

9.3.13.1 $LiIO_3$

The salt exists in two crystalline modifications.[18,46,47] Vaporization of an aqueous solution below 60°C yields hexagonal crystals, above 90°C it yields the tetragonal form; the area between it yields both. At 260°C the hexagonal transforms to the tetragonal with a volume change of 3.2 cm^3mol^{-1}. The latter decomposes on melting. The decomposition has been studied by Bousquet and David.[48] The overall process is the formation of the paraperiodate:

$$5\ LiIO_3 = Li_5IO_6 + I_2 + 9/2\ O_2 \qquad (a)$$

analogous to the reactions of the alkaline earth iodates. The reaction proceeds a little faster in an O_2 than N_2 atmosphere. The authors, on the basis of numerous subsidiary experiments, propose the following mechanism to account for the overall reaction (a).

$$LiIO_3 = LiI + 3/2\ O_2 \quad \text{(I)}$$

$$2\ LiI = 1/2\ O_2 = Li_2O + I_2 \quad \text{(II)}$$

$$6\ Li_2O + 4\ I_2 = Li_5IO_6 + 7\ LiI \quad \text{(IIIa)}$$

$$5\ Li_2O + I_2 + 7/2\ O_2 = 2\ Li_5O_6 \quad \text{(IIIb)}$$

$$2\ LiIO_3 + 3\ LiI = Li_5O_6 + 2\ I_2 \quad \text{(IV)}$$

with I and IV constituting the major steps.

Enthalpy increments of $LiIO_3$ and Li_5IO_6 have been measured[49] up to 623 and 867 K, respectively. The experimental values were used to generate enthalpy and entropy increments for the two salts. Since entropy data are lacking, it was not possible to calculate equilibrium constants for the decomposition reactions.

A density 4.502 gcm^{-3} at 305 K was reported for this salt.[48]

9.3.13.2 Li_2IO_4

This unusual compound, in which the valence of iodine is +6, between iodate (+5) and periodate (+7), results from the thermal decomposition of $Li_2H_3IO_6$.[50] Detailed X-ray, infrared, and magnetic susceptibility measurements were used to establish the chemical individuality of the salt which is stable to about 450°C.

9.3.13.3 Li_5IO_6

The salt sublimes at 730°C. At 850°C it decomposes by the reaction[47,51]

$$2\ Li_5IO_6 = 5\ Li_2O + I_2 + 7/2\ O_2$$

Phase transitions of $LiIO_3$ are shown in Table 9.11, thermodynamic data at 298.15 K in Table 9.12, and high-temperature enthalpy and entropy increments for $LiIO_3$ and Li_5IO_6 in Table 9.13.

9.3.14 Magnesium

9.3.14.1 $Mg(IO_3)_2$

Very little information about this salt is available. According to Unezawa et al.,[52] the anhydrous salt, prepared by dehydrating the 4-hydrate is stable up to 575°C. Above this temperature it decomposes to the oxide. No thermodynamic data, not even at 298.15 K, are listed in the standard compilations. Enthalpy increments were measured calorimetrically by Sanyai and Nag.[14]

9.3.14.2 $Mg(IO_4)_2$

This salt has been reported as an intermediate in the decomposition of its 8-hydrate,[54] stable up to near 525°C. Above this temperature decomposition to $Mg(IO_3)_2$ occurs.

9.3.14.3 $Mg_2I_2O_9$

This salt is probably not stable in the anhydrous state. Varhely and Kekedy[11] report the dehydration of a 12-hydrate, but their TGA curve is not horizontal and MgO is formed gradually. A more careful examination by Dratovsky and Koslerova[54] indicates the starting material to be the 13- hydrate. Heating leads to the successive formation of the 2.5-hydrate, $MgIO_4.1/2H_2O$, and MgO.

Enthalpy and entropy increments for $Mg(IO_3)_2$ are given in Table 9.14.

9.3.15 Manganese

9.3.15.1 $Mn(IO_3)_2$

Minovici and Kollo[55] reported that the salt is stable up to 200°C, but according to Duval[56] it increases up to 2% in weight on being heated to 340°C due to uptake of oxygen. Apparently the iodate is partially transformed to the periodate on heating. Decomposition of both salts is complete at 875°C, the only solid phase remaining is Mn_3O_4. However, other oxides (MnO, MnO_2) have also been reported as products in the range 555 to 835°C.[11] The oxide formed is expected to depend on the partial pressure of oxygen in the system.

9.3.15.2 $Mn_3(IO_5)_2$

Preparation of this compound has been reported.[57] The pink crystals decomposed above 12°C. Since the product was stated to be manganese hydroxide, it is questionable whether the salt was really anhydrous.

The density of $Mn(IO_3)_2$ is 4.934 gcm^{-3}.[58] Wagman et al.[1] lists ΔHf° = –669 kJ mol^{-1}, S° = 264 J $deg^{-1}mol^{-1}$. Data for the oxides can be found in Chapter 2.

9.3.16 Mercury

Several iodates and periodates of both Hg(I) and Hg(II) have been known for a long time. The older literature is summarized by Gmelin[59] and Mellor.[60] However, for more than 60 years very little work has been done on these compounds. No thermodynamic data are listed by Wagman et al.[1] A brief summary relating to thermal decomposition is given below.

9.3.16.1 $Hg_2(IO_3)_2$

The salt is stable to near 175°C. Above 500 K, decomposition results in the formation of HgI_2, which sublimes.[59]

9.3.16.2 $Hg(IO_3)_2$

Heating results in the formation of HgI_2 and O_2.[59]

9.3.16.3 $Hg_2(IO_4)_2$

The salt is stable to near 100°C. Decomposition begins near 130°C and is complete at 200°C.[59]

9.3.16.4 $Hg_5(IO_6)_2$

The salt can be heated to at least 100°C without weight loss.[61] There is disagreement concerning stability above this temperature. Duval[56] finds continuous weight loss above 200°C, whereas Varhely and Kekedy,[11] also using TGA, find no weight loss up to 500°C. The decomposition reactions have not been studied in detail, but Hg_2I_2, HgI_2, HgO, and Hg have been reported as products.[59]

9.3.16.5 $Hg_4I_2O_9$

Preparation of this salt has been reported.[62,63] Its mode of decomposition is not known, but it is probably similar to that of the other mercury periodates.

9.3.17 Nickel

9.3.17.1 $Ni(IO_3)_2$

This salt may exist in two modifications: green and yellow. The latter appears to be the form reported by Meusser.[64] Martinez-Cros and Le Boucher[65] report distinctly different densities for the two forms. Pechkovskii and Sofronova[66] find from DTA and TGA that the salt is stable up to 487°C and decomposes above this temperature by the reaction

$$Ni(IO_3)_2 = NiO + I_2 + 5/2\ O_2$$

9.3.17.2 $Ni_3(IO_5)_2$

The preparation of a greenish-yellow amorphous salt of this composition was described by Kimmins.[39]

9.3.17.3 $Ni_2I_2O_4$

The black crystalline salt is stable to at least 100°C.[39]

The density of the two modifications of $Ni(IO_3)_2$ is given in Table 9.15, and thermodynamic data in Table 9.16.

9.3.18 Plutonium

9.3.18.1 $Pu(IO_3)_4$

The preparation of submicrogram quantities of the yellow salt were reported more than 50 years ago,[67] but its properties were not studied.

9.3.19 Potassium

9.3.19.1 KIO_3

A melting point of 833 K has been reported.[68,69]

The decomposition of the salt has been studied experimentally by both a dynamic and a static method[70] between 655 and 736 K. Over this temperature range the reaction:

$$KIO_3 = KI + 3/2\ O_2$$

required 50 to 600 days to reach equilibrium. Using the *gef* values for K in Stull and Prophet[71] and C_p for KIO_3 determined by Ahlberg and Latimer,[72] equilibrium constants for the reaction were calculated (Table 9.19). Calculated values were about 0.3 more negative in log K than experimental values, with the same $\Delta H°$.

The KIO_3-KI system exhibits a eutectic at 730 K.[73] Above this temperature a liquid phase is present. If the system is one component, however, $p(O_2)$ will be a unique function of temperature. This will be true as long as there is a solid phase present (see Chapter 1).

It is interesting to note that Duval[74] found no weight loss when KIO_3 was heated *rapidly* to 500°C, the temperature at which the equilibrium pressure exceeds 1 atm. As is also evident from the long equilibration times, the decomposition is very sluggish. There is also some indication that there is also a secondary decomposition to the oxide:

$$4\ KIO_3 = 2\ K_2O + 2\ I_2 + 5\ O_2$$

9.3.19.1.1 Periodates

Several anhydrous periodates are known. Their description is complicated by the fact that some authors describe these salts in terms of the formula $nK_2O.m\ I_2O_7$, although (1) I_2O_7 probably does not exist as a discrete compound, and (2) the I_2O_7 group apparently does not exist in the crystal structure of periodates. The main reason for the system seems to be that I_2O_7 can be regarded as the (hypothetical) anhydride of HI_2O_4 (HIO_4), and hydrates of the salts can then be represented on the ternary I_2O_7-K_2O-H_2O diagram (see Bianco et al.[75]).

9.3.19.2 KIO_4

The formula of this salt is frequently given as $K_2I_2O_8$, although the existence of tetrahedral IO_4 groups in the solid is well established.[76] The salt is stable to at least 260 to 290°C[74,75] and perhaps 330°C.[76] The decomposition reaction is

$$KIO_4 = KIO_3 + 1/2\ O_2$$

No thermodynamic values have been measured. Decomposition kinetics have been studied in some detail,[77–80] with some of the studies being concerned with the effect of additives and irradiation on the course of decomposition. Various equations were used to represent different stages of the reaction. The activation energy is 209 kJ, consistent with the breaking of an I–O bond and the formation of atomic oxygen as the rate-determining step.

9.3.19.3 $K_4I_2O_9$ ($2\ K_2O.I_2O_7$)

The anhydrous salt is stable to 445°C.[75] Its crystal structure has been determined.[80] The $I_2O_9^{4-}$ ion is a distinct entity that can be represented as shown in Figure 9.7, but its decomposition reaction is not known, although, like all potassium periodates, the final product is KI.

9.3.19.4 K_3IO_5 ($3\ K_2O.I_2O_7$)

The salt was prepared by dehydration of its hydrates.[75] It is stable to 440°C, but its decomposition reaction has not been studied, nor are its properties known.

9.3.19.5 K_5IO_6 ($5\ K_2O.I_2O_7$)

The orthoperiodate is stable up to 570°C,[75] but no further information is available.

Densities are shown in Table 9.17, thermodynamic data at 298.15 K in Table 9.18, and data for the thermal decomposition $KIO_3 = KI + 3/2\ O_2$ in Table 9.19.

9.3.20 Radium

9.3.20.1 $Ra(IO_3)_2$

The salt has been reported,[81] but no high-temperature properties have been measured.

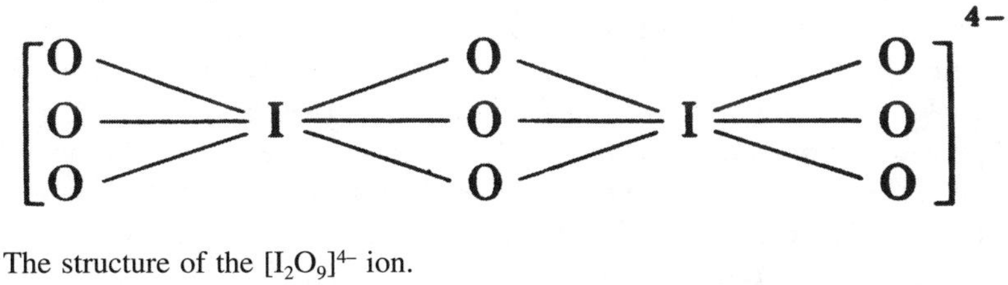

FIGURE 9.7 The structure of the $[I_2O_9]^{4-}$ ion.

9.3.21 Rare Earths

9.3.21.1 Iodates

During the early 1970s researchers at Bell Laboratories studied these salts, both hydrated and anhydrous, because of their interest in the electrical, magnetic, and structural properties of the compounds. Syntheses were also studied and the results were compared with previously published results.[82,83] Densities of all the anhydrous rare-earth iodates, derived from crystallographic data,[83a] are listed in Table 9.20. Extensive crystallographic data of these salts are also provided, as is a comparison with previous work cited in our earlier publication.[2]

Further work has been reported since 1980. Roy, Ghosh, and Nag[84] have studied phase transitions for the iodates of La, Pr, and Nd. These are listed in Table 9.21. Note that S° for these compounds is not known. The thermal decomposition of hydrated $La(IO_3)_3$ has also been studied.[85] The final H_2O is driven off at 250°C, and the iodate then decomposes to the periodate at 560°C:

$$10\ La(IO_3)_3 = 2\ La_5(IO_6)_3 + 12\ I_2 + 27\ O_2$$

This decomposition has also been reported for other lanthanide iodates[86,87] at temperatures between 400 and 500°C.

The periodate then decomposes to the oxide at 715 to 720°C.

9.3.21.2 Periodates

The orthoperiodates, $M_5(IO_6)_3$, that are formed by the thermal decomposition of iodates (see above) are stable to near 1000°C, but their decomposition has not been studied in detail. They decompose to the oxides. The mesoperiodates of La and Pr have been reported.[88] $LaIO_5$ is stable to near 420°C, $PrIO_5$ to 400°C. Decomposition goes to the oxides:

$$2\ LaIO_5 = La_2O_3 + I_2 + 7/2\ O_2$$

$$6\ PrIO_5 = Pr_6O_{11} + 3\ I_2 + 19/2\ O_2$$

Thermodynamic data are listed in Table 9.22.

9.3.22 Rubidium

9.3.22.1 $RbIO_3$

The salt decomposes according to

$$RbIO_3 = RbI + 3/2\ O_2$$

Bousquet and Remy[27] determined the melting point, ΔH_m and ΔS_m from the phase diagram of the $RbIO_3$-RbI system. The eutectic is at 777 K. Consequently, any decomposition above this temperature will result in the formation of a liquid phase, but the O_2 pressure will remain a unique function of temperature, as long as a solid phase is present and the system is one component.

Enthalpy increments of $RbIO_3$ were measured calorimetrically by Bousquet, Perachon, and Remy[89] up to 707 K. The resultant data were fitted to a power series in T which was then used to generate enthalpy and entropy increments at equal temperature intervals. Since only the lack of an S° (298.15 K) value for $RbIO_3$ prevented the calculation of *gef* values, it was estimated by comparison with other alkali metal halates. The estimate agrees with similar ones made by Bousquet and Remy.[70] C_p values of RbI estimated by Kelley[26] were used to generate enthalpy and entropy increments. In this way equilibrium constants for the decomposition reaction were calculated that are probably good to within an order of magnitude. These calculations indicate that decomposition in air becomes appreciable between 600 and 700 K, well below the melting point.

Bousquet and Remy[90] have carried out a similar calculation with values that differ only slightly from those used in this work. They also measured the equilibrium pressure of O_2 by both static and dynamic methods between 597 and 740 K. Under static conditions, 25 to 200 days (at various temperatures) were required to reach a steady pressure. Agreement between both sets of calculations and experimental values is quite good.

The melting point is 863 K. ΔH and ΔS for the phase change are 8.8 kJ mol^{-1} and 10.2 $Jdeg^{-1}mol^{-1}$, respectively.

9.3.22.2 $RbIO_4$

Older references are given by Gmelin.[59] No recent work has been reported. By analogy with $CsIO_4$ the decomposition probably proceeds according to

$$RbIO_4 = RbIO_3 + 1/2\ O_2$$

Densities are in Table 9.23; thermodynamic data for the decomposition reaction are in Table 9.24.

9.3.23 Silver

9.3.23.1 $AgIO_3$

Little work has been done on the thermal decomposition of this salt. It is apparently stable to near 405°C. Above this temperature decomposition yields AgI.[91–93]

9.3.23.2 $AgIO_4$

The salt decomposes near 200°C, and explosively at a lower temperature in a vacuum. The products are $AgIO_3$ and AgI.[94]

9.3.23.3 $AgIO_5$

The thermal decomposition of this salt has been studied by several researchers. The products reported seem to vary with the temperature. The lowest temperature is reported by Curti and Coggiola,[95] who find formation of $AgIO_3$ near 175°C. However, both Varhely and Kekedy[18] and Pacesova, Rathanova, and Rosicky[94] find the salt to be stable to near 425 to 445°C. Since $AgIO_3$ is unstable at this temperature, decomposition proceeds directly to the iodide:

$$2\ AgIO_5 = 2\ AgI + 5\ O_2$$

9.3.23.4 Ag_5IO_6

This salt has been reported to be stable at least to 60°C.[94] Both it and Ag_3IO_5 are black and give the same X-ray pattern.[95] However, the existence of the salt seems to be beyond question, particularly since the reaction

$$4\ Ag_3IO_5 = 2\ Ag_5IO_6 + 2\ AgIO_3 + O_2$$

in the presence of base has been studied.[95]

9.3.23.5 $Ag_4I_2O_9$

There is some question as to whether the pure anhydrous salt has been prepared. The 3-hydrate has been synthesized, but attempts to remove H_2O by a variety of procedures always leads to the loss of some O_2 along with the water.[95–97]

Densities of $AgIO_3$ and $AgIO_5$ are listed in Table 9.25, thermodynamic data at 298.15 K in Table 9.26.

9.3.24 Sodium

9.3.24.1 $NaIO_2$

This salt has been prepared by treating $NaIO_3$ with Pb in a nitrogen atmosphere at 420 to 425°C.[98] Some optical properties are reported.

9.3.24.2 $NaIO_3$

The salt decomposes according to

$$NaIO_3 = NaI + 3/2\ O_2$$

Equilibrium pressures were measured by Bousquet and Remy[99] by two different methods between 593 and 900 K, but the results exhibit considerable scatter. The same authors also calculated equilibrium pressures from thermodynamic data[90] up to 593 K. We have calculated equilibrium values up to 700 K, since C_p values for $NaIO_3$ are only available up to this temperature.[90] For NaI we have accepted the evaluation of Stull and Prophet,[71] although the value of $\Delta Hf°$ (298.15 K) differs somewhat from the newer value[1] of –481.788 kJ.

Since $NaIO_3$ decomposes only slowly, the melting point obtained by a short extrapolation of the $NaIO_3$-NaI phase diagram is probably accurate to within 1°C.

If the compound is heated fairly rapidly, as in DTA measurements,[100] the salt can be melted without appreciable decomposition, perhaps to near 460°C. There is some evidence that, at least under nonequilibrium conditions, about 28% of the $NaIO_3$ decomposes to the oxide:

$$4\ NaIO_3 = 2\ Na_2O + 2\ I_2 + 5\ O_2$$

9.3.24.3 Na_2IO_4

This salt, in which iodine is in a +6 oxidation state, results from heating $Na_2H_3IO_6$ at 120 to 180°C.[101] Using X-ray diffraction, magnetic susceptibility, and infrared absorption, the chemical individuality of the salt was established; i.e., it is not a mixture of $NaIO_3$ and Na_3IO_5. The compound is stable up to 370°C, but its decomposition products were not identified.

9.3.24.4 $NaIO_4$

The thermal expansion of the salt is quite different along the a and c axes. Between 25 and 120°C the expansion coefficients are represented by[102]

$$\alpha_a = 36.57 \times 10^{-6} + 3.89 \times 10^{-8}t + 14.19 \times 10^{-11}t^2$$

$$\alpha_c = 49.49 \times 10^{-6} + 18.42 \times 10^{-8}t - 9.22 \times 10^{-11}t^2$$

Heating of $NaIO_4$ at 290 to 300°C leads to the formation of $NaIO_3$:[103,104]

$$NaIO_4 = NaIO_3 + 1/2\ O_2$$

No high-temperature thermodynamic data for the periodate are available.

9.3.24.5 Na_5IO_6

As with the alkaline earths, this is the stablest of the periodates. It decomposes above 700°C according to Zintl and Morawietz:[105]

$$2\ Na_5IO_6 = 5\ Na_2O + I_2$$

Densities of $NaIO_3$ and $NaIO_4$ are given in Table 9.27, transitions in Table 9.28, thermodynamic data at 298.15 K in Table 9.29, and thermodynamics of the decomposition of $NaIO_3$ in Table 9.30.

9.3.25 Strontium

9.3.25.1 $Sr(IO_3)_2$

The salt is stable to at least 550°C. Above 600°C it forms the periodate:[11,106]

$$5\ Sr(IO_3)_2 = Sr_5(IO_6)_2 + 4\ I_2 + 9\ O_2$$

gef values for the iodate were calculated from m, ΔHf°, and S° values in Wagman et al.[1] and the enthalpy increments determined by David et al.[107] However, a lack of S° (298.15 K) for the paraperiodate prevents the calculation of the equilibrium constant for the above reaction.

9.3.25.2 $Sr_5(IO_6)_2$

The salt is stable to near 930°C. At about 1000°C it decomposes to the oxide at an appreciable rate:[52]

$$2\ Sr_5(IO_6)_2 = 10\ SrO + 2\ I_2 + 7\ O_2$$

Enthalpy and entropy increments are based on the calorimetric measurements of David et al.[107] The lack of S° prevents further calculation.

9.3.25.3 SrI_2O_9

The dimesoperiodate may only exist as a hydrate, since heating it results in the formation of $SrIO_4$ at 230°C:[14]

$$SrI_2O_9.n\ H_2O = 2\ SrIO_4 + n\ H_2O + 1/2\ O_2$$

On further heating this decomposes to SrI_2O_7.

The density of $Sr(IO_3)_2$ is 5.045 gcm^{-1} at 288 K.[13]

Thermodynamic data for $Sr(IO_3)_2$ and $Sr(IO_6)_2$ at 298.15 K are given in Table 9.31, high-temperature thermodynamic data in Table 9.32.

9.3.26 Thallium

9.3.26.1 $Tl(IO_3)_3$

Two studies agree that the salt decomposes with the simultaneous evolution of I_2 and O_2:[8,108]

$$Tl(IO_3)_3 = TlIO_3 + I_2 + 5/2\ O_2 \text{ to } 380\text{–}420°C$$

$$5TlIO_3 = Tl_5IO_6 + 2\ I_2 + 9/2\ O_2 \text{ to } 420\text{–}440°C$$

$$Tl_5IO_6 = 2\ Tl_2O + TlI + 2\ O_2 \text{ to } 500\text{–}520°C$$

In addition to these reactions, Udupa[109] proposed a decomposition of the iodate at 480 to 600°C:

$$7\ TlIO_3 = 3\ Tl_2O_3 + TlI + 3\ I_2 + 6\ O_2$$

9.3.26.2 Tl_5IO_6

The preparation and X-ray spectrum of this salt have been reported.[109]
Thermodynamic data are given in Table 9.33.

9.3.27 Thorium

9.3.27.1 $Th(IO_3)_4$

This salt probably exists as an anhydrous compound up to near 300°C. However, since it is very hygroscopic, it has not yet been prepared in the anhydrous state at ambient temperature, but only by the thermal dehydration of a hydrate.[110]

9.3.28 Titanium

9.3.28.1 $Ti(IO_3)_4$

Preparation of the anhydrous salt by the dehydration of a dihydrate at 100°C has been reported.[111]
The density at 305 K (determined pycnometrically) is 4.080 gcm^{-3}.[13]

9.3.29 Yttrium

9.3.29.1 $Y(IO_3)_3$

The anhydrous salt can be prepared by the dehydration of its 4-hydrate below 100°C. It then decomposes, beginning near 470°C, according to the overall reaction.[112]

$$10\ Y(IO_3)_3 = 2\ Y_5(IO_6)_3 + 12\ I_2 + 27\ O_2$$

The periodate decomposes to the oxide near 1000°C, but this reaction has not been studied in detail.

9.3.29.2 YIO_5

The anhydrous salt can be prepared by the thermal decomposition of $YH_2IO_5.3H_2O$.[113] It is stable up to 250°C and then decomposes to the oxide:

$$2\ YIO_5 = Y_2O_3 + I_2 + 7/2\ O_2$$

9.3.30 Zinc

9.3.30.1 $Zn(IO_3)_2$

The salt is stable up to about 520°C. Above this temperature it decomposes according to Varhely and Kekedy.[114]

$$Zn(IO_3)_2 = ZnO + I_2 + 5/2\ O_2$$

9.3.30.2 $Zn(IO_5)_2$

Preparation of the anhydrous compound has been reported,[115,116] but its properties have not yet been studied.

The density of $Zn(IO_3)_2$ is 5.064 gcm^{-3} at 298 K.[117] No enthalpies and entropies are available, but Wagman et al.[1] gives ΔG° at 298.15 K = –433.70 kJ mol^{-1}.

9.3.31 Zirconium

9.3.31.1 $Zr(IO_3)_4$

The crystal structure of the anhydrous salt has been determined.[118] The density of the tetragonal prisms is 5.00 gcm^{-3} at ambient temperature.

TABLE 9.1
Classification of the Periodates

Periodic Acids

As Hydrates of I_2O_7	Acid	As Hydrates of HIO_4	Prefix
$I_2O_7.H_2O$	HIO_4	HIO_4	meta-
$I_2O_7.2H_2O$	$H_4I_2O_9$	$2HIO_4.H_2O$	di-, meso-
$3I_2O_7.7H_2O$	$H_7I_3O_{14}$	—	tri-
$I_2O_7.4H_2O$	$H_8I_2O_{11}$	$2HIO_4.3H_2O$	diortho-
$I_2O_7.5H_2O$	H_5IO_6	$HIO_4.2H_2O$	ortho-

Periodates

Salts[a]	As $M_2O.I_2O_7$	Prefix
$MIO_4(M_2I_2O_8)$	$M_2O.I_2O_7$	meta-
$M_4I_2O_9$	$2M_2O.I_2O_7$	dimeso-, pyro-
M_3IO_5	$3M_2O.I_2O_7$	meso-
M_2HIO_5		
$M_6I_2O_{11}$	$4M_2O.I_2O_7$	diortho-
$M_3H_2IO_6$		
$M_2H_3O_6$		

[a] Formulas written for a monovalent cation.

TABLE 9.2
Thermodynamic Data (298.15 K) of Ba-I Salts

Compound	ΔHf° kJ mol^{-1}	S° J mol^{-1} deg^{-1}	Reference
$Ba(IO_3)_2$	–1027.2	249.4	1
$Ba_5(IO_6)_2$	–3952	—	1

TABLE 9.3
High-Temperature Thermodynamic Data[2]

T(K)	$Ba(IO_3)_2$ *gef* J deg^{-1}	BaO *gef* J deg^{-1}	$Ba_5(IO_6)_2$ $H_T - H_{298}$ J	$S_T - S_{298}$ J deg^{-1}
298.15	–294.4	–70.3	0	0
400	–256.7	–72.1	42873	123.38
500	–271.8	–75.8	90523	229.56
600	–289.4	–79.9	140746	321.11
700		–85.2	190615	398.05
800		–88.4	237212	460.37

TABLE 9.4
Thermodynamic Data (298.15 K) of Ca-I Salts

Compound	ΔHf° kJ mol^{-1}	S° J mol^{-1} deg^{-1}	Reference
$Ca(IO_3)_2$	–1002.5	230.	1
$Ca_5(IO_6)_2$	–3828	453.6	1, 2

TABLE 9.5
$5\ Ca(IO_3)_2 = Ca_5(IO_6)_2 + 4\ I_2(g) + 9\ O_2(g)$

T K	Δ*gef* J deg^{-1}	G°/T	K	p_{tot}(atm)
298.15	–2190	2619	1.56E – 137	6.99E – 11
400.00	–2187	1398	9.40E – 74	4.48E – 6
500.00	–2180	687.6	1.21E – 36	3.19E – 3
600.00	–2173	216.7	4.79E – 12	2.50E – 1
700.00	–2167	–118.7	1.59E + 6	5.56

TABLE 9.6
$Ca_5(IO_6)_2 = 5\ CaO + I_2(g) + 7/2\ O_2(g)$

T K	Δ*gef* J deg^{-1}	G°/T	K	p_{tot}(atm)
298.15	–723.2	1466.7	2.45E – 77	1.61E – 17
400.00	–721.7	910.6	2.73E – 48	4.58E – 11
500.00	–716.9	588.9	1.74E – 31	2.48E – 7
600.00	–711.2	376.9	2.05E – 20	7.16E – 5
700.00	–706.4	226.3	1.52E – 12	4.01E – 3
800.00	–702.1	114.1	1.10E – 6	8.05E – 2
900.00	–696.9	28.6	3.22E – 2	0.79
1000.00	–692.8	–39.8	120.5	4.92

TABLE 9.7
Thermodynamic Data (298.15 K) of Cs-I Salts

Compound	ΔHf° kJ mol⁻¹	S° J mol⁻¹ deg⁻¹	Reference
$CsIO_3$	–525.95	167	2
$CsIO_4$	—	130	2
CsI	–345.82	123.05	1

TABLE 9.8
$CsIO_3 = CsI + 3/2\ O_2(g)$

T K	Δ*gef* J deg⁻¹	G°/T	K	p_{tot} (atm)
298.15	–270.55	325.78	9.62E – 18	4.52E – 12
400.00	–271.80	172.69	9.53E – 10	9.68E – 7
500.00	–276.04	81.57	5.49E – 5	1.44E – 3
600.00	–276.60	19.73	9.32E – 2	2.05E – 1
700.00	–279.25	–25.26	20.8	7.57

TABLE 9.9
Thermodynamic Data (298.15K) of $Cu(IO_3)_2$

Compound	ΔHf° kJ mol⁻¹	S° J mol⁻¹ deg⁻¹	Reference
$Cu(IO_3)_2$	–389.5	218	122

TABLE 9.10
Thermodynamic Data (298.15 K) of $Pb(IO_3)_2$

Compound	ΔHf° kJ mol⁻¹	S° J mol⁻¹ deg⁻¹	Reference
$Pb(IO_3)_2$	–495.4	313.0	1

TABLE 9.11
Transitions of Li-I Salts

Phase Change	T(K)	ΔH kJ/mol	ΔS J/(mol*K)	Reference
$LiIO_3$, hexagonal → tetragonal	533	2.2	4.2	46
$LiIO_3$, tetragonal → liquid	723	—	—	23

TABLE 9.12
Thermodynamic Data (298.15 K) of Li-I Salts

Compound	ΔHf° kJ mol^{-1}	S° J mol^{-1} deg^{-1}	Reference
$LiIO_3$	–503.38	—	1
Li_5IO_6	–2044	—	123

TABLE 9.13
Enthalpy and Entropy Increments of $LiIO_3$ and Li_5IO_6

	$LiIO_3$		Li_5IO_6	
T(K)	$H_T - H_{298.15}$ kJ mol^{-1}	$S_T - S_{298.15}$ J deg^{-1} mol^{-1}	$H_T - H_{298.15}$ kJ mol^{-1}	$S_T - S_{298.15}$ J deg^{-1} mol^{-1}
400	10.98	31.54	25.76	74.08
500	22.94	58.18	51.02	130.43
600	36.09	82.11	76.27	176.48
700			105.71	215.41
800			126.78	249.13

TABLE 9.14
Enthalpy and Entropy Increments of $Mg(IO_3)_2$ [2]

T(K)	$H_T - H_{298.15}$ kJ mol^{-1}	$S_T - S_{298.15}$ J deg^{-1} mol^{-1}
400	21.84	63.19
500	53.96	110.98
600	64.67	150.03
700	86.08	183.05
800	107.50	211.65

TABLE 9.15
Density of $Ni(IO_3)_2$

Phase	T(K)	d	Reference
$Ni(IO_3)_2$(yellow)	298	5.02	65
$Ni(IO_3)_2$(green)	298	4.61	65

TABLE 9.16
Thermodynamic Data (298.15 K) of $Ni(IO_3)_2$

Compound	ΔHf° kJ mol^{-1}	S° J mol^{-1} deg^{-1}	Reference
$Ni(IO_3)_2$	–489.1	213.1	1

TABLE 9.17
Densities of K-I Salts

Phase	T(K)	d	Reference
KIO_3	305	3.93	111
KIO_4	288	3.618	119
$K_4I_2O_9$	Ambient	3.7	124

TABLE 9.18
Thermodynamic Data (298.15 K) of K-I Salts

Compound	ΔHf° kJ mol^{-1}	S° J mol^{-1} deg^{-1}	Reference
KIO_3	–501.37	151.46	1
KIO_4	–467.23	176.	1
KI	–327.9	106.38	1

TABLE 9.19
$KIO_3 = KI + 3/2\ O_2(g)$

T K	Δ*gef* J deg^{-1}	G°/T J deg^{-1}	K	p_{tot}(atm)
298.15	–262.48	348.39	6.33E – 19	7.37E – 13
400.00	–262.19	193.13	8.16E – 11	1.84E – 11
500.00	–261.54	102.73	4.31E – 6	2.65E – 6
600.00	–260.81	42.74	5.86E – 3	5.86E – 3
700.00	–260.14	0.043	0.995	1.00
800.00	–259.63	–31.97	4.67	46.7

TABLE 9.20
Densities of the Rare-Earth Iodates (calculated, Abrahams et al.[83a])

Rare Earth	d(gcm^{-3})
Ce	5.35
Pr	5.39
Nd	5.44
Sm	5.58
Eu	5.65
Gd	5.68
Tb	5.75
Dy	5.86
Ho	5.92
Er	5.95
Tm	6.03
Yb	6.03
Lu	6.12
Y	5.26

TABLE 9.21
Enthalpy Changes for Phase Transitions of Rare-Earth Iodates[84]

Compound	Transition (°C)	ΔH (kJ mol^{-1})
$La(IO_3)_3$	352 (endothermic)	20.75
	355 (endothermic)	20.75
	380 (exothermic)	4.39
	484 (endothermic)	10.25
$Pr(IO_3)_3$	356 (exothermic)	2.09
	371 (exothermic)	19.46
	412 (exothermic)	14.02
$Nd(IO_3)_3$	350 (exothermic)	1.46
	372 (exothermic)	12.97
	418 (exothermic)	10.08

TABLE 9.22
ΔHf°(298.15K) of Rare-Earth Iodates[1]

Rare Earth	ΔHf° (kJ mol^{-1})
Lu	–1339
Yb	–1347
Tm	–1372
Er	–1381
Ho	–1381
Dy	–1377
Tb	–1364
Gd	–1368
Eu	–1290.3
Sm	–1381
Nd	–1389
Pr	–1396.6
Ce	–1389
La	–1397

TABLE 9.23
Density of Rb-I Salts

Phase	T(K)	d (g cm^{-3})	Reference
$RbIO_3$	305	4.471	119
$RbIO_4$	289	3.918	111

TABLE 9.24
$RbIO_3 = RbI + 3/2\ O_2$

T K	Δgef J deg⁻¹	G°/T J deg⁻¹	K	$p(O_2)$ atm
298.15	–264.88	346.20	8.25E – 19	8.80E – 13
400.00	–264.46	191.02	1.05E – 10	2.23E – 11
500.00	–263.65	100.73	5.47E – 6	3.10E – 7
600.00	–262.71	40.94	2.73E – 3	3.75E – 4
700.00	–261.89	–1.62	1.21	1.14

TABLE 9.25
Density of Ag-I Salts

Phase	T(K)	d (g cm^{-1})	Reference
$AgIO_3$	293	5.525	13
Ag_5IO_6	Ambient	1.884	98a

TABLE 9.26
Thermodynamic Data (298.15 K)

Compound	ΔHf° kJ mol^{-1}	S° J mol^{-1} deg^{-1}	Reference
$AgIO_3$	–171.1	149.4	1
AgI	–61.84	115.5	1

TABLE 9.27
Density of Na-I salts

Phase	T(K)	d (g cm^{-3})	Reference
$NaIO_3$	293	4.277	13
$NaIO_4$	289	3.865	127

TABLE 9.28
Transition of $NaIO_3$

Phase Change	T(K)	ΔH kJ/mol	ΔS J/(mol*K)	Reference
$NaIO_3$, c→liq.	695	35.1	50.9	128

TABLE 9.29
Thermodynamic Data (298.15 K) of Na-I Salts

Compound	ΔHf° kJ mol^{-1}	S° J mol^{-1} deg^{-1}	Reference
$NaIO_3$	–481.788	135.1	1
$NaIO_4$	–429.28	163	1
NaI	–287.78	98.53	1

TABLE 9.30
$NaIO_3 = NaI + 3/2\ O_2(g)$

T K	Δgef J deg⁻¹	ΔG°/T J deg⁻¹	K	$p(O_2)$ atm
298.15	–270.91	409.47	4.09E – 22	5.51E – 15
400.00	–283.42	223.71	2.06E – 12	1.63E – 8
500.00	–148.58	115.45	9.32E – 7	9.43E – 5
600.00	–293.72	44.37	4.81E – 3	2.83E – 2
700.00	–194.55	–4.76	1.77	1.46

TABLE 9.31
Thermodynamic Data (298.15 K) of Sr-I Salts

Compound	ΔHf° $kJ\ mol^{-1}$	S° $J\ mol^{-1}\ deg^{-1}$	Reference
$Sr(IO_3)_2$	–1019.2	234	1
$Sr_5(IO_6)_2$	–3975	—	129

TABLE 9.32
High-Temperature Thermodynamic Data of Strontium Compounds[2]

T	$Sr(IO_3)_2$ *gef*	SrO *gef*	$Sr_5(IO_6)_2$	
K	$J\ deg^{-1}$	$J\ deg^{-1}$	$H_T - H_{298.15}$ kJ	$S_T - S_{298.15}$ $J\ deg^{-1}$
298.15	–243.3	–54.3	0	0
400	–242.0	–56.2	42.39	121.60
500	–257.2	–59.7	89.89	227.32
600	–275.2	–63.7	140.48	319.42
700	–294.0	–67.9	191.44	397.92
800		–71.9	240.06	462.82

TABLE 9.30
Thermodynamic Data (298.15 K) of Tl-I Salts

Compound	ΔHf° $kJ\ mol^{-1}$	S° $J\ mol^{-1}\ deg^{-1}$	Reference
$TlIO_3$	–267.4	176.6	1
TlI	–123.8	127.6	1

REFERENCES

1. Wagman, D. D., Evans, W. H., Parker, V. B., Schumm, R. H., Bailey, S. M., Churney, K. L., and Nuttall, R. L., The NBS tables of chemical thermodynamic properties, *J. Phys. Chem. Ref. Data*, 2, Suppl. 2, 1982.
2. Stern, K. H., High temperature properties and decomposition of inorganic salts, part 4. Oxysalts of the halogens, *J. Phys. Chem. Ref. Data,* 3, 481, 1974.
3. Harmelin, M. and Duval, C., *Compt. Rend.,* 260, 2361, 1965.
4. Dasent, W. E. and Waddington, T. C., *J. Chem. Soc.,* 2429, 1960.
5. Brown, R. G., Denning, J., Hallett, A., and Ross, S. D., Spectrochim. Acta 26A, 963, 1970.
6. Siebert, H., *Z. Anorg. Chem.,* 273, 21, 1953.
7. Helmholtz, L., *J. Am. Chem. Soc.,* 59, 2036, 1937.
8. Ivanov, B. N., Rybina, V. I., Venkovskii, N. U., Kaziev, G. A., and Zaitsev, B. F., *Russian J. Inorg. Chem.*, 26, 1730, 1981.
9. Duval, C., *Anal. Chim. Acta*, 20, 263, 1959.
10. Pechkovskii, V. V. and Sofronova, A. V., *Russian J. Inorg. Chem.*, 11, 828, 1966.
11. Varhely, C. and Kekedy, E., *Studia Ser. Chem. Univ. Babes-Bolyai,* 7, 11, 1962.
12. David, J., Mathurin, D., and Thevenot, F., *Bull. Soc. Chim. France,* 2508, 1970.
13. *International Critical Tables*, McGraw-Hill, New York, 1926.
14. Sanyai, G. S. and Nag, K., *J. Inorg. Nucl. Chem.,* 39, 1127, 1977.
15. Arnikar, H. J. and Patnaik, S. J., *Radiochem. Radioanal. Lett.,* 32, 45, 1978.
16. Frit, B., Tanguy, B., and Hagenmuller, P., *Bull. Soc. Chim. France,* 2190, 1966.
17. Neckers, J. W. and Hernandez, R. A., *J. Indian Chem. Soc.,* 30, 675, 1953.
18. Unezawa, T., Ninimiya, Y., and Tatuoka, S., *J. Appl. Cryst.*, 3, 417, 1970.
19. Pacesova, L. and Bohanesova, E., *Coll. Czech. Comm.*, 33, 664, 1968.
20. Bousquet, J. and Remy, J. C., *Bull.Soc. Chim.*, 214, 1964.
21. Peltier, S. and Duval, C., *Anal. Chim. Acta,* 1, 345, 1947.
22. Bousquet, J., Audidier, J. D., and Vermande, P., *Bull. Soc. Chim.,* 1564, 1966.
23. Bousquet, J. and David, J. C., *Bull. Soc. Chim.,* 3169, 1965.
24. Breusov, O. N., Kashina, N. L., and Vedizade, N. A., *Russ. J. Inorg. Chem.,* 15, 316, 1970.
25. Wheeler, H. L., *Z. Anorg. Chem.,* 2, 437, 1892.
26. Kelley, K. K., *U.S. Bur. Mines Bull.*, 584, 1960.
27. Bousquet, J. and Remy, J. C., *Bull. Soc. Chim.,* 3169, 1965.
28. Arend, H., Granicher, H., and Hofman, R., *Helv. Phys. Acta,* 43, 484, 1970.
29. Bousquet, J. and Vermande, P., *Bull. Soc. Chim.,* 214, 1964.
30. Breusov, O. N., Kashina, N. L., and Revzina, T. V., *Russ. J. Inorg. Chem.,* 15, 316, 1970.
31. Nissan, K., Shiever, J. W., and Prescott, B. E., *J. Solid State Chem.*, 7, 186, 1973.
32. Duval, C., *Inorganic Thermogravimetric Analysis*, Elsevier, Amsterdam, 1953.
33. Abraham, S. C., Sherwood, R. C., and Bernstein, J. L., *J. Solid State Chem.,* 8, 274, 1973.
34. Lumme, P. and Lumme, H., *Suomen Kemistilehti,* 35B, 129, 1962.
35. Mathers, F. C. and Schluederberg, C. G., *J. Am. Chem. Soc.,* 30, 211, 1908.
36. Ensalin, F., *Z. Anorg. Allgem. Chem.*, 250, 199, 1942.
37. Endredy, A., *Z. Anorg. Allgem. Chem.*, 217, 53, 1934.
38. Music, S., Simmons, G. W., and Leidheise, H., *Radiochem. Radioanal. Chem. Lett.,* 48, 227, 1981.
39. Kimmins, C. W., *J. Chem. Soc.,* 148, 1899.
40. Seifer, G. B., *Russ. J. Inorg. Chem.,* 3, 2309, 1958.
41. Duval, C., *Anal. Chim. Acta,* 4, 159, 1950.
42. Ray, P. R. and Saha, H., *Z. Anorg. Allgem. Chem.,* 217, 376, 1934.
43. Willard, H. H. and Thompson, J. J., *J. Am. Chem. Soc.,* 56, 1828, 1934.
44. Nasanen, R. and Uggla, R., *Suomen Kemistilehti,* 31, 225, 1958.
45. Staritzky, E. and Walker, D., *Anal. Chem.,* 28, 914, 1956.
46. Liebertz, J., *Z. Phys. Chem. (NF),* 67, 94, 1969.
47. Robertson, D. S. and Roslington, J. M., *J. Phys. D,* 4, 1582, 1971.
48. Bousquet, J. and David, J. C., *Bull. Soc. Chim. France,* 3169, 1965.
49. Bousquet, J., David, J. C. , Diot, M., and Guillon, A., *Bull. Soc. Chim. France,* 2390, 1968.

50. Dratovsky, M., *Coll. Czech. Comm.*, 29, 579, 1964.
51. Hagenmuller, P., Rault, M., and Tanguy, B., *Compt. Rend.*, 249, 2324, 1959.
52. Unezawa, T., Ninimiya, Y., and Tatuaka, S., *J. Appl. Cryst.*, 3, 417, 1970.
53. Bousquet, J. and Perachon, G., *Compt. Rend.*, 266C, 951, 1968.
54. Dratovsky, M. and Koslerova, J., *Coll. Czech. Comm.*, 31, 4375, 1966.
55. Minovici, S. and Kollo, C., *Chim. and Ind. (Paris)*, 8, 449, 1922.
56. Duval, C., *Anal. Chim. Acta*, 4, 159, 1950.
57. Raychoudhury, P. C., *J. Indian Chem. Soc.*, 18, 576, 1941.
58. Abrahams, S. C., Sherwood, R. C., Bernstein, J. L., and Nassau, K. J., *Solid State Chem.*, 7, 205, 1973.
59. *Gmelin's Handbuch der Anorganischen Chemie*, 8th ed., Verlag Chemie, Weinheim, Germany.
60. Mellor, J. W., *A Comprehensive Treatise of Inorganic and Theoretical Chemistry*, Longmans, Green, New York, 1924.
61. Willard, H. H. and Thompson, J. J., *Ind. Eng. Chem.*, 3, 398, 1931.
62. Sahney, R. C., Aggarwal, S. L., and Singh, S. M., *J. Indian Chem. Soc.*, 23, 177, 1946.
63. Singh, S. and Singh, A., *J. Indian Chem. Soc.*, 29, 134, 1952.
64. Meusser, A., *Ber. d. Chem. Ges.*, 34, 2432, 1901.
65. Martinez-Cros, J. and LeBoucher, L., *Ann. Soc. Espan. de Fis. y Quim.*, 33, 229, 1935.
66. Pechkovskii, V. V. and Sofronova, A. V., *Russ. J. Inorg. Chem.*, 11, 826, 1966.
67. Cunningham, B. B. and Werner, L. B., *J. Am. Chem. Soc.*, 71, 1521, 1949.
68. Bousquet, J. and Dode, M., *Compt. Rend.*, 238, 87, 1950.
69. Ray, R. D. and Dayal, V., *Trans. Faraday Soc.*, 32, 742, 1936.
70. Bousquet, J. and Rermy, J. G., *Bull. Soc. Chim.*, 3430, 1967.
71. Stull, D. R. and Prophet, H., *JANAF Thermochemical Tables*, 2nd ed., NSRDS-NBS 37, 1971.
72. Ahlberg, J. E. and Latimer, W. M., *J. Am. Chem. Soc.*, 56, 857, 1934.
73. Bousquet, J. and Dode, M., *Compt. Rend.*, 230, 87, 1950 .
74. Duval, C., *Anal. Chim. Acta*, 15, 223, 1956.
75. Bianco, P., Sabbah, R., and Perinet, G., *Bull. Soc. Chim.*, 3437, 1967.
76. Simon, J., Gal, S., and Erdey, L., *Acta Chim. Acad. Sci. Hung.*, 66, 175, 1970.
77. Phillips, B. R. and Taylor, D., *J. Chem. Soc.*, 5583, 1963.
78. Arnikar, H. J. and Patnaik, S. K., *Radiochem. Radioanal. Lett.*, 53, 367, 1982.
79. Furuichi, R., Ishii, T., Yamanaka, Z., and Shimokawabe, M., *Thermochim. Acta*, 53, 367, 1982.
80. Brehler, B., Jacobi, H., and Siebert, H., *Z. Anorg. Chem.*, 362, 301, 1968.
81. Polesickij, A. and Talmacev, P., *Dokl. Akad.*, Sci. USSR III, 102, 1936.
82. Nassau, K., Shiever, J. W., Prescott, B. E., and Couper, A., *J. Solid State Chem.*, 11, 314, 1974.
83. Nassau, K., Shiever, J. W., and Prescott, B. E., *J. Solid State Chem.*, 14, 122, 1975.
83a. Abrahams, S. C., Bernstein, J. L., and Nassau, K., *J. Solid State Chem.*, 14, 122, 1975.
84. Roy, A., Ghosh, B. P., and Nag, K., *Thermochim. Acta*, 47, 105, 1981.
85. Tarasova, G. N., Vinogradov, E. N., and Kudinov, I. B., *Russian J. Inorg. Chem.*, 30, 315, 1985.
86. Hajek, B. and Hradilova, J., *J. Less Common Metals*, 23, 217, 1971.
87. Zintl. E. and Morawietz, W., *Z. Anorg. Chem.*, 236, 372, 1938; ibid. 245, 20, 1940.
88. Lokio, A. H. and Kyrki, J. R., *Suomen Kemistilehti*, B44, 334, 1971.
89. Bousquet, J., Perachon, G., and Remy, J. C., *Bull. Soc. Chim. France*, 240, 1967.
90. Bousquet, J. and Remy, J. C., *Bull. Soc. Chim. France*, 3430, 1967.
91. Dupuis, T. and Duval, C., *Compt. Rend.*, 229, 51, 1949.
92. Greensfelder, B. S. and Latimer, W. H., *J. Am. Chem. Soc.*, 53, 3813, 1931.
93. Kireev, A., *Zh. Obsch. Khim.*, 16, 1199, 1946.
94. Pacesova, L., Rathanova, H., and Rosicky, *J. Coll. Czech. Chem. Comm.*, 33, 3670, 1968.
95. Curti, R. and Coggiola, M., *Gazz. Chim. Ital.*, 71, 847, 1941.
96. Jonas, L. and Pacesova, L., *Coll. Czech. Comm.*, 31, 4718, 1966.
97. Partington, J. R. and Bahl, R. K., *J. Chem. Soc.*, 1091, 1934.
98. Harmelin, M. and Duval, C., *Compt. Rend.*, 260, 2461, 1965.
98a. Ferrari, A., Braibanti, A., and Tiripicchio, A., *Gazz. Chim. Ital.*, 52, 22, 1962.
99. Bousquet, J. and Remy, J. C., *Bull. Soc. Chim. France*, 705, 1963.
100. Erdey, L., Simon, J., and Gal, S., *Talanta*, 15, 653, 1968.
101. Dratovsky, M., *Russian J. Inorg. Chem.*, 8, 1276, 1963.

102. Deshpande, V. T., Suryanarayana, S. V., and Pawar, R. R., *Acta Cryst.,* A24, 398, 1968.
103. Dratovsky, M. and Turnwald, J., *Coll. Czech. Chem. Comm.*, 25, 2503, 1960.
104. Arnikar, H. J. and Patnaik, S. K., *Radiochem. Radioanal. Chem. Lett.,* 32, 45, 1978.
105. Zintl, E. and Morawietz, W., *Z. Anorg. Chem.,* 236, 372, 1940.
106. Duval C., *Inorganic Thermogravimetric Analysis*, Elsevier, Amsterdam, 1953.
107. David, J., Mathurin, D., and Thevenot, F., *Bull. Soc. Chim. France,* 2508, 1970.
108. Udupa, M. R., *Indian J. Chem.*, 19A, 1014, 1981.
109. Busev, A. T., Guseinzade, S. M., and Verdizade, N. A., *J. Anal. Chem. USSR*, 24, 914, 1969.
110. Dupuis, T. and Duval, C., *Anal. Chim. Acta*, 3, 589, 1949.
111. Ray, P. R. and Saha, H., *Z. Anorg. Chem.,* 208, 100, 1932.
112. Yakunina, G. M., Alekseenko, L. A., and Serebrennikov, V. V., *Russ. J. I Inorg. Chem.*, 14, 1414, 1969.
113. Kyrki, J. R. and Lokio, A. H., *Suomen Khemi.*, B44, 105, 1971.
114. Varhely, C. and Kekedy, E., *Studia Ser. Chem. Univ. Babes-Bolyai,* 7, 11, 1962.
115. Fabre, R. and Tomesco, T. G., *Bull. Soc. Chim. Romania*, 18A, 59, 1937.
116. Sirokomskii, V. S. and Melamed, S. T., *Zavodskaya Lab.*, 16, 398, 1950.
117. Martines-Cros, J., *Ann. Espan.*, 34, 72, 1936.
118. Larson, A. C. and Cromer, D. T., *Acta Cryst.*, 14, 128, 1961.
119. Barker, T. V., *J. Chem. Soc.,* 15, 1908.
120. Bousquet, J. and Remy, J., *Bull. Soc. Chim.,* 3433, 1967.
121. Bousquet, J., Perachon, G., and Remy, J., *Bull. Soc. Chim.,* 240, 1967.
122. Gedansky, L. M., Pearce, P. J., and Hepler, L. G., *Canad. J. Chem.,* 48, 1770, 1970.
123. Shidlovskii, A. A. and Voskresenskii, A. A., *Russ. J. Phys. Chem.*, 39, 810, 1965.
124. Brehler, B., Jacobi, H., and Siebert, H., *Z. Anorg. Chem.,* 362, 301, 1968.
125. Ray, R. D. and Dayal, V., *Trans. Faraday Soc.,* 32, 742, 1936.
126. Bousquet, J. and Remy, J., *Bull. Soc. Chim.,* 3430, 1967.
127. Wheeler, H. L., *Z. Anorg. Chem.,* 2, 437, 1892.
128. Bousquet, J., Perachon, G., and Remy, J., *Bull. Soc. Chim.,* 240, 1967.
129. Bousquet, J., David, J. C., Mathurin, D., and Perachon, G., *Bull. Soc. Chim.,* 3991, 1968.